Design
of Wood
Structures

Design of Wood Structures

by

Donald E. Breyer

Professor of Civil Engineering
and Engineering Technology

California State Polytechnic
University, Pomona

with
contributions
by

John A. Ank

Senior Structural
Engineer

C. F. Braun Engineers
and Constructors, Inc.

McGraw-Hill Book Company

New York   St. Louis   San Francisco   Auckland   Panama   São Paulo   Hamburg
Bogotá   Singapore   Johannesburg   London   Sydney   Tokyo   Paris
Madrid   Mexico   Montreal   New Delhi   Toronto

Library of Congress Cataloging in Publication Data
Breyer, Donald E
  Design of wood structures.

  Includes index.
  1. Building, Wooden.   I. Ank, John A., joint
author.   II. Title.
TA666.B74   694'.1   80–10226
ISBN 0–07–007671–5

1234567890   KPKP   89876543210

The editors for this book were Jeremy Robinson and Joseph
Williams, the designer was Mark E. Safran, and the production
supervisor was Paul A. Malchow. It was set in Baskerville by
The Kingsport Press.

Printed and bound by The Kingsport Press.

Contents

v

CONTENTS

10. Shearwalls   329

11. Nailed and Stapled Connections   359

12. Bolts, Lag Screws, and Timber Connectors   391

CONTENTS

Preface

The purpose of this book is to introduce engineers, technologists, and architects to the design of wood structures. It is designed to serve either as a text for a course in timber design or as a reference for systematic self-study of the subject.

The book will lead the reader through the complete design of a wood structure (except for the foundation). The sequence of the material follows the same general order that it would in actual design:

1. Design loads (vertical and lateral)
2. Design for vertical loads (beams and columns)
3. Design for lateral loads (horizontal diaphragms and shearwalls)
4. Connection design (including the overall tying together of the vertical- and lateral-force-resisting systems).

The need for such an overall approach to the subject became clear from experience gained in teaching timber design at the undergraduate level.

This text pulls together the design of the various elements into a single reference. A large number of practical design examples are provided throughout the text. Because of their wide usage, buildings naturally form the basis of the majority of these examples. However, the principles of member design and diaphragm design have application to other structures (such as concrete formwork and falsework).

This book relies on practical, current industry literature as the basis for structural design. This includes publications of the National Forest Products Association, International Conference of Building Officials, and the American Plywood Association.

In the writing of this text, an effort has been made to conform to the spirit and intent of the reference documents. The interpretations are those of the authors and are intended to reflect current structural design practice. The material presented is suggested as a guide only, and final design responsibility lies with the structural engineer.

If this book is used as a text for a formal course, an Instructor's Manual is available from the author in care of California State Polytechnic University, 3801 W. Temple Blvd., Pomona, CA 91768.

It would be greatly appreciated if readers would notify the authors of any errors or ambiguities which may be found in the text or examples. Comments can be directed to the author at the address given above.

A number of individuals and organizations have contributed suggestions and reviews of the text during its writing. A special thanks goes to Warren Bower, Chairman of the Wood Subcommittee of the Structural Engineers Association of California. Appreciation is also expressed to Robert A. Hewett of the National Forest Products Association; Donald R. Watson of the International Conference of Building Officials; Thomas E. Brassell and Russell P. Wibbens of the American Institute of Timber Construction; William A. Baker, James R. Elliott, and John R. Tissell of the American Plywood Association; Harlow Richardson, formerly with APA; and William R. Bloom of Steinbrugge and Thomas, Inc. Structural Engineers.

Acknowledgment and appreciation are also given to Don Wood of Wood Engineering for his contribution to the initial part of the book, and to Dr. Ronald L. Carlyle of California State Polytechnic University for his preparation of Appendix E. Thanks is also given to Russell W. Krivchuck and T. Verne Phillips for their assistance in the preparation of the Instructor's Manual.

Finally, a great many thanks go to my wife Cecelia (Cid) for her typing of the manuscript.

DONALD E. BREYER

Nomenclature

Organizations

AITC American Institute of Timber Construction
333 West Hampden Avenue
Englewood, CO 80110

APA American Plywood Association
1119 A Street
P.O. Box 2277
Tacoma, WA 98401

AWPI American Wood Preservers Institute
1651 Old Meadow Road
McLean, VA 22101

BOCA Building Officials and Code Administrators
17926 South Halsted
Homewood, IL 60430

ICBO International Conference of Building Officials
5360 South Workman Mill Road
Whittier, CA 90601

NFPA National Forest Products Association
1619 Massachusetts Avenue, N.W.
Washington, DC 20036

SBCC Southern Building Code Congress International
5200 Montclair Road
Birmingham, AL 35213

WWPA Western Wood Products Association
1500 Yeon Building
Portland, OR 97204

Additional addresses are given in Appendix F, *References*.

Publications

NDS National Design Specifications for Wood Construction (Ref. 2)

PDS Plywood Design Specification (Ref. 14.1)

TCM Timber Construction Manual (Ref. 3)

UBC Uniform Building Code (Ref. 1)

WWUB Western Woods Use Book (Ref. 4)

Units

ft foot, feet

ft² square foot, square feet

in. inch, inches

in.² square inch, square inches

k 1000 lb (kip, kilopound)

ksi k per square inch (k/in²)

pcf pounds per cubic foot (lb/ft³)

plf pounds per lineal foot (lb/ft)

psf pounds per square foot (lb/ft²)
psi pounds per square inch (lb/in²)

sec second

Abbreviations

Adj. adjusted

Allow. allowable

BLF bearing length factor

B&S beam and stringer

c.c. center to center

cg center of gravity

CUF condition-of-use factor

DF Douglas fir

DL dead load (lb, k, lb/ft, k/ft, psf)

EMC equilibrium moisture content

FBD free body diagram

FDL floor dead load (lb, k, lb/ft, k/ft, psf)

FLL floor live load (lb, k, lb/ft, k/ft, psf)

FS factor of safety

FSP fiber saturation point

Glulam structural glued-laminated timber

IP inflection point (point of reverse curvature and point of zero moment)

J&P joist and plank

LDF load duration factor

LF light framing

LFRS lateral-force-resisting system

LL live load (lb, k, lb/ft, k/ft, psf)

MC moisture content

MPF metal plate factor

NA neutral axis

o.c. on center

OM overturning moment (ft-lb, ft-k)

PII panel identification index (plywood)

$\mathcal{R}$ plate

P&T post and timber

RDL roof dead load (lb, k, lb/ft, k/ft, psf)

RLL roof live load (lb, k, lb/ft, k/ft, psf)

RM resisting moment (ft-lb, ft-k)

S4S dressed lumber (surfaced four sides)

SL snow load (lb, k, lb/ft, k/ft, psf)

SV shrinkage value

Tab. tabulated

T&G tongue and groove

TL total load (lb, k, lb/ft, k/ft, psf)

Trib. tributary

Symbols

A area (in.², ft²)

A_g gross cross-sectional area of a tension member or column (in.²)

A_n net area of a tension member or column (in.²)

A_1 gross cross-sectional area of the main member in a connection

A_2 Sum of the gross cross-sectional areas of the side members in a connection (in.²)

b width of a horizontal diaphragm (ft), length of shearwall (ft), width of a beam (in.)

C seismic design response spectrum value, compressive force (lb, k)

C_F size factor for bending stress

C_f form factor for bending stress

C_k limit on C_s separating the intermediate and long unbraced beam (allowable bending stress) formulas

C_p response spectrum coefficient for the seismic forces on a portion of a structure

C_s slenderness factor for laterally unbraced beams

c distance from the neutral axis of a beam to the extreme outside fibers (in.)

D dimension of structure parallel to a seismic force (ft)

D_1 depth of a wood member at some initial MC (in.)

D_2 depth of a wood member at some final MC (in.)

d beam depth or column width (in.), bolt diameter (in.), distance (ft, in.)

d_e effective depth of a beam at a connection for horizontal shear stress calculation (in.)

d_x width of column parallel to the y axis; used to calculate the slenderness ratio about the x axis (in.)

d_y width of column parallel to the x axis; used to calculate the slenderness ratio about the y axis (in.)

d' reduced depth of a beam at a notched end (in.)

d_1 shank diameter of a lag screw (in.)

d_2 pilot hole diameter for the threaded portion of a lag screw (in.)

E modulus of elasticity (psi)

e eccentricity (in.)

F_b allowable bending stress (psi)

F_b' allowable bending stress adjusted for stability (psi)

f_b actual bending stress (psi)

F_c allowable compressive stress parallel to grain (psi)

F_c' allowable compressive stress parallel to grain adjusted for stability (psi)

f_c actual compressive stress parallel to grain (psi)

$F_{c\perp}$ allowable compressive stress perpendicular to grain (psi)

$f_{c\perp}$ actual compressive stress perpendicular to grain (psi)

F_n allowable compressive stress at an angle to the grain (psi)

f_n actual compressive stress at an angle to the grain (psi)

F_t allowable tension stress parallel to grain (psi); additional seismic force applied at the top level (lb, k)

f_t actual tension stress parallel to grain (psi)

F_v allowable horizontal shear stress (psi)

f_v actual horizontal shear stress (psi)

f_v' reduced actual horizontal shear stress (psi)

F lateral force (lb, k)

F_2 seismic force generated by W_2 (lb, k)

F_2' seismic force generated by W_2' (lb, k)

F_R seismic force generated by W_R (lb, k)

F_R^t seismic force generated by W_R^t (lb, k)

F_p seismic load on a portion of a structure (lb, k, lb/ft, psf)

F_x seismic horizontal force at level x (lb, k)

h height of a shearwall (ft); height of a level above the base of a structure (ft); height (or depth) of a beam (in.)

I moment of inertia (in.^{4}); seismic occupancy importance coefficient

J coefficient to adjust for P-Δ effect in beam-columns

K seismic framing coefficient; reduction factor for connection design accounting for the number of fasteners in a row; column design—slenderness ratio limit separating intermediate and long column formulas; multiplying factor for DL to account for the effect of creep in beam deflection calculations

L length of column or beam span (ft, in.)

L_c length of cantilever overhang (ft, in.)

L_e effective unbraced length of a beam (ft, in.)

L_u unbraced length of a beam (ft, in.)

L_x unbraced length of a column considering buckling about the x axis

L_y unbraced length of a column considering buckling about the y axis

l length of bolt in the main member of a connection (in.)

l_b bearing length parallel to grain (in.)

M moment (in.-k, ft-k)

N number of stories (seismic loading)

n allowable load on bolt, lag screw, or timber connector at an angle to grain (lb, k)

P load or force (lb, k)

p allowable load on bolt, lag screw, or timber connector parallel to grain (lb, k)

P_L allowable lateral load on a nail driven into the side grain of the holding member (lb)

P_r reduced allowable load for a row of fasteners in a connection (lb, k)

P_s sum of the allowable loads for a number of fasteners in a row (lb, k)

P_w allowable withdrawal load on a nail driven into the side grain of the holding member (lb)

Q statical moment of an area for calculation of shear stress in a beam $(A'\bar{y})$

q allowable load on bolt, lag screw, or timber connector perpendicular to grain (lb, k)

R reaction (lb, k)

R_1 seismic load generated by the DL of a wall that is parallel to the seismic force (lb, k)

r radius of gyration (in.)

S section modulus (in.^{3}); site—structure resonance coefficient (seismic)

T tension force (lb, k) period of vibration (sec)

T_s period of soil system (sec)

v unit shear in horizontal diaphragm or shearwall (lb/ft)

V total seismic base shear (lb, k); shear in a diaphragm or shearwall (lb, k); shear in a beam (lb, k)

W total load (lb, k); total dead load for seismic load calculation (lb, k)

W_1 DL of a 1-ft-wide strip (lb/ft, k/ft)

W_2 total DL tributary to the second-floor level (lb, k)

W_2' portion of W_2 that generates a seismic load in the second-floor horizontal diaphragm (lb, k)

W_R total DL tributary to roof level (lb, k)

W_R' portion of W_R that generates a seismic load in the roof horizontal diaphragm (lb, k)

W_p weight of a portion of a structure (lb, k, lb/ft, k/ft, psf)

w uniformly distributed load (lb/ft, k/ft, psf, ksf)

Z seismic zone coefficient

Δ deflection (in.)

Δ_s shrinkage (in.)

θ angle of load or stress to direction of grain; slope of roof

Wood Buildings and Design Criteria

1.1 Introduction

There are probably more buildings constructed with wood than any other structural material. Many of these buildings are single-family residences, but many larger apartment buildings as well as commercial and industrial buildings also use wood framing.

The widespread use of wood in the construction of buildings has both an economic and an esthetic basis. The ability to construct wood buildings with a minimal amount of equipment has kept the cost of wood-frame buildings competitive with other types of construction. On the other hand, where architectural considerations are important, the beauty and warmth of exposed wood is difficult to match with other materials.

Wood-frame construction has evolved from a method used in primitive shelters into a major field of structural design. However, in comparison with the amount of time devoted to steel and reinforced-concrete design, timber design is not given sufficient attention in most colleges and universities.

This book is designed to introduce the subject of timber design as applied to wood-frame building construction. Although the discussion centers around building design, the concepts also apply to the design of other types of wood-frame structures. Final responsibility for the design of a building rests with the structural engineer. However, this book is written to introduce the subject to a broad audience. This includes engineers, engineering technologists, architects, and

1

others concerned with building design. A background in statics and strength of materials is required in order to adequately follow the text. Most wood-frame buildings are made up of statically determinate members, and the ability to analyze these types of trusses, beams, and frames is also necessary.

1.2 Types of Buildings

There are various types of framing systems that can be used in wood buildings. The most common type of wood-framed construction uses a system of horizontal diaphragms and shearwalls to resist lateral loads, and this book deals specifically with the design of this basic type of building. The building code classifies shearwall buildings as *box systems,* and the distinctions between these and other systems are given in Chap. 3.

Other types of building systems, such as wood (glulam) arches, are beyond the scope of this book. It is felt that the designer should first have a firm understanding of the behavior of basic shearwall buildings and the design procedures that are applied to them. With a background of this nature, the designer can acquire from currently available sources (e.g., Ref. 3) the design techniques for other systems.

The basic box system can be constructed entirely from wood components. See Fig. 1.1. Here the *roof, floors, and walls* use wood framing. The calculations necessary to design these structural elements are illustrated throughout the text in comprehensive examples.

Fig. 1.1 Two-story wood-framed building. *(Photo by Mike Hausmann.)*

In addition to buildings that use only wood components, another common type of construction makes use of wood components with some other type of structural material. Perhaps the most common mixture of structural materials is in buildings that use *wood roof and floor systems and concrete tilt-up or masonry (concrete block or brick) shearwalls.* See Fig. 1.2. This type of construction is very common, especially in one-story commercial and industrial buildings.

This construction is economical for small buildings, but its economy increases as the size of the building increases. Trained crews can erect large areas of *panelized* roof systems in short periods of time.

Design procedures for the wood components that are used in buildings with concrete or masonry walls are also illustrated throughout this book.

Fig. 1.2a Foreground: Office portion of wood-framed construction. Background: Warehouse with concrete tilt-up walls and a wood roof system. (*Photo by Mike Hausmann.*)

Fig. 1.2b Building with reinforced concrete block walls and a wood roof system with plywood sheathing. (*Photo by Mark Williams.*)

The connections between wood and concrete or masonry elements are particularly important, and are treated in considerable detail.

This book covers the *complete* design of a wood-frame *box*-type building from the roof level down to (but not including) the foundation. In a complete building design, *vertical and lateral* loads must be considered, and the design procedures for both types of loads are covered in detail.

Wind loads and seismic (earthquake) loads are the two lateral loads that are normally taken into account in the design of a building. In recent years the design for lateral loads has become a significant portion of the design effort. The reason for this is an increased awareness of the effects of lateral loads. In addition, the building codes have substantially increased the magni-

3

tudes of the required design loads, especially in areas of high seismic risk. The San Fernando earthquake of 1971 was a major factor in causing codes to increase their design load requirements.

1.3 Building Codes and Design Criteria

Cities and counties across the United States typically adopt a building code in order to ensure public welfare and safety. Most local governments use one of the three *model codes* as the basic framework for their local building code. The three major model codes are the

1. *Uniform Building Code* (Ref. 1)
2. *Basic Building Code* (Ref. 32)
3. *Standard Building Code* (Ref. 33)

The *Building Code Requirements for Minimum Design Loads in Buildings and Other Structures* (Ref. 34) serves as the basis for some of the loading criteria in several of the model codes and a number of local codes.

Generally speaking, the *Uniform Building Code* is used in the western portion of the United States, the *Basic Building Code* in the north, and the *Standard Building Code* in the south. The model codes are revised and updated periodically, usually on a 3-year cycle.

Taken as a group, the model codes set forth the building code requirements for the large majority (roughly 90 percent) of the United States. Certain large cities write their own building codes, and those that use one of the model codes may enact legislation which modifies the model code in some manner.

In writing this design text, it was considered desirable to use one of the model building codes to establish the loading criteria and certain allowable stresses. The *Uniform Building Code* (UBC) is used throughout the text for this purpose. The UBC was selected because it is the most widely used of the three model codes, and, because of its use in the western states, it reflects recent seismic design trends.

It was noted previously that the trend in earthquake-resistant design is toward larger design loads. In addition, the UBC emphasizes the need to tie the building together to withstand the dynamic motions generated in an earthquake.

Throughout the text reference is made to the *Code* and the *UBC*. When references of this nature are used the design criteria are taken from the 1979 edition of the *Uniform Building Code*.

Although specific design criteria have been taken from the UBC, these criteria will normally meet or exceed the requirements of other codes. In order to make the text useful to users of other codes, many of the UBC tables are reproduced in Appendix D. By comparing the design values of another code with the criteria in Appendix D, the designer will be able to determine quickly whether or not the two are in agreement. Appendix D will also be a helpful cross-reference in checking future editions of the UBC against the values used in this text.

The other main reference used in this book is the *National Design Specification for Wood Construction* (Ref. 2). This pamphlet is commonly referred to as the *NDS*, and it is widely known and used by the design profession. The NDS is the basic authority on structural lumber and wood connections. Practically speaking, the NDS is a basic tool of timber design, and the designer should become acquainted with this document from the beginning. The reader should have a copy of the NDS to adequately follow this text.

The 1977 edition of the NDS is referenced throughout this book. As with most building codes, the NDS is revised periodically (usually on a 3- to 5-year cycle). When an NDS table is referenced in this book, the title of the table is also given. This will allow the designer to use future editions of the NDS with this book even if the tables are renumbered.

1.4   Relationship between the Code and the NDS

In this book the NDS is used as the basis for determining the allowable stresses in wood members and the allowable loads on connections. The Code is used for design load criteria and diaphragm and shearwall design.

The designer may question the combined use of these two design criteria (the NDS and the Code). Why not simply base the discussion on the Building Code?

The answer is simple. Portions of the timber design requirements of *most codes* are based on all or part of the NDS. Thus, the NDS criteria are commonly accepted. Much of the information contained in the wood design section of the UBC comes from the NDS and its supplement. Additional portions of the NDS are incorporated into the UBC standards (UBC Standard 25–17).

Occasionally the allowable stresses in lumber are revised, and these changes usually appear first in the NDS. They are normally incorporated in a later edition of the Code. In this respect it should be noted that many of the revisions that have taken place in recent years represent a design approach that is more conservative (i.e., on the side of greater safety).

It should also be noted that, once enacted into law, the *building code* is the *legal authority*. The designer should cross-check between the Code and the NDS criteria. If there are conflicting criteria in the two references, prudent design would follow the more conservative approach. Thus, the Code criteria represent minimum design requirements, and the designer may use a more conservative approach. An attempt has been made to follow this procedure throughout the text.

1.5 Organization of the Text

The text has been organized to present the complete design of a wood-framed building in an orderly manner. The subjects covered are presented roughly in the order that they would be encountered in the design of a building.

In a building design, the first items that need to be determined are the design loads. The Code design-load requirements for vertical and lateral loads are reviewed in Chap. 2, and the distribution of these loads in a building with wood framing is described in Chap. 3.

After the distribution of the loads has been determined, attention is turned to the design of the elements that support these loads. As noted previously, there are basically two systems that must be designed, one for *vertical loads* and one for *lateral loads.*

The vertical-load-carrying system is considered first. In a wood-frame building this system is basically composed of beams and columns. Chapters 4 and 5 cover the characteristics and design properties of these wood members. Chapter 6 then outlines the design procedures for beams, and Chap. 7 treats the design methods for columns and members subjected to combined axial and bending.

As one might expect, there are some parts of the vertical-force-carrying system that are also a part of the lateral-force-resisting system. The sheathing for wood roof and floor systems is one such element. The sheathing distributes the vertical loads to the supporting members, and it also serves as the *skin* or *web* of the horizontal diaphragm for resisting lateral loads. The most widely used structural sheathing material in wood-frame construction is plywood. Chapter 8 introduces the grades and properties of plywood and essentially serves as a transition from the vertical- to the lateral-force-resisting system. Chapters 9 and 10 deal specifically with the lateral-force-resisting system. In the typical box-type buildings covered in this text, the lateral-force-resisting system is made up of a diaphragm that spans horizontally between vertical shear-resisting elements known as shearwalls.

After the design of the main elements in the vertical- and lateral-force-resisting systems, attention is turned to the design of the connections. The importance of proper connection design cannot be overstated, and design procedures for various types of wood connections are outlined in Chaps. 11 through 13.

Chapter 14 concludes the text with a treatment of the anchorage requirements between horizontal and vertical diaphragms. Basically the anchorage requirements ensure that the horizontal and vertical elements in the building are adequately tied together.

1.6 Structural Calculations

This book contains a large number of practical design examples and sample calculations. An attempt has been made to use a consistent set of symbols and abbreviations throughout the text. Comprehensive lists of symbols and abbreviations, and their definitions, follow the table of contents. A number of the symbols and abbreviations are unique to this book, but where possible, they are in agreement with those accepted in the industry. The units of measure used in the main body of the text are the U.S. customary units. The abbreviations for these units are also summarized after the table of contents. Factors for converting to SI metric units are included in Appendix E.

Where possible an expression for a calculation is first given in generalized terms (i.e., a formula is first stated), the numerical values are then substituted into the expression, and finally the result of the calculation is given. In following this pattern the designer should be able to readily follow the sample calculation.

It should be noted that the conversion from pounds (lb) to kips (k) is often made without a formal notation. This is common practice and should be of no particular concern to the reader. For example, the calculations below illustrate the axial load capacity of a tension member

$$\text{Allow. } T = F_t A$$
$$= (1200 \text{ lb/in.}^2)(20 \text{ in.}^2)$$
$$= 24.0 \text{ k}$$

where T = tensile force
F_t = allowable tensile stress
A = cross-sectional area

The following illustrates the conversion for the above calculations which will normally be done mentally

$$\text{Allow. } T = F_t A$$
$$= (1200 \text{ lb/in.}^2)(20 \text{ in.}^2)$$
$$= (24,000 \text{ lb}) \left(\frac{1 \text{ k}}{1000 \text{ lb}} \right)$$
$$= 24.0 \text{ k}$$

The appropriate number of significant figures used in calculations should be considered by the designer. Until recently, most of the types of calculations illustrated in this book were performed with the aid of a slide rule. In that case, three or possibly four significant figures was all that could be obtained, and generally this degree of *precision* is sufficient.

A return to the "good old days" of the slide rule is not being advocated. However, when structural calculations are done on a calculator there is a tendency to present the results with too many significant figures. Variations in loading and material properties make the use of a large number of significant figures inappropriate. A false degree of *accuracy* is implied when the stress in a wood member is calculated to the third decimal point.

As an example, consider the bending stress in a wood beam. If the calculated stress as shown on the calculator is 1278.356 · · · psi, it is reasonable to

7

report 1280 psi in the design calculations. Rather than representing sloppy work, the latter figure is more realistic in presenting the degree of accuracy of the problem.

Although the calculations for problems in this text were performed on a calculator, intermediate and final results are generally presented with three or four significant figures.

1.7 Detailing Conventions

With the large number of examples included in this text, the sketches are necessarily limited in detail. For example, a number of the building plans are shown without doors or windows. However, each sketch is designed to illustrate certain structural design points, and the lack of full details should not detract from the example.

One common practice in drawing wood structural members is to place an X in the cross-section of a *continuous* wood member. On the other hand, a *noncontinuous* wood member is shown with a single diagonal line in cross section. See Fig. 1.3.

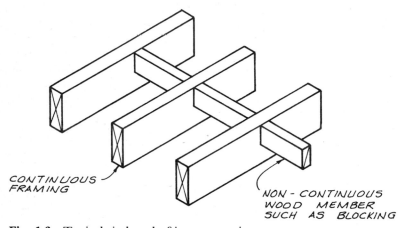

CONTINUOUS FRAMING

NON - CONTINUOUS WOOD MEMBER SUCH AS BLOCKING

Fig. 1.3 Typical timber drafting conventions.

1.8 Fire-Resistive Requirements

Building codes place restrictions on the materials of construction based on the occupancy (i.e., what the building will house), area, height, number of occupants, and a number of other factors. The choice of materials affects not only the initial cost of a building, but the recurring cost of fire insurance premiums as well.

The fire-resistive requirements are of great importance to the building designer. This topic can be a complete subject in itself and is beyond the scope of this book. However, several points that affect the design of wood buildings will be mentioned here to alert the designer.

Wood (unlike steel and concrete) is a combustible material, and certain *types of construction* (defined by the Code) do not permit the use of combustible materials. There are arguments for and against this type of restriction, but these limitations do exist.

Generally speaking, the *unrestricted use* of wood is allowed in buildings of limited floor area. In addition, the height of these buildings is limited to one, two, or three stories, depending upon the occupancy.

Wood is also used in another type of construction that is known as *heavy timber*. Experience and fire endurance tests have shown that the tendency of a wood member to ignite in a fire is affected by its cross-sectional dimensions. In a fire, large-size wood members form a protective coating of char which insulates the inner portion of the member. Thus large wood members may continue to support a load in a fire long after an uninsulated steel member has collapsed because of the elevated temperature. This is one of the arguments used against the restrictions placed on "combustible" building materials. (It should be noted that properly insulated steel members can perform adequately in a fire.)

The minimum cross-sectional dimensions required in order to qualify for the heavy timber fire rating are set forth in building codes. As an example, the UBC states that the minimum cross-sectional dimension for a wood column is 8 in. Different minimum dimensions apply to different types of wood members, and the Code should be consulted for these values. Limits on maximum allowable floor areas are much larger for wood buildings with heavy timber members, compared with buildings without wood members of sufficient size to qualify as *heavy timber*.

1.9   Industry Organizations

There are a number of organizations that are actively involved in promoting the proper design and use of wood and related products. These include the model building code groups as well as a number of industry-related organizations. The names and addresses of some of these organizations are listed after the table of contents. Others are included in the list of references, Appendix F.

Design Loads

2.1   Introduction

The calculation of design loads for buildings is covered in Chaps. 2 and 3. Chapter 2 deals primarily with Code-required design loads and how these loads are calculated and modified for a specific building design. Chapter 3 is concerned with the distribution of these design loads throughout the structure.

In ordinary building design, one normally distinguishes between two major types of loads: (1) *vertical (gravity) loads* and (2) *lateral loads*. Although certain members may function only as vertical-load carrying members, or only as lateral-load-carrying members, quite often members may be subjected to a combination of vertical and lateral loads. For example, a member may function as a beam when subjected to vertical loads and as an axial-load-carrying member under lateral loads (or vice versa).

Regardless of how a member functions, it is convenient to classify design loads into these two main categories. Vertical loads offer a natural starting point. Little introduction to gravity loads is required. "Weight" is something with which most people are familiar, and the design for vertical loads is often accomplished first. The reason for starting here is twofold. Gravity loading is an ever-present load, and quite naturally it has been the basic, traditional design concern. Second, in the case of lateral seismic loads, it is necessary to know the magnitude of the vertical loads before the earthquake forces can be estimated.

11

It should also be noted that the design of structural framing members usually follows the reverse order in which they are constructed in the field. That is to say, design starts with the lightest framing member on the top level and proceeds downward, and construction starts at the bottom with the largest members and proceeds upward.

Design loads are the subject of UBC Chap. 23. It is suggested that the reader accompany the remaining portion of this chapter with a review of Chap. 23 of the Code. For convenience, the Code load tables are reproduced in Appendix D.

2.2 Vertical Loads—Dead

Vertical loads are classified as either *dead* or *live* loads. Dead loads include the weight of all materials which are permanently attached to the structure. In the case of a wood roof or wood floor system, this would include the weight of the roofing or floor covering, the sheathing, framing, insulation, ceiling (if any), and any other permanent materials such as piping or automatic fire sprinklers.

Another dead load which must be included, but one that is easily overlooked (especially on a roof), is that of mechanical or air-conditioning equipment. Often this type of load is supported by two or three beams or joists side by side which are the same size as the standard roof or floor framing members. See Fig. 2.1. The alternative is to design special larger (and deeper) beams to carry these isolated equipment loads.

The magnitude of dead loads for various construction materials can be found in a number of references. A fairly complete list of weights is given in Appendix B, and additional tables are given in Refs. 3 and 12.

Because most building dead loads are estimated as uniform loads in terms of pounds per square foot (psf), it is often convenient to convert the weights of framing members to these units. For example, if the weight per lineal foot of a wood framing member is known, and if the center-to-center spacing of parallel members is also known, the dead load in psf can easily be determined by simply dividing the weight per lineal foot by the center-to-center spacing. For example, if 2×12 beams weighing 4.3 lb/ft are spaced at 16 in. o.c., the equivalent uniform load is 4.3 lb/ft $\div$ 1.33 ft = 3.2 psf. A table showing these equivalent uniform loads for typical framing sizes and spacings is given in Appendix A.

It should be pointed out that in a wood structure, the dead load of the *framing members* usually represents a fairly minor portion of the total design load. For this reason a small error in estimating the weights of framing members (either lighter or heavier) typically has a negligible effect on the final member choice. *Slightly* conservative (larger) estimates are usually used for design.

The estimation of the dead load of a structure requires some knowledge of the methods and materials of construction. A "feel" for what the unit dead loads of a wood-frame structure should total is readily developed after exposure to several buildings of this type. The dead load of a typical wood floor or roof system typically ranges between 7 and 20 psf, depending on the materials of construction, spans, and whether a ceiling is suspended below

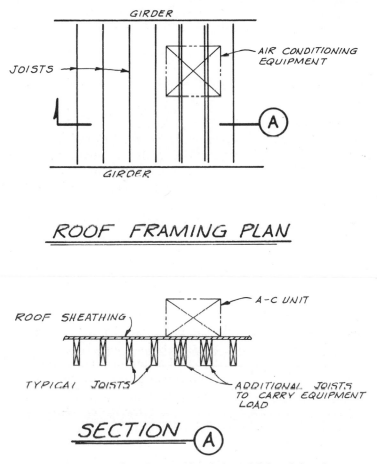

Fig. 2.1 Support of equipment loads by additional framing.

the floor or roof. For wood wall systems, values might range between 4 and 20 psf, depending on stud size and spacing and the type of wall sheathings used (e.g., ⅜-in. plywood weighs approximately 1 psf whereas ⅞-in. stucco weighs 10 lb per square foot of wall surface area). Typical load calculations provide a summary of the makeup of the structure. See Example 2.1.

The dead load of a wood structure that differs substantially from the typical ranges mentioned above should be examined carefully to ensure that the various individual dead load components are in fact correct. It pays in the

13

EXAMPLE 2.1 **Sample DL Calculation Summary**

Roof Dead Loads

Roofing (5-ply with gravel)	= 6.5 psf
½-in. plywood (3 psf × ½-in.)	= 1.5
Framing (estimate 2 × 12 at 16 in. o.c.) =	3.2
Insulation	= 0.5
Suspended ceiling (acoustical tile)	= 2.0
Roof DL	= 13.7
	Say RDL = 14.0 psf

Floor Dead Loads

Floor covering (lightweight concrete 1½-in. at 100 lb/ft³)	= 12.5
1⅛-in. plywood (3 psf × 1⅛-in.)	= 3.4
Framing (estimate 4 × 12 at 4 ft-0 in. o.c.) =	2.5
Ceiling supports (2 × 4 at 24 in. o.c.)	= 0.7
Ceiling (½-in. drywall, 5 psf × ½-in.)	= 2.5
Floor DL	= 21.6 psf
Movable partition load*	= 20.0
	= 41.6 psf
	Say FDL = 42.0 psf

* Uniform partition loads are required when the location of partitions is subject to change.

long run to stand back several times during the design process and ask, "Does this figure seem reasonable compared with typical values for other similar structures?"

Before moving on to another type of loading, the concept of the *tributary area* of a member should be explained. The area that is assumed to load a given member is known as the tributary area. For a beam or girder, this area can be calculated by multiplying the *tributary width* times the span of the member. See Example 2.2.

When the load to a member is uniformly distributed, the load per foot can readily be determined by taking the unit load in psf times the tributary width (lb/ft² × ft = lb/ft).

The concept of tributary area will play an important role in the calculation of many types of loads.

2.3 Vertical Loads—Live

Vertical live loads are those gravity loads which are not permanently applied, such as people, furniture, construction workers, construction equipment, contents, and so on. Building codes typically specify the minimum roof and floor live loads that must be used in the design of a structure. For example, UBC Table 23A specifies unit floor live loads in psf for use in the design of floor systems. The Code design requirements are usually thought to represent minimum performance criteria. Thus, if the designer has knowledge that the

---

EXAMPLE 2.2   Tributary Areas

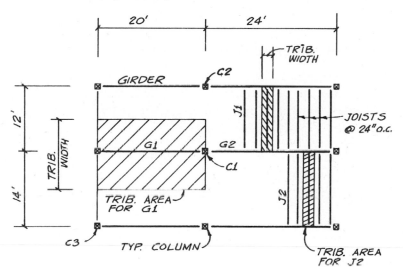

ROOF or FLOOR FRAMING PLAN

Fig. 2.2

In many cases a uniform spacing of members is used throughout the framing plan. This example is designed to illustrate the *concept* of tributary areas rather than typical framing layouts. See Fig. 2.2.

Tributary Area Calculations

$$\text{Trib. } A = \text{trib. width} \times \text{span}$$

Joist J1	Trib. $A = 2 \times 12 = 24$ ft²
Joist J2	Trib. $A = 2 \times 14 = 28$ ft²
Girder G1	Trib. $A = (1\frac{3}{2} + 1\frac{1}{2})20 = 260$ ft²
Girder G2	Trib. $A = (1\frac{3}{2} + 1\frac{1}{2})24 = 312$ ft²
Column C1	Trib. $A = (1\frac{3}{2} + 1\frac{1}{2})(20\frac{1}{2} + 24\frac{1}{2}) = 286$ ft²
Exterior column C2	Trib. $A = (1\frac{3}{2})(20\frac{1}{2} + 24\frac{1}{2}) = 132$ ft²
Corner column C3	Trib. $A = (1\frac{1}{2})(20\frac{1}{2}) = 70$ ft²

15

actual load will exceed these minimum values, the higher values must be used.

In the case of both roof live loads and floor live loads, the *tributary area of the member under design consideration* is taken into account. The idea that the tributary area should be considered in determining the magnitude of the *unit* live load (psf)—not just *total* load—is as follows:

If a member has a small tributary area, it is likely that a fairly high *unit* live load could be imposed over that relatively small surface area.

On the other hand, as the tributary area becomes large, it is less likely that this large area will be uniformly loaded by the same high unit load considered in the design of a member with a small tributary area.

Therefore the consideration of the tributary area in determining the *unit* live load has to do with the probability that high unit loads are likely to occur over small areas, but that these unit loads will probably not occur over large areas.

The Code considers the tributary area of the *member being designed* when specifying both roof and floor design live loads. The approach, however, is somewhat different. In the case of roof live loads, the Code takes the tributary area directly into account when specifying the live load. For floor live loads, the Code specifies the *basic* unit load depending on the occupancy of the structure (i.e., what it houses). The determination of the floor live load as a function of the tributary area is then left to the designer (the Code, however, provides the method to be used in the determination of this load).

2.4   Roof Live Load

The Code specifies minimum unit live loads that are to be used in the design of a roof system. The live load on a roof is usually applied for a relatively short period of time during the life of a structure. This fact is normally of no concern in the design of structures other than wood buildings. However, as will be shown in subsequent chapters, the length of time for which a load is applied to a *wood structure* does have an effect on the load capacity.

Roof live loads are specified to account for the miscellaneous loads that may occur on a roof. These include loads that are imposed during the construction of the building including the roofing process. Roof live loads that may occur after construction include reroofing operations, air-conditioning and mechanical equipment installation and servicing, and, perhaps, loads caused by fire-fighting equipment. Wind loads and snow loads are not normally classified as live loads, and they are covered separately.

Unit roof live loads can be obtained from UBC Table 23C. Method 1 in this table gives directly the unit roof live loads by taking into account the tributary area of the member being designed. The larger the tributary area, the lower the unit roof live load. Therefore, each time the design of a new member is undertaken, the first step should be the calculation of the tributary area that the member is assumed to support.

Method 2 in UBC Table 23C is a more recent addition to the Code and allows the designer to *calculate* the unit roof live load as a function of tributary area. This approach provides a continuous range of live loads whereas Method 1 provides incremental changes in live loads. These two methods are independent and are not to be combined. Method 2 calculates the percent reduction in the basic roof live load as the smallest of the following three values:

1. $R = r(A - 150)$

2. $R = 23.1 \left(1 + \dfrac{DL}{RLL}\right)$

3. $R =$ maximum permitted reduction given in UBC Table 23C

where $R =$ reduction in percent
$r =$ rate of reduction (percent per square foot over 150 ft²) given in UBC Table 23C
$A =$ tributary area of the roof member under consideration
$DL =$ roof dead load
$RLL =$ tabulated roof live load from UBC Table 23C

EXAMPLE 2.3   Calculation of Live Roof Loads

Determine the uniformly distributed roof loads for the members in the building shown in Fig. 2.3, assuming that the roof is flat (minimum slope for drainage only). Roof DL = 8 psf.

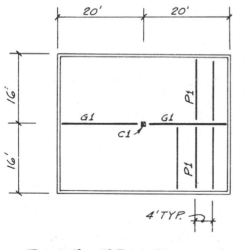

ROOF FRAMING PLAN

Fig. 2.3

Tributary Areas

Purlin	Pl:	$A = 4 \times 16 = 64$ ft²
Girder	Gl:	$A = 16 \times 20 = 320$
Column	Cl:	$A = 16 \times 20 = 320$

UBC Table 23C

METHOD 1:

a. Purlin.

$$\text{Trib. } A = 64 \text{ ft}^2 < 200$$
$$\therefore \text{RLL} = 20 \text{ psf}$$
$$w = (\text{DL} + \text{LL}) \text{ (trib. width)}$$
$$= (8 + 20)4 = 112 \text{ lb/ft}$$

b. Girder.

$$\text{Trib. } A = 320; \ 200 < 320 < 600$$
$$\therefore \text{RLL} = 16 \text{ psf}$$
$$w = (8 + 16)16 = 384 \text{ lb/ft}$$

c. Column.

$$\text{Trib. } A = 320; \text{ same as girder}$$
$$\therefore \text{RLL} = 16 \text{ psf}$$
$$P = (8 + 16)320 = 7680 \text{ lb}$$

METHOD 2:

a. Purlin.

$$\text{Trib. } A = 64 \text{ ft}^2 < 150$$
$$\therefore \text{ reduction not allowed}$$
$$w = (\text{DL} + \text{LL}) \text{ (trib. width)}$$
$$= (8 + 20)4 = 112 \text{ lb/ft}$$

b. Girder.

$$\text{Trib. } A = 320 > 150$$
$$R = r(A - 150)$$
$$= 0.08(320 - 150)$$
$$= 13.6 \text{ percent} \qquad \text{(governs)}$$

$$R = 23.1(1 + \text{DL/LL})$$
$$= 23.1(1 + 8/20) = 32.3 \text{ percent}$$

$$R = 40 \text{ percent (UBC Table 23C)}$$

The smallest of the three values is taken as the reduction of RLL:
$$\text{RLL} = 20(1.00 - 0.136)$$
$$= 17.3 \text{ psf}$$
$$w = (8 + 17.3)16 = 404 \text{ lb/ft}$$

c. Column.

$$\text{Trib. } A = 320 \qquad \text{same as girder}$$
$$\therefore \text{RLL} = 17.3 \text{ psf}$$
$$P = (8 + 17.3)320 = 8090 \text{ lb}$$

A second criterion that UBC Table 23C uses in establishing roof live loads is the slope or pitch of the roof. Consideration of roof slope follows similar reasoning to that which is used in the case of tributary areas. In evaluating the effects of slope, it is reasoned that on a roof that is relatively flat, fairly high *unit* live loads are likely to occur. Conversely, on steeply pitched roofs, the unit live load to be expected will be much less. Both Method 1 and Method 2 take roof slope into account. An example will illustrate the application of both methods. See Example 2.3.

EXAMPLE 2.4 Combined DL + LL on Sloping Roof

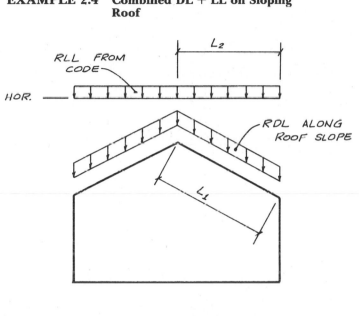

END ELEVATION

Fig. 2.4

The total roof load (DL + LL) can be obtained either as a distributed load along the roof slope or as a load on a horizontal plane. The lengths (L_1 and L_2) on which the loads are applied must be considered.

Equivalent total roof loads (DL + LL):
  Load on horizontal plane:

$$w_{TL} = w_{DL}(L_1/L_2) + w_{LL}$$

  Load along roof slope:

$$w_{TL} = w_{DL} + w_{LL}(L_2/L_1)$$

It should be pointed out that the unit live loads specified in the Code are applied on a horizontal plane. Therefore, roof live loads on a flat roof can be added directly to the roof dead load. In the case of a sloping roof, the dead load would probably be estimated along the sloping roof; the roof live load, however, would be on a horizontal plane. In order to be added together, the roof dead load or live load must be converted to a load along a length consistent with the load to which it is added. Note that both the dead load and the live load are gravity loads and they both, therefore, are *vertical* (not inclined) vector resultant forces. See Example 2.4.

2.5 Snow Load

Snow load is another type of gravity load that primarily loads roof structures. In addition, certain types of floor systems, including balconies and decks, may be subjected to snow loads.

The magnitude of snow loads can vary greatly over a relatively small geographical area. For this reason the UBC simply refers the designer to the local building official for the design snow load.

As an example of how snow loads can vary, the design snow load in a certain mountainous area of southern California is 100 psf, but approximately 5 miles away at the same elevation, the snow load is only 50 psf. This emphasizes the need to be aware of local conditions.

Snow loads can be extremely large. For example, a basic snow load of 240 psf is required in an area near Lake Tahoe. It should be noted that the specified snow loads are on a horizontal plane (similar to roof live loads). Unit snow loads (psf), however, are not subject to the tributary area reductions that can be used for roof live loads.

The slope of the roof has a substantial effect on the magnitude of the design snow load. The Code provides a method by which the basic snow load, obtained from the local building official, may be reduced, depending on the slope of the roof. No reduction is allowed for roofs with slopes less than 20 degrees or for snow loads of 20 psf or less. The UBC provides the following method of reducing the design snow load

$$R_s = \frac{SL}{40} - \frac{1}{2}$$

where R_s = reduction in snow load in psf per degree of roof slope over 20 degrees

SL = total snow load in psf

The use of this reduction and the combination of loads on sloping and horizontal planes is illustrated in Example 2.5. This example also illustrates the effects of using a load on a horizontal plane in design calculations.

The Code requires that the potential accumulation of snow at roof valleys, parapets, and other uneven roof configurations be considered. The method of accounting for this buildup of snow is not specified by the UBC.

EXAMPLE 2.5   Reduction of Snow Loads

Determine the total design dead load plus snow load for the rafters in the building shown in Fig. 2.5a. Determine the design shear and moment for the rafters if they are spaced 4 ft-0 in. o.c. Roof DL has been estimated as 10 psf along the roof, and the basic snow load is given as 75 psf on a horizontal plane.

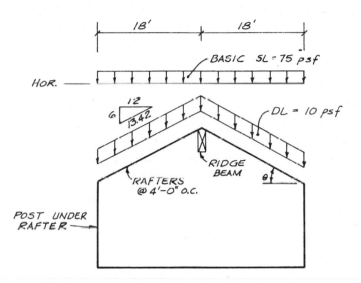

$$TYPICAL \quad SECTION$$

Fig. 2.5a

Snow Load

Check reduction:

$$\text{Roof slope} = \theta = \tan^{-1}\left(\frac{6}{12}\right) = 26.6 \text{ degrees} > 20 \text{ degrees}$$

∴ Reduction can be used.

$$R_s = \frac{SL}{40} - \frac{1}{2} = \frac{75}{40} - \frac{1}{2}$$

$$= 1.38 \text{ psf/degree slope over 20 degrees}$$

Reduced snow load:

$$SL = 75 - 1.38(26.6 - 20) \approx 66 \text{ psf}$$

Total Loads

Total loads can be calculated either along the roof or on a horizontal plane. Both methods are shown.

21

Along roof (left rafter):

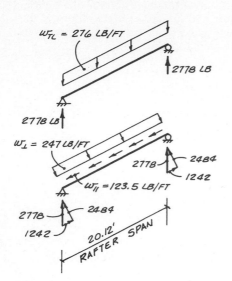

Horiz. plane (right rafter):

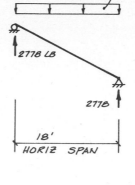

Fig. 2.5b

TL = DL + SL

$$= 10 + 66 \left(\frac{18}{20.12}\right)$$

$$= 69 \text{ psf}$$

$$w = 69 \times 4 = 276 \text{ lb/ft}$$

Use component of total load that is normal to roof and rafter span parallel to roof.

$$V = \frac{wL}{2} = \frac{0.247(20.12)}{2}$$

$$= 2.48 \text{ k}$$

$$M = \frac{wL^2}{8} = \frac{0.247(20.12)^2}{8}$$

$$= 12.5 \text{ ft-k}$$

TL = DL + SL

$$= 10 \left(\frac{20.12}{18}\right) + 66$$

$$= 77.2 \text{ psf}$$

$$w = 77.2 \times 4 \approx 309 \text{ lb/ft}$$

Use total vertical load and the projected horizontal span.

$$V = \frac{wL}{2} = \frac{0.309(18)}{2}$$

$$= 2.78 \text{ k} \qquad \text{(conservative)}$$

$$M = \frac{wL^2}{8} = \frac{0.309(18)^2}{8}$$

$$= 12.5 \text{ ft-k}$$

NOTE: The horizontal plane method is commonly used in practice to calculate design values for sloping beams such as rafters. This approach is convenient and gives equivalent design moments and conservative values for shear compared with the inclined span analysis. (By definition *shear* is an internal force *perpendicular* to the longitudinal axis of a beam. Therefore, the calculation for shear for the left rafter in this example is theoretically correct.)

Reference 34 provides comprehensive information on design snow loads, including a different method for accounting for snow load reductions for roof slope *and* a comprehensive treatment of increased snow loads as a function of roof configuration. The *Basic Building Code* and the *Standard Building Code* use snow load criteria based on ANSI A58.1.

2.6 Floor Live Loads

As noted earlier, floor live loads are specified in UBC Table 23A. These loads are based on the occupancy or use of the building. Typical *occupancy or use* floor live loads range from a minimum of 40 psf for residential structures to much larger values, say 250 psf for heavy storage facilities.

These Code unit live loads are for members supporting small tributary areas. A small tributary area is defined as an area of 150 ft² or less. From the previous discussion of tributary areas, it will be remembered that the magnitude of the *unit* live load can be reduced as the size of the tributary area increases.

It should be pointed out that no reduction is permitted where live loads exceed 100 psf or in areas of public assembly. Reductions are not allowed in these cases because an added measure of safety is desired in these critical structures. In warehouses with high storage loads, and in areas of public assembly (especially in emergency situations), it is possible for high unit loads to be distributed over large surface areas.

However, for the majority of wood-framed structures, reductions in floor live loads will be allowed. For these structures, the smallest value of R as given by the following three criteria represents the percent reduction in floor live load:

1. $R = r(A - 150)$

2. $R = 23.1 \left(1 + \dfrac{DL}{FLL}\right)$

3. $R = 40$ percent maximum for horizontal members or vertical members receiving load from one level only

    $R = 60$ percent maximum for other vertical members

where r = reduction rate equal to 0.08 percent per square foot of tributary floor area over 150 ft²

A = tributary floor area of the member under consideration

DL = floor dead load

FLL = tabulated floor live load from UBC Table 23A

The calculation of reduced floor live loads is illustrated in Example 2.6.

In addition to basic floor unit live loads, UBC Table 23A also provides special alternate concentrated floor loads. Whichever combination of loads, (DL + FLL) or (DL + special concentrated load), produces the more critical situation is the one which is to be used in sizing the framing members.

23

EXAMPLE 2.6   Reduction of Floor Live Loads

Determine the total axial force required for the design of the interior column in the floor framing plan shown in Fig. 2.6. The structure is an apartment building with a floor DL of 10 psf and, from UBC Table 23A, a tabulated floor live load of 40 psf. Assume roof loads are not part of this problem, and the load is received from one level.

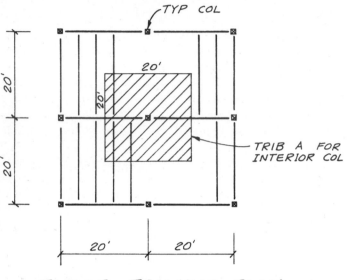

FLOOR FRAMING PLAN

Fig. 2.6

Floor Live Load

$$\text{Trib } A = 20 \times 20 = 400 \text{ ft}^2 > 150$$

∴ FLL can be reduced:

$$R = r(A - 150) = 0.08(400 - 150)$$
$$= 20 \text{ percent} \qquad \text{(governs)}$$

$$R = 23.1 \left(1 + \frac{DL}{FLL}\right) = 23.1 \left(1 + \frac{10}{40}\right)$$
$$= 28.9 \text{ percent}$$

$$R = 40 \text{ percent} \qquad \text{(vertical members)}$$

The smallest of the three values is taken as the reduction of FLL:

$$FLL = 40(1.00 - 0.20) = 32 \text{ psf}$$

Total Load

$$TL = DL + FLL = 10 + 32 = 42 \text{ psf}$$
$$P = 42 \times 400 = 16.8 \text{ k}$$

Concentrated floor loads can be distributed over an area 2½ ft square. Their purpose is to account for miscellaneous nonstationary equipment loads which may occur. In sizing framing members, it will be found that the majority of the designs will be governed by the uniform loads. However, both the concentrated loads and the uniform loads should be checked.

2.7 Deflection Criteria

The above discussion of gravity loads provides a general summary of vertical design load criteria. Certainly any special loads in addition to those mentioned must be taken into account by the designer. Refer to UBC Table 23B for special loads required by the Code.

The Code establishes deflection limitations that are not to be exceeded under certain gravity loads. The deflection criteria is given in UBC Tables 23D and 23E and applies to floor members and roof members that support plastered ceilings. The same limitations apply to both types of members. These fairly restrictive deflection limits are to ensure user comfort and to prevent excessive cracking of plaster ceilings.

The question of user comfort is tied directly to the confidence that occupants have regarding the safety of a structure. It is possible for a structure to be very safe with respect to satisfying stress limitations, but it may deflect under load to such an extent as to render it unsatisfactory.

Excessive deflections can occur under a variety of loading conditions. For example, user comfort is essentially related to deflection caused by live loads only. The Code therefore requires that the deflection under live load be calculated. This deflection should be less than or equal to the span length divided by 360 ($\Delta_{LL} \leq L/360$).

Another loading condition that relates more to the cracking of plaster and the creation of an unpleasant visual situation is that of total load deflection (i.e., dead load plus live load). For this case, the actual deflection is controlled by the limit of the span divided by 240 ($\Delta_{(KDL+LL)} \leq L/240$).

Notice that in the second criterion above, the calculated deflection is to be under K times the DL plus the LL. K in this calculation is the Code's attempt to reflect the tendency of various structural materials to creep under sustained load.

For steel members, K is taken as zero, which indicates no tendency to creep (at least under normal temperature conditions). Reinforced concrete, on the other hand, can creep, but this behavior is reduced by the presence of compression steel in bending members. K is defined by an expression in UBC Table 23E that is a function of the amount of compression steel in relation to the amount of tension steel. Finally, the tendency of wood beams to creep is affected by the moisture content (see Chap. 4) of the member. The dryer the member, the less the deflection under sustained load. Thus, for seasoned lumber, a K factor of 0.5 is used; for unseasoned wood, K is

25

taken as 1.0. Seasoned lumber here is defined as wood having a moisture content of less than 16 percent at the time of construction, and it is further assumed that the wood will be subjected to dry conditions of use (as in most covered structures). Although the K factor is included in the Code, many designers take a conservative approach and simply use the full dead load (i.e., $K = 1.0$) in the check for deflection under DL + LL in wood beams.

The question of what deflection limitations should be applied to members not covered by the Code criteria should also be considered. For example, what deflection limits should be used in checking roof members that do not support plastered ceilings?

In the *Timber Construction Manual* (Ref. 3), the American Institute of Timber Construction (AITC) recommends that the live load deflection be limited to $L/240$ and that the total load deflection be limited to $L/180$. These larger allowable deflections seem reasonable for this less critical situation. The deflection of members in other possible critical situations should be evaluated by the designer. Members over large glazed areas and members which affect the alignment or operation of special equipment are examples of two such potential problems.

In recent years, there has been an increasing concern about the failure of roof systems associated with excessive deflections on *flat* roof structures caused by the entrapment of water. This type of failure is known as a ponding failure, and represents a progressive collapse caused by the accumulation of water on a flat roof. The initial deflection allows water to become trapped. This trapped water, in turn, causes additional deflection. A vicious circle is generated which can lead to failure if the roof structure is too flexible.

Ponding failures may be prevented by proper design. The simplest method is to provide adequate drainage together with a positive slope (even on essentially flat roofs) so that an initial accumulation of water is simply not possible. AITC recommends a minimum roof slope of ¼ in./ft to guard against ponding. An adequate number and size of roof drains must be provided to carry off this water unless, of course, no obstructions are present.

However, in lieu of providing the minimum ¼-in./ft slope, ponding can be prevented by designing a sufficiently stiff and strong roof structure so that water cannot accumulate in sufficient quantities to cause a progressive failure. This is accomplished by imposing additional deflection criteria and by designing for increased stresses and deflections. These increased stresses are obtained by multiplying calculated actual stresses under service loads by a magnification factor. The magnification factor is a number greater than 1.0 and is a measure of the sensitivity of a roof structure to accumulated (pond) water. It is a function of the total design roof load (DL + LL) and the weight of ponding water.

Because the first method of preventing ponding is the more direct, positive, and less costly method, it is recommended for most typical designs. Where the minimum slope cannot be provided for drainage, the roof structure should be designed, as described above, for ponding. Because this latter approach is not the more common solution, the specific design criteria are not included here. The designer is referred to Ref. 3 for these criteria and a detailed example of their application.

There are several methods that can be used in obtaining the recommended ¼-in./ft roof slope. The most obvious solution is to place the supports for framing members at different elevations. These support elevations (or the *top of sheathing,* abbreviated TS, elevations) should be clearly shown on the roof plan.

A second method which can be used in the case of glulam construction is to provide *additional* camber (see Chap. 5) so that the ¼-in./ft slope is built into supporting members. It should be emphasized that this slope camber is in addition to the camber provided to account for long-term (dead-load) deflection.

2.8 Lateral Loads

The subject of lateral loads can easily fill several volumes. Wind and seismic loads have been the topics of countless research projects, and complete texts deal with the calculation of these loads. Interest in the design for earthquake effects has increased substantially in light of the experience obtained in the San Fernando earthquake of 1971 and other well-documented earthquakes.

The design criteria included in the Code will be summarized in the remainder of this chapter. The calculation of lateral loads for typical buildings using shearwalls (box systems) is covered in Chap. 3.

Some consideration should be given as to what loads will act concurrently. For example, it is extremely unlikely that the maximum seismic load and the maximum wind load will act simultaneously. Consequently, the Code simply requires that the horizontal seismic load *or* the wind load be used (in combination with other appropriate loads) in design. Of course, the loading which creates the more critical condition is the one which must be used.

Similarly, the Code does not require that roof live loads (loads which act relatively infrequently) be considered simultaneously with wind or seismic. In areas subjected to snow loads, all or part (depending on local conditions) of the snow load must be considered simultaneously with lateral loads.

2.9 Wind Loads

The UBC takes a simplified, but normally adequate, approach in accounting for wind load effects. The Code establishes *basic* horizontal wind pressures for various locations in the country (UBC Chap. 23, Fig. 4). In UBC Table 23F, this *basic* wind pressure is adjusted for various height zones. The height zones are: less than 30 feet, 30 to 49 feet, 50 to 99 feet, and so on. See Example 2.7.

EXAMPLE 2.7   Wind Loads

Sketch the wind pressure diagram for the building in Fig. 2.7 if the *basic* wind pressure is 20 psf (obtained from UBC Chap. 23, Fig. 4).

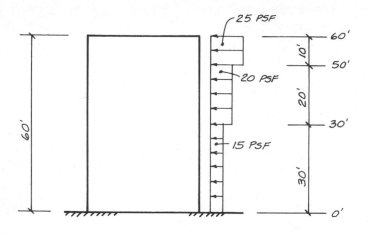

ELEVATION

Fig. 2.7

The adjustment for height zones is given in UBC Table 23F. A step loading results.
   NOTE: Most wood buildings are considerably less than 60 ft in height. The above building is used simply to illustrate the Code loading criteria.

In certain areas, high local winds may not be adequately taken into account by the Code's *basic* wind loading. In these circumstances, the designer may want to arbitrarily increase the magnitude of the design wind pressure, or the local building official may require that certain areas use higher wind loads. For example, some isolated areas of Southern California experience *locally* high "Santa Ana" winds, and the local building department should be consulted for the design wind loads in these areas.

   The wind loads described above can act either entirely as a pressure on the windward side of the building or as a suction on the leeward side, or the load may be divided in some proportion between the windward and the leeward sides. The critical condition in the above distribution must be used in the design. The horizontal wind pressure is applied to the horizontal projection of the building on a vertical plane. See Example 2.8.

28

EXAMPLE 2.8   Wind on Sloping Roofs

The building in this example is less than 30 ft in height. The *basic* wind pressure is 30 psf and is adjusted to 25 psf by UBC Table 23F.

Wind Load for Lateral Force System

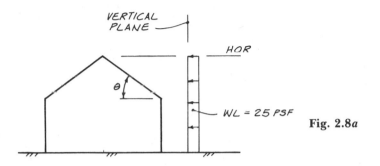

Fig. 2.8*a*

For lateral force design of a building, use the horizontal load applied to the projected profile of the building (Fig. 2.8*a*) regardless of the slope of the roof θ.

Wind Load for Roof Structure

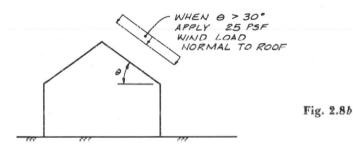

Fig. 2.8*b*

In high-wind-load areas the roof structure may be critical under wind loads. When roof slope θ exceeds 30 degrees, apply wind load perpendicular to roof slope (Fig. 2.8*b*). When θ ≤ 30 degrees, apply horizontal wind pressure to projection of roof as in Fig. 2.8*a*.

In the design of roofs, buildings with mildly sloping roofs simply use the *horizontal* load. However, when the roof slope exceeds 30 degrees, the roof must be designed to withstand an inward wind pressure acting *normal* to the roof. The pressure to be used is the same as the horizontal pressure for the height zone under consideration.

In addition to the wind pressures discussed above, the question of uplift may be important. The term *uplift* is used to describe two different situations.

One use of the term uplift has to do with a wind force that acts normal to the surface of a roof (see Example 2.9). As the wind passes over a structure, a suction force can be generated. The magnitude of this uplift load depends on a number of factors, but the UBC provides a very simple design value. The uplift is taken as a percentage of the horizontal wind pressure. The

---

EXAMPLE 2.9   Wind Uplift Forces

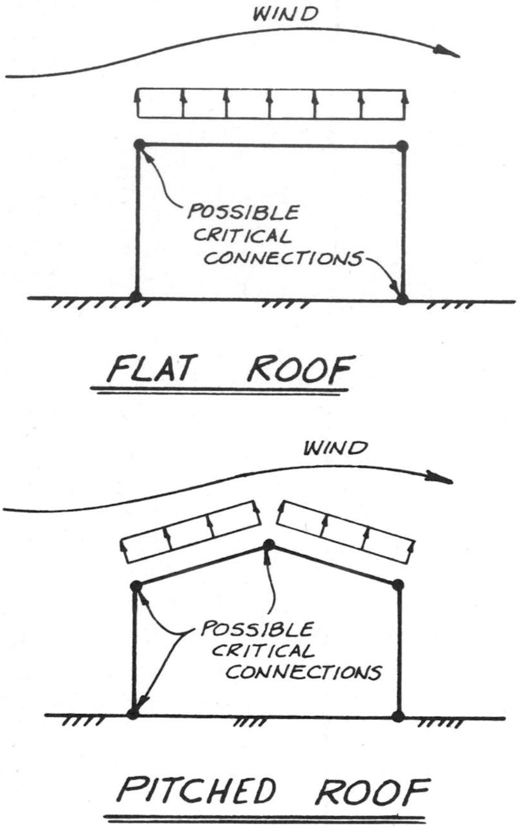

Fig. 2.9

Wind passing over a building can create a suction known as uplift. Uplift forces are applied over the entire roof area and act perpendicular to the surface of the roof.

Uplift Forces
Enclosed structure:

$$\text{Uplift} = 0.75 \times \text{Horizontal wind load}$$

Unenclosed structure:

$$\text{Uplift} = 1.25 \times \text{Horizontal wind load}$$

percentage depends on whether the building is enclosed or unenclosed. An enclosed structure is one that has a perimeter of solid walls. These "solid" walls can have openings, but openings must be protected by door or window assemblies. The uplift force for an unenclosed structure is applied to the overhangs of an enclosed building.

If the dead load of the roof structure exceeds this uplift force, little is required in the way of design for uplift. However, in the cases of unenclosed structures and structures with light dead loads (these often go hand in hand), design for uplift may affect member sizes. Connections and footing sizes are the items that typically require special consideration even if member sizes are not affected. For example, connections are normally designed for gravity (vertically downward) loads. For high uplift loads, the connection may need to be modified to act in tension. It may be necessary to connect a roof beam to a column, or a column to a footing, to transmit the net uplift force from the member on top to the supporting member below. In fact, it may be necessary to size column footings to provide an adequate dead load to counter the uplift.

The second use of the term uplift relates to the moment stability of the structure when subjected to horizontal loads. The net overturning moment OM is the difference between the gross OM and the resisting moment RM. See Example 2.10.

The Code requires that ⅔RM be greater than the OM. In other words, a factor of safety FS of ³⁄₂ or 1.5 is required for stability for wind. Notice that in this stability check, an overestimation of dead load tends to be unconservative (normally an overestimation of loading is considered conservative). In order to obtain the design OM, two-thirds of the RM is subtracted from the gross OM. It should also be pointed out that up to this point, the DL being used in the calculation of RM does not include the weight of the foundation.

Now, if the design OM is a positive value (i.e., the gross OM is more than ⅔RM), the structure will have to be tied to the foundation. The design OM can be replaced by a couple (T and C). The tension force T must be developed by the connection to the foundation. This tension force is also known as the design *uplift* force. If the design OM is negative (i.e., the gross OM is less than or equal to ⅔RM), there would be no uplift problem. Should an uplift problem occur, the DL of the foundation plus the DL of the building must be sufficient to counteract the gross OM.

The preceding discussion of overturning and the required factor of safety of 1.5 for stability applies to lateral wind loading. A similar analysis must be used for lateral seismic loading. The UBC does not require a FS of 1.5 for overturning for seismic loads, but it is conservative to use this factor of safety in design. For uniformity and ease of application, a FS of 1.5 will be used in this book for overturning due to *wind* or *seismic* loads.

One other point about wind loading should be noted at this time. The Code permits allowable stresses to be increased by a factor of one-third if

EXAMPLE 2.10   Uplift—Overall Moment Stability

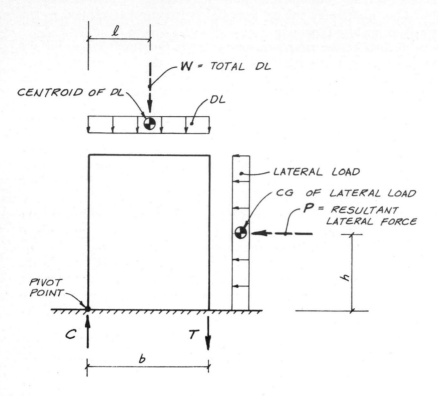

SHEARWALL

Fig. 2.10

$$\text{Gross overturning moment} = \text{OM} = P(h)$$
$$\text{Resisting moment} = \text{RM} = W(l)$$

Required factor of safety for overall stability:

$$\text{Req'd FS} = \tfrac{3}{2} = 1.5$$

∴ For no uplift force T the following criteria must be satisfied:

$$\text{Gross OM} \leqslant \tfrac{2}{3}\text{RM}$$

If this criterion is not satisfied, the net OM is the difference between the gross OM and the RM:

$$\text{Net OM} = \text{gross OM} - \text{RM}$$

The design OM, however, must reflect the required FS and is obtained as follows:

$$\text{Design OM} = \text{gross OM} - \tfrac{2}{3}\text{RM}$$

This moment can then be resolved into a couple (T and C):

$$\text{Uplift force} = T = \frac{\text{design OM}}{b}$$

The design uplift force T is to be used for the design of the connection of the shearwall to the foundation. A subsequent stability check which includes the foundation weight in the resisting moment must satisfy the criterion

$$\text{Gross OM} \leqslant \tfrac{2}{3}\text{RM}$$

---

the stresses are caused by wind (or seismic) loads acting alone, or in combination with other loads (such as dead load or floor live load).

This increase is mentioned here because it is permitted for all structural materials (wood, steel, concrete, masonry, and soil). Other allowable stress increases based on the duration of loading are applied to wood only, and are therefore covered in later chapters which deal specifically with wood.

The logic behind the one-third increase in allowable load values for the loading conditions described is that these are basically short-term loads. It is reasoned that loads of this nature are not as critical as those that are sustained for long periods of time. A somewhat lower factor of safety is the Code's method of taking this less critical condition into account. (Actually in wood structures, the factor of safety remains constant, because wood has the unique ability to support higher loads for short periods of time. See Chap. 4.)

The UBC is currently considering a possible revision to the wind-loading requirements. A more comprehensive method for calculating design wind loads may be found in Refs. 3 and 34.

2.10 Seismic Loads—Introduction

Courses in structural dynamics and earthquake engineering deal at length with the subject of seismic loads. The purpose of Secs. 2.10 to 2.13 is to introduce some basic concepts concerning this subject, and to review the Code seismic design criteria.

The treatment of seismic loads given here is somewhat lengthy in comparison with the discussion of other types of loads. Many designers have a good understanding of gravity and wind loads, but the loads created in an earthquake may not be as widely known. For this reason a fairly complete introduction to seismic loading is included.

Techniques are available for the theoretical analysis of structures subjected to ground motion records from previous earthquakes. These solutions, however, are typically reserved for large, multistory buildings, and are not applied to small shearwall-type buildings. For this reason, rectangular box-type structures (defined in UBC Chap. 23) are usually designed according to empirical Code seismic loads.

The Structural Engineers Association of California (SEAOC) pioneered the work in this area, and their lateral force requirements (Ref. 26) have been adopted by most of the model building codes. These design loads have undergone major revisions in recent years, and they are included in the 1979 UBC.

From structural dynamics it is known that a number of different forces act on a structure during an earthquake. These forces include inertia forces, damping forces, elastic forces, and an equivalent forcing function (mass times ground acceleration). The theoretical solution involves the addition of individual responses of a number of "modes" of vibration. Each mode is described by an equation of motion which includes a term reflecting each of the forces mentioned above.

Rather than attempting to define all of the forces acting during an earthquake, the Code takes a simplified approach, and one that is particularly easy to visualize. The earthquake force is treated as an *inertial* problem only. Before the start of an earthquake, a building is in static equilibrium (i.e., it is at rest). Suddenly, the ground moves and the structure attempts to remain stationary. The key to the problem is, of course, the length of time during which the movement takes place. If the ground displacement takes place very slowly, the structure would simply ride along quite peacefully. However, because the ground movement occurs quickly, the structure lags behind and "seismic" forces are generated. See Example 2.11.

These equivalent static loads are applied at the story levels of the building (i.e., at roof and floor elevations). It should be noted that no such simplified loads are truly "equivalent" to the complicated combination of forces generated during an earthquake. However, it is felt that the Code design loads give *reasonable* building designs when the structure is designed to *elastically* resist the specified Code loads. The coefficients in the Code loads reflect many of the important dynamic aspects of the problem.

During an earthquake, vertical ground movement can create vertical loads in addition to the horizontal loads discussed above. However, the vertical components are usually smaller than the horizontal, and the structure typically has much more inherent strength vertically than horizontally. For these reasons, normal design practice for earthquakes considers horizontal forces only.

The method used to calculate these Code horizontal story forces is to first calculate the total *base shear* (the total horizontal force acting at the base of the building, V). Once the base shear has been determined, the appropriate percentages of this total force are calculated and assigned to the various story levels throughout the height of the structure. These story forces are given the symbol F_x (the force at level x), and a special extra force F_t may be applied at the top level. The sum of the F_x forces and F_t must equal the total base shear, V. See Fig. 2.12a. The formulas used to calculate F_x and F_t are examined in Sec. 2.12.

Before the Code expressions used to calculate these forces are examined, it should be noted that the story forces are shown to increase with increasing

EXAMPLE 2.11   Building Subjected to Earthquake

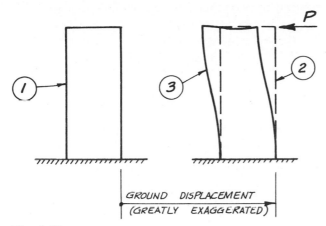

Fig. 2.11

1.   Original static position of the building before earthquake
2.   Position of building if ground displacement occurs very slowly (i.e., in a static manner).
3.   Deflected shape of building because of "dynamic" effects caused by rapid ground displacement

The force P in Fig. 2.11 is an "equivalent static" design load provided by the Code and can be used in lieu of a more complicated dynamic analysis. This is common practice in ordinary wood building design.

height above the base of the building. The magnitude of the story forces depends on the mass (dead load) distribution throughout the height of the structure. However, if the dead load is equally distributed to each story level, the distribution provided by the Code formula for F_x will be essentially triangular (i.e., maximum at the roof level and decreasing linearly to the ground level).

The reason for this distribution is that the Code bases its forces on the fundamental mode of vibration of the structure. The fundamental mode is also known as the first mode of vibration, and it is the significant mode for most structures.

In order to develop a feel for the above distribution, the dynamic model used to theoretically analyze buildings should briefly be discussed. See Figure 2.12b. In this model, the mass (weight) *tributary* to each story is assigned to that level. In other words, the weight of the floor and the tributary wall loads halfway between adjacent floors is assumed to be concentrated or

35

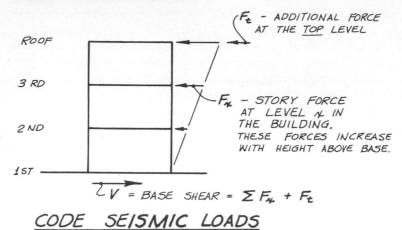

CODE SEISMIC LOADS

Fig. 2.12a Seismic force distribution: Distribution of Code seismic loads.

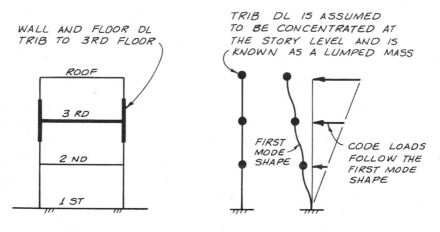

TRIB DL'S   LUMPED MASSES

Fig. 2.12b Seismic force distribution: Code loads follow fundamental mode.

"lumped" at the floor level. In analytical studies, this model simplifies the solution of the dynamic problem.

Now, with lumped mass defined, the concept of a mode shape can be explained. A mode shape is a simple *displacement pattern* that occurs as a structure moves when subjected to a dynamic force. If the appropriate percentages of all of the modes of vibration are added together, the complex motion of the complete structure can be described. The first mode shape is defined as the displacement pattern where *all* lumped masses are on one side of the reference axis. Higher mode shapes will show masses on both sides of the vertical reference axis.

The point of this discussion is to explain why the F_x story forces increase with increasing height above the base. To summarize, the fundamental or first mode is the critical displacement pattern (deflected shape). The first mode shape shows all masses on one side of the vertical reference axis. Greater displacements and accelerations occur *higher* in the structure, and the F_x story forces follow this distribution.

2.11 Seismic Load—Base Shear Calculation

The horizontal base shear is calculated from an expression which is essentially inertial in form:

$$F = Ma = \left(\frac{W}{g}\right) a = W\left(\frac{a}{g}\right)$$

where F = inertia force

M — mass
W = weight
a = acceleration
g = acceleration of gravity

The Code form of this expression is somewhat modified. The (a/g) term is replaced by a "seismic coefficient." The Code formula is

$$V = (ZIKCS)W$$

Each of the terms in the above expression is defined as follows

V = *Base Shear.*   The total horizontal seismic force assumed to act at the base of the structure (Fig. 2.12a).

W = *Weight of Structure.*   The total weight of the structure which is assumed to contribute to the development of seismic forces. For most structures, this weight is simply taken as dead load. However, in structures where a large percentage of the live load is likely to be present at any given time, it is reasonable to include at least a portion of this live load in the value of W. The Code specifies that in storage warehouses, W must include at least 25 percent of the floor live load. Other live loads are not covered specifically by the Code, and the designer must use judgment.

Roof live loads need not be included in the calculation of W, but the Code does require that the snow load be included if it exceeds 30 psf. The local building official may allow some reduction in the amount of snow load that is included in W based on the duration of the snow load. This reduction can be as high as 75 percent.

ZIKCS = *Seismic Coefficient.*   The product of these five coefficients is referred to as the seismic coefficient and represents the (a/g) quantity in the basic inertia expression. Actually, the terms are *judgment coefficients* which reflect a number of the important factors that are known to affect the magnitude of seismic forces.

Z = *Zone Coefficient*.   The zone coefficient accounts for the relative seismic risk of various areas in the country. Four seismic risk zones are defined, and these are shown on a map in the Code (UBC Chap. 23, Figs. 1, 2, and 3). Zones are established from geologic studies and from intensities of previous earthquakes using the modified Mercalli scale. The probable frequency of occurrence is not considered in the assignment of seismic risk zones. The value of Z ranges from a maximum of $Z = 1$ for zone 4 down to a minimum of $Z = \frac{3}{16}$ for zone 1. Obviously zone 4 represents the area of highest seismisity.

I = *Occupancy Importance Coefficient*.   This coefficient has recently been introduced into the base shear expression and is a result of failures which occurred in the San Fernando earthquake. The I coefficient provides that certain *essential facilities* be designed for higher seismic forces than other types of structures.

Essential facilities are defined as those which must remain functional for emergency, postearthquake operations. These include hospitals, certain communication centers, fire-fighting stations, and other buildings intended to house disaster-related services.

For a structure that is classified as an essential facility, $I = 1.5$. Nonessential facilities use $I = 1.0$. Certain other structures that are not defined as "essential," but are important from the standpoint that a large number of people may be present at a single location are assigned an intermediate value for I. For example, an assembly building housing more than 300 occupants is to be designed using $I = 1.25$. Values of I are given in UBC Table 23K.

K = *Framing Coefficient*.   The framing coefficient is a judgment factor that depends on the type of lateral-force-resisting system that is used in the structure. Certain systems have performed better than others in previous earthquakes. K reflects this past performance, and, to some extent, it evaluates the energy-absorbing capability of the system.

The energy-absorbing capacity (damping) is related to the ductility of a structure. Ductility refers to the ability of a structure to deform under load without rupture. Hence, structures that are built with a lateral-force-resisting system that can undergo extensive deformations (especially into the plastic range) are assigned lower framing coefficients.

For example, the minimum value of K is 0.67 and is assigned to structures with ductile moment-resisting space frames. These structures are three-dimensional moment-resisting systems which carry the entire lateral load and must be specially designed to provide ductility (see UBC Sec. 2626 for reinforced concrete and Sec. 2722 for structural steel requirements).

At the other end of the scale is the maximum value of K, which is 1.33 for buildings (other types of structures can have higher values). The Code assigns this maximum value to "box" systems. Box systems resist lateral loads either by shearwall action or by braced frames (i.e., vertical truss action).

For the most part, wood structures, or structures that use wood roof and floor systems, fall into this category.

Intermediate values of K may be used in the design of buildings with other types of framing systems (UBC Table 23I). For example, glulam arch buildings may qualify for a K factor of 1.0 since this type of structure is not specifically covered by other K factors.

However, care should be used in establishing K factors for wood structures other than $K = 1.33$. The problem develops when a wood arch system resists lateral loads in one direction, and shearwalls or diagonal braces are used in the other direction. Some designers quite naturally use $K = 1.0$ for the arch design and $K = 1.33$ for the shearwall or bracing design. The Code does not specifically preclude such an interpretation of its criteria, but the Commentary to the SEAOC *Recommended Lateral Force Requirements and Commentary* (Ref. 26) states that where a K factor of 1.33 is required in one direction, it shall also be used in the other direction. This certainly establishes a more severe design criteria.

$C = $ *Design Response Spectrum Value.* In order to discuss this coefficient, the period of vibration of a structure must be defined. Assume that a one-story building has its mass tributary to the roof level, assigned or "lumped" at that level. See Fig. 2.13.

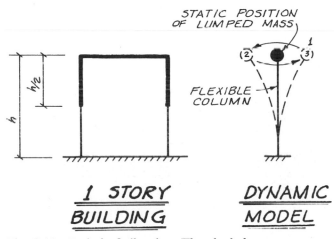

1 STORY BUILDING

DYNAMIC MODEL

Fig. 2.13 Period of vibration. The shaded area represents the tributary wall DL and roof DL, which is assumed to be concentrated at the roof level.

The dynamic model then becomes a flexible column with a single concentrated mass at its top. Now, if the mass is given some initial horizontal displacement (point 1) and then released, it will oscillate back and forth (i.e., from 1 to 2 to 3). This movement with no externally applied load is termed "free vibration." The period of vibration, T, of this structure is defined as the

39

length of time (seconds) that it takes for one complete cycle of free vibration. The period is a characteristic of the structure (a function of mass and stiffness), and it is a value that can be calculated from dynamic theory.

When the multistory building of Fig. 2.12 was discussed (Sec. 2.10), the concept of the fundamental mode of vibration was defined. Characteristic periods are associated with all the modes of vibration. The *fundamental period* can be defined as the length of time (seconds) that it takes for the first or fundamental mode (deflected shape) to undergo one cycle of free vibration (Fig. 2.12*b*). The fundamental period can be calculated from theory, or the Code's simple, normally conservative method of estimating T can be used.

With the concept of period of vibration now defined, the idea of a *response spectrum* can be introduced. In a study of structural dynamics, it has been found that structures which have the same *period* and the same amount of *damping*, have essentially the same response to a given *earthquake record*.

Damping is the resistance to motion provided by the internal friction of building materials as the molecules forming the material are forced across one another. It is a property of the type of building construction and materials used.

**EXAMPLE 2.12   Typical Theoretical Response
Spectrum**

The response spectrum name comes from the fact that *all building periods* are summarized on one graph (for a *given earthquake record* and a *given percentage of critical damping*). Figure 2.14 shows the complete *spectrum* of building periods. The curve shifts upward or downward for different amounts of damping.

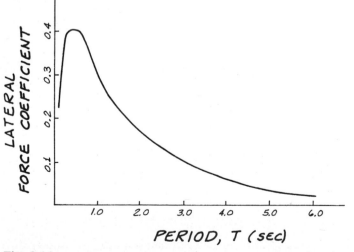

Fig. 2.14

Earthquake records are obtained from strong-motion instruments known as accelerographs which are triggered during an earthquake. Time histories of ground acceleration are obtained and serve as the basis of theoretical solutions. UBC Chap. 23 requires that accelerographs be installed in certain buildings over six stories in height and in all buildings over ten stories in height.

Computer solutions of the dynamic problem for a number of different buildings with various periods are used to generate a response spectrum. A response spectrum is defined as a plot of maximum response (acceleration, velocity, displacement, or equivalent static force) vs. period of vibration. See Example 2.12. Once the response spectrum has been generated by the computer, it can be used to determine the effects of an earthquake on other buildings. The information required to obtain values from the response spectrum is simply the period of the structure.

It should be pointed out that a large number of earthquake records are available and that each record can be used to generate a family of theoretical response spectra for buildings with different damping characteristics. The Code, however, provides only *one response spectrum curve* and takes the question of *damping* into account in the *framing coefficient K*. The code response spectrum value C is given by the expression $C = 1/15\sqrt{T}$ and need not exceed 0.12. See Fig. 2.15.

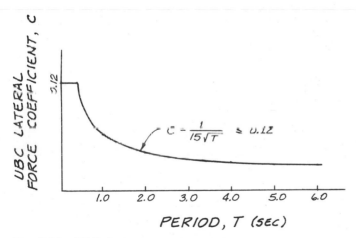

Fig. 2.15 UBC design response spectrum.

The simplified Code formulas for estimating the period are as follows:

$T = 0.10N$ for buildings with ductile moment-resisting frames

$T = \dfrac{0.05 h_n}{\sqrt{D}}$ for other types of buildings

41

where N = number of stories

h_n = height above base to nth or uppermost level (ft)

D = dimension of structure parallel to the applied forces (ft)

These estimates are, in most cases, conservative values of the fundamental period. From Fig. 2.15 it can be seen that a conservative value of period would underestimate the actual period. In other words, within limits, the shorter the period, the greater the response. The theoretical response spectra curve of Fig. 2.14 shows a definite hump or resonance effect that occurs where the period of the structure and the period of the earthquake are close. The Code design curve, however, makes no attempt to consider a decreased response for a structure with a very short period.

S = *Site—Structure Resonance Coefficient.* This is another coefficient that was recently included in the base shear expression. The S coefficient is the result of studies that have shown a significant influence of local soil conditions on the seismic motion that reaches a structure on the surface of the earth. Local soil layers may amplify the response of a building to earthquake motions originating in bedrock.

This amplification is taken into account by considering the geotechnical profile (soil layers) beneath the structure. Soil layers to bedrock, or to a depth of 500 ft (maximum), are considered.

It is perhaps more difficult to visualize, but a number of soil layers have a characteristic period T_s similar to the period of vibration of a structure, T. Greater structural damage is likely to occur when the fundamental period of the structure is close to the period of the underlying soil. Thus, buildings with short periods tend to suffer greater damage when they are supported on shallow, short-period soil deposits. Conversely, tall multistory buildings tend to suffer greater damage when they are located on long-period deposits. In these cases a quasi-resonance effect between the structure and the underlying soil develops. See Example 2.13.

The site—structure resonance coefficient takes the above behavior into account by specifying a large value of S when T and T_s are close. On the other hand, a small value of S is specified when T and T_s are substantially different. See Fig. 2.16b. As can be seen, S lies in the range

$$1.0 \leq S \leq 1.5$$

Two different expressions are used to define S, depending on the value of T/T_s.

In the calculation of S, T must be substantiated by calculations or comparisons, and T_s must be obtained from a soils (geotechnical) report. Without substantiation, the maximum value of 1.5 for S can be used for any structure, and this will be the approach taken in this book.

CS = *Combined Response Spectrum and Site—Structure Resonance Coefficient.* In order that the combined effect of C times S not become excessively large, the Code provides that the product CS need not exceed 0.14.

EXAMPLE 2.13   Site—Structure Resonance Effect

When the building period T and the soil period T_s are nearly equal, a resonance effect is generated. The S coefficient attempts to take this behavior into account. Consider soil layers down to bedrock or to a depth of 500 ft. See Fig. 2.16.

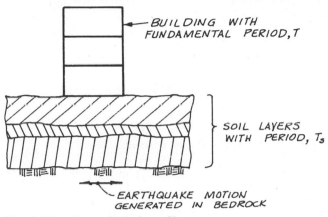

Fig. 2.16a   Geotechnical profile.

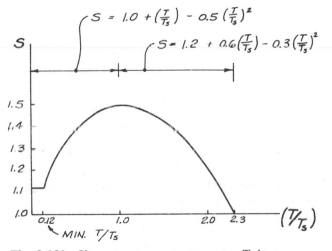

$$S = 1.0 + \left(\frac{T}{T_s}\right) - 0.5\left(\frac{T}{T_s}\right)^2$$

$$S = 1.2 + 0.6\left(\frac{T}{T_s}\right) - 0.3\left(\frac{T}{T_s}\right)^2$$

Fig. 2.16b   Site—structure resonance coefficient.

2.12   Seismic Load—Distribution of Forces

The reasoning behind the distribution of seismic forces provided by the Code was given in Sec. 2.10. The general distribution was described, and it was seen that the shape of the first mode served as the basis for this distribution.   **43**

The method used to actually calculate these distributed forces (F_x and F_t) is covered in this section.

The horizontal forces are assumed to be concentrated at the story levels in much the same manner that the masses tributary to a level are "lumped" or assigned to a particular story height. The force at any level, x (Fig. 2.12a), is calculated by the following expression:

$$F_x = \frac{(V - F_t)\, w_x h_x}{\sum\limits_{i=1}^{n} w_i h_i}$$

where F_x = horizontal force at story level x

$\qquad V$ = total base shear

$\qquad F_t$ = additional force applied to the top level

$\qquad w_x, w_i$ = tributary weight assigned to story level x and i

$\qquad h_x, h_i$ = height above the base of the building to level x and i, ft

As mentioned earlier, if the weights assigned to each level are equal, a triangular distribution of F_x forces results from use of this formula. If the weights are not all equal, some variation from the straight-line distribution will result, but the trend will follow the first-mode shape. Accelerations, and correspondingly, inertia forces $(F = Ma)$ increase with increasing height above the base.

The item which remains to be defined is the F_t force. The F_t force is an additional force which is applied at the top level of the structure. The purpose of this force is the Code's method of accounting for whip action in tall, slender buildings and to allow for the effects of the higher modes (i.e., other than the first mode) of vibration. F_t is given by the following expression, which need not exceed 25 percent of the total base shear:

$$F_t = (0.07\,T)V \qquad \leq 0.25V$$

and

$$F_t = 0 \qquad \text{where } T \leq 0.7 \text{ seconds}$$

For many wood-frame buildings, $F_t = 0$ according to this criterion.

The final step in the process of distributing seismic loads is to consider the distribution of the story force F_x at a given level. Once the total story force has been determined, it is distributed at that level in proportion to the mass (DL) distribution of that level. See Example 2.14.

The purpose behind this distribution relates back to the idea of an inertial

---

EXAMPLE 2.14 **Distribution of Forces at Story Level x**

Transverse and Longitudinal Directions Defined:
The direction of a lateral load applied to a building is typically described as being in the transverse or longitudinal direction. These terms are interpreted as follows:

Transverse lateral load is parallel to the short dimension of the building.

Longitudinal lateral load is parallel to the long dimension of the building.

Buildings are designed for seismic loads applied independently in both the transverse and longitudinal directions.

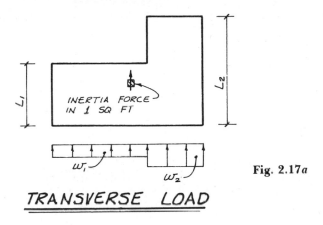

Fig. 2.17*a*

TRANSVERSE LOAD

Each square foot of DL can be visualized as generating its own inertial force. If all of the inertial forces generated by these unit areas are summed in the transverse direction, the loads w_1 and w_2 are in proportion to the lengths L_1 and L_2, respectively.

The sum of the distributed seismic loads w_1 and w_2 equals the total transverse story force F_{xt}.

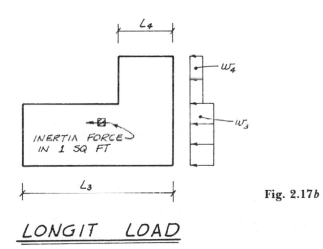

Fig. 2.17*b*

LONGIT LOAD

Here L_3 and L_4 are measures of the distributed loads w_3 and w_4. The sum of these distributed seismic loads equals the total longitudinal story force F_{xl}.

NOTE: The distribution of inertial forces generated by the DL of the walls is considered in Chap. 3.

force. If it is visualized that each square foot of dead load has a corresponding inertial force generated by an earthquake, then the distributions shown in the sketches become clear.

If each square foot of area has the same DL, then the distributed seismic load is in direct proportion to the length of the floor or roof parallel to the load in question. Hence the magnitude of the distributed load is large where the dimension of the floor or roof parallel to the load is large. Correspondingly, the distributed seismic load is small where the dimension parallel to the load is small.

The calculation of seismic forces and their distributions will be clarified by the numerical examples which are given in Chap. 3.

2.13 Seismic Loads—Forces on Portions of Buildings

The seismic forces which have been discussed up to this point are those assumed to be developed in the structure when it acts as a *unit* in response to an earthquake. However, when individual elements of the building are considered separately, it may be necessary to consider different seismic effects. The reason for this is that certain elements which are attached to the structure respond dynamically to the motion of the *structure* rather than to the motion of the *ground*. Resonance between the structure and the attached element may occur.

The Code provides a force F_p which is an "equivalent static" force for various elements (portions) of a structure. By including the F_p forces, the Code takes into account the possible response to the motion of the structure and the consequences involved if they collapse or fail. The force on a portion of the structure is given by the following expression, which uses a number of the coefficients used in the base shear expression:

$$F_p = (ZIC_p)W_p$$

Here (ZIC_p) is the seismic coefficient. The coefficients Z and I have been defined previously. W_p is simply the dead load of the portion of the structure being considered. C_p is a response coefficient that is given as a constant for various elements (components) of a building. Values of C_p are given in UBC Table 23J. An example will demonstrate the calculation of a force on a portion of a building. See Example 2.15.

EXAMPLE 2.15   Seismic Loads Normal to Wall

Determine the seismic design load normal to the wall for the building shown in Fig. 2.18. The wall spans vertically between the floor and the roof. The wall is constructed of reinforced brick masonry that weighs 90 psf. Compare the seismic load to the wind load of 20 psf to determine the critical lateral design load (wind and seismic are not considered simultaneously). $Z = 1.0$ and $I = 1.0$.

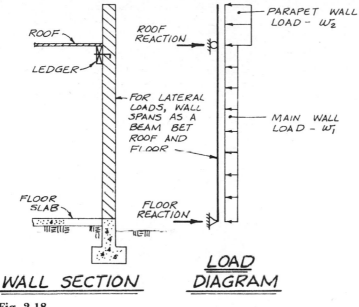

WALL SECTION   LOAD DIAGRAM

Fig. 2.18

NOTE: Loads w_1 and w_2 can act normal to the wall in either direction (i.e., inward or outward).

Load to main wall:

$$F_p = (ZIC_p)W_p$$
$$= (1.0)(1.0)(0.3)\,W_p$$

$$= 0.3\,W_p$$

Since W_p is in psf, w can be calculated directly as a uniformly distributed load.

$$w_1 = 0.3\,W_p = 0.3(90)$$
$$w_1 = 27 \text{ psf} > 20 \text{ psf wind}$$

∴ Seismic governs.

Load to parapet wall:

$$F_p = (ZIC_p)W_p$$
$$= (1.0)(1.0)(0.8)\,W_p$$
$$= 0.8\,W_p$$
$$w_2 = 0.8\,W_p = 0.8(90)$$
$$w_2 = 72 \text{ psf} > 20$$

∴ Seismic governs.

2.14 Problems

2.1   *Given:* The house framing section shown in Fig. 2.A.

Find: a. Roof DL in psf on a horizontal plane.
   b. Wall DL in psf of wall surface area.
   c. Wall DL in lb/ft of wall.
   d. Basic (i.e., consider roof slope but not trib. area) unit roof LL in psf.
   e. Basic unit roof LL in psf if the slope is changed to $3/12$.

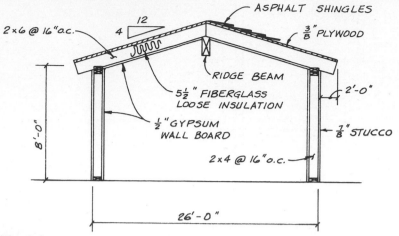

Fig. 2.A

2.2   *Given:*   The house framing section shown in Fig. 2.B. Note that a roofing square is equal to 100 ft².

   Find: *a.* Roof DL in psf on a horizontal plane.
   b. Ceiling DL in psf.
   c. Basic (i.e., consider roof slope but not trib. area) unit roof LL in psf.

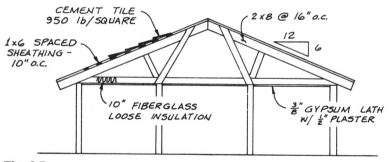

Fig. 2.B

2.3   *Given:*   The building framing section shown in Fig. 2.C.

   Find: *a.* Roof DL in psf.
   b. Second floor DL in psf.
   c. Basic unit roof LL in psf.

2.4   *Given:*   The roof framing plan of the industrial building shown in Fig. 2.D. Roof slope is ¼ in./ft. General construction:

Roofing—5-ply felt
Sheathing—½-in. plywood

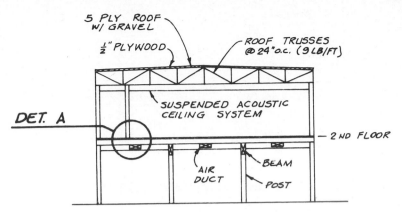

DET. A

SECTION

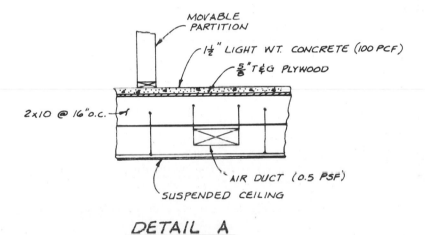

DETAIL A

Fig. 2.C

Subpurlin—2 × 4 at 24 in. o.c.
Purlin—4 × 14 at 8 ft-0 in. o.c.
Girder—6¾ × 33 at 20 ft-0 in. o.c.

Assume loads are uniformly distributed on supporting members.

Find: a. Average DL of entire roof in psf.
 b. Trib. DL to subpurlin in lb/ft.
 c. Trib. DL to purlin in lb/ft.
 d. Trib. DL to girder in lb/ft.
 e. Trib. DL to column C1 in k.
 f. Basic unit roof LL in psf.

49

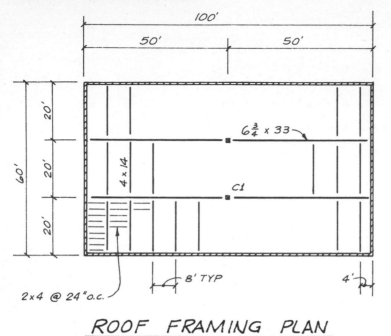

ROOF FRAMING PLAN

Fig. 2.D

2.5 *Given:* Figure 2.A. The ridge beam spans 18 ft-0 in.

Find: a. Trib. area to the ridge beam.
 b. Unit roof LL in psf using UBC Method 1.
 c. Unit roof LL in psf using UBC Method 2.
 d. Trib. width to the beam and the uniform RLL on the beam using Method 1 and Method 2.

2.6 *Given:* A roof similar to Fig. 2.A with $3/12$ roof slope. The ridge beam spans 20 ft-0.

Find: a. Trib. area to the ridge beam.
 b. Unit roof LL in psf using UBC Method 1.
 c. Unit roof LL in psf using UBC Method 2.
 d. Trib. width to the beam and the uniform RLL on the beam using Method 1 and Method 2.

2.7 *Given:* Figure 2.B and a basic snow load SL of 60 psf.

Find: a. Slope of roof in degrees from horizontal.
 b. Reduced SL in psf on a horizontal plane.

2.8 *Given:* A roof similar to Fig. 2.B with $8/12$ slope and a basic snow load SL of 100 psf.

Find: a. Slope of roof in degrees from horizontal.
 b. Reduced SL in psf on a horizontal plane.

2.9   *Given:*   The roof structure in Fig. 2.D.

   Find: *a.* Unit RLL in psf for
   1. 2×4 subpurlin (use Method 1 or 2).
   2. 4×14 purlin (use Method 2).
   3. $6\frac{3}{4} \times 33$ glulam beam (use Methods 1 and 2).
   b. Using the unit RLLs from *(a)*, determine the uniformly distributed LLs in lb/ft for each of the members.

2.10   *Given:*   The roof structure in Fig. 2.D and a 25-psf snow load.

   Find: *a.* Uniformly distributed SL in lb/ft for
   1. 2×4 subpurlin.
   2. 4×14 purlin.
   3. $6\frac{3}{4} \times 33$ glulam beam.
   b. Trib. SL to column C1 in k.

2.11   *Given:*   The building in Fig. 2.C.

   Find: Second floor basic (i.e., consider occupancy but not trib. areas) unit LL and concentrated loads for the following uses:
   a. Offices
   b. Light storage
   c. Retail store
   d. Apartments
   e. Hotel restrooms
   f. School classroom

2.12   A column supports only loads from the second floor of an office building. The trib. area to the column is 220 ft² and the DL is 35 psf.

   Find: *a.* Basic floor LL in psf.
   b. Reduced FLL in psf.
   c. Total load to column in k.

2.13   A beam supports the floor of a classroom in a school building. The beam spans 25 ft and the trib. width is 15 ft. DL = 20 psf.

   Find: *a.* Basic floor LL in psf.
   b. Reduced FLL in psf.
   c. Uniformly distributed total load to the beam in lb/ft.

2.14   *Given:*   UBC Table 23A.

   Find: Four occupancies where the unit floor LL can not be reduced. List the occupancy and the corresponding unit FLL.

2.15   *Given:*   UBC beam deflection criteria.

   Find: The allowable deflection limits for the following members. Consider both LL only and total load TL. Neglect creep effects.
   a. Floor beam with 20-ft span.
   b. Roof rafter that supports a plaster ceiling below. Span = 14 ft.

2.16   *Given:*   The *Timber Construction Manual* (Ref. 3) beam deflection recommendations.

   Find: The allowable deflection limits for the following roof beams. Consider

both LL and TL. Creep adjustments do not apply.
   a. Roof rafter that supports a gypsum board ceiling below. Span = 14 ft.
   b. Roof beam supporting an acoustic suspended ceiling. Span = 40 ft.

2.17 *Given:* A two-story enclosed building 26 ft-0 in. high located in Nevada.

   Find: The following UBC wind load requirements:
   a. The basic wind pressure from UBC Fig. 4.
   b. The design wind pressure based on the height of the building.
   c. The upward design wind pressure normal to the surface of the roof.
   d. The upward wind pressure on the roof overhangs.

2.18 *Given:* UBC seismic design load requirements.

   Find: a. The maximum numerical value that the product CS need not exceed.
   b. The maximum numerical value of C.
   c. The minimum value of S.
   d. The value of S to be used when there is not sufficient information to calculate a value.

2.19 *Given:* UBC seismic design load requirements.

   Find: a. The definition of period of vibration and the Code methods for estimating the fundamental period.
   b. How does period of vibration affect seismic loads?
   c. Describe the effects of the interaction of the soil and structure on seismic loads.
   d. What is damping, and how does it affect seismic loading? Do the Code criteria take damping into account?

2.20 *Given:* UBC seismic design load requirements.

   Find: a. The expression for the distribution of the base shear over the height of the building.
   b. Describe the distribution of the base shear over the height of the building and why it is used.
   c. The section of the UBC where the terms for the distribution expression are defined.

2.21 A two-story building has a box system for resisting lateral loads. The building is located in San Francisco and will be used as an "essential" hospital.

   Find: The following seismic design values:
   a. The value of Z
   b. The value of I
   c. The value of K
   d. V in terms of W if $CS = 0.14$.

2.22 A 10-story office building in Los Angeles has a ductile moment-resisting space frame.

   Find: The following seismic design values:
   a. The value of Z
   b. The value of I
   c. The value of K
   d. V in terms of W if $CS = 0.12$

2.23   A one-story wood-framed office building is in seismic zone 4.

Find: The following seismic design values:
  a. The value of Z.
  b. The value of I.
  c. The value of K.
  d. V in terms of W if $CS = 0.14$.

2.24   *Given:*   UBC seismic design requirements.

Find: *a.* The expression for the lateral force on an element of a building.
  b. The section in the Code where the terms of the expression are defined.
  c. Briefly state why a different expression is used for the seismic force on the building acting as a unit and for individual elements of a building.

Behavior of Structures under Load

3.1 Introduction

The loads required by the Code for designing a building were described in Chap. 2. Chapter 3 deals primarily with the transfer of these loads from one member to another throughout the structure. The distribution of *vertical loads* in a typical wood-frame building follows the traditional "post-and-beam" concept. This subject is briefly covered at the beginning of the chapter.

The distribution of *lateral loads* may not be as evident as the distribution of vertical loads. The majority of Chap. 3 deals with the transfer of lateral loads from the point of origin, through the building, and into the foundation. This subject is first introduced by reviewing the three basic types of lateral-force-resisting systems used in conventional rectangular-type buildings.

Shearwalls and horizontal diaphragms make up the box system used in most wood-frame buildings (or buildings with a combination of wood framing and concrete or masonry walls). The chapter concludes with two detailed examples of lateral load calculations in box buildings.

3.2 Structures Subject to Vertical Loads

The behavior of framing systems (post-and-beam type) under vertical loads is relatively straightforward. Sheathing (decking) spans between the most closely spaced beams; these short-span beams are given

55

various names: stiffeners, rafters, joists, subpurlins. The reactions of these members in turn cause loads on the next set of beams in the framing system; these next beams may be referred to as beams, joists, or purlins. Finally, reactions of the second set of beams impose loads on the largest beams in the system. These large beams are known as girders. The girders, in turn, are supported by columns. See Example 3.1.

EXAMPLE 3.1   Typical Post-and-Beam Framing

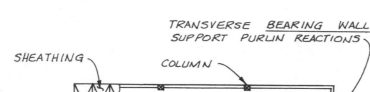
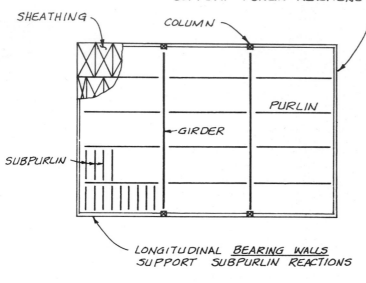

FRAMING   PLAN

Fig. 3.1

1. Sheathing spans between subpurlins
2. Subpurlins span between purlins
3. Purlins span between girders
4. Girders span between columns

Subpurlins and purlins are also supported by bearing walls. *Bearing walls* are defined as walls that support vertical loads in addition to their own weight.

When this framing system is used for a roof, it is often constructed as a *panelized* system. Panelized roofs typically use glulam girders spaced at 18 to 24 ft on center, sawn lumber purlins at 8 ft on center, subpurlins at 24 in. on center, and plywood sheathing. The name of the system comes from the fact that 8-ft-wide roof *panels* are prefabricated and then lifted onto preset girders using forklifts. See Fig. 3.2. The speed of construction and erection

Fig. 3.2 Panelized roof system installed with forklift. *(Photo by Mike Hausmann.)*

makes panelized roof systems very economical. Panelized roofs are widely used on large one-story commercial and industrial buildings.

Although the loads to successively larger beams are a result of reactions from lighter members, for structural design the loads on beams in this type of system are often assumed to be uniformly distributed. In order to obtain a feel for whether this approach produces conservative values for shear and moment, it is suggested that a comparison be made between the values of shear and moment obtained by assuming a *uniformly distributed load* and those obtained by assuming *concentrated loads* from lighter beams. The actual loading probably falls somewhere between the two conditions described. See Example 3.2.

EXAMPLE 3.2   Beam Loading Diagrams

Figure 3.3 shows the girder from the building in Fig. 3.1. The load to the girder can be considered as a number of concentrated reaction loads from the purlins. However, a more common design practice is to assume that the load is uniformly distributed. The uniformly distributed load is calculated as the unit load times the tributary width to the girder. As the number of concentrated loads increases, the closer the loading approaches the uniform load case.

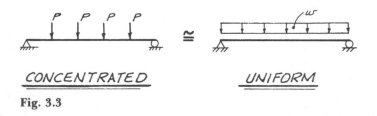

CONCENTRATED   UNIFORM

Fig. 3.3

Regardless of the type of load distribution that is used, it should be remembered that it is the *tributary area of the member being designed* (Sec. 2.3) which is used in establishing unit live loads, rather than the tributary area of the lighter members which impose the load. This concept is often confusing when it is first encountered.

As an example, consider the design load for the girder in Fig. 3.1. Confusion may occur when the *unit* live load for the girder (based on a large tributary area) turns out to be less than the *unit* live load used in the design of the purlin. Obviously, the reaction of the purlin (using the higher live load) must be supported by the girder. Why is the lower live load used for design?

The reasoning is this. The girder must be capable of supporting *individual* reactions from purlins calculated using the larger unit live load (obtained using the tributary area of the purlin). However, when the entire tributary area of the girder is considered loaded, the smaller unit live load may be used (this was discussed in detail in Chap. 2). Of course, each connection between the purlin and the girder must be designed for the higher unit live load, but not all purlins are subjected to this higher load simultaneously.

The spacing of members and the spans used depend on the function and purpose of the building. Closer spacing and shorter spans require smaller member sizes, but short spans require closely spaced columns or bearing walls. The need for clear, unobstructed space must be considered when the framing system is first established. Once the layout of the building has been established, dimensions for framing should be chosen which result in the best utilization of materials. For example, the standard size of a sheet of plywood is 4 ft by 8 ft, and a joist spacing should be chosen which fits this basic module. Spacings of 16 in., 24 in., and 48 in. o.c. (o.c. = on center

58

and c.c. = center to center) all provide supports at the edge of a sheet of plywood.

Certainly an unlimited number of framing systems can be used, and the choice of the framing layout should be based on a consideration of the requirements of a particular structure. Several other examples of framing arrangements are shown in Fig. 3.4a, b, and c. These are given to suggest possible

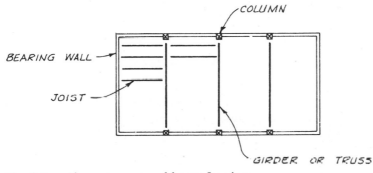

Fig. 3.4a Alternate post-and-beam framing.

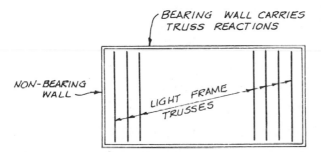

Fig. 3.4b Light frame trusses.

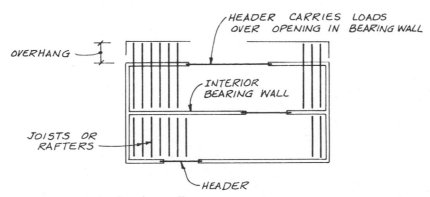

Fig. 3.4c Interior bearing walls.

arrangements and are not intended to be a comprehensive summary of framing systems.

It should be noted that in the framing plans, a break in a member represents a simple end connection. For example, in Fig. 3.4a the lines representing joists are broken at the girder. If a continuous joist is to be shown, a solid line with no break at the girder would be used. This is illustrated in Fig. 3.4c where the joist is continuous at the rear wall overhang. Such points may seem obvious, but a good deal of confusion results if they are not recognized.

3.3 Structures Subject to Lateral Loads

The behavior of structures under lateral loads usually requires some degree of explanation. In covering this subject, the various types of lateral-force-resisting systems (LFRSs) used in ordinary rectangular buildings should be clearly distinguished. See Example 3.3. These LFRSs include

1. Rigid frame
2. Vertical truss (braced frame)
3. Shearwall

EXAMPLE 3.3   **Basic Lateral Force Resisting System (LFRS)**

Rigid Frame

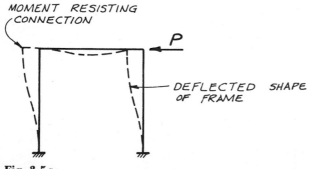

Fig. 3.5a

Resistance to lateral loads is provided by *bending* in the column and girders of the rigid frame members.

Vertical Truss (Braced Frame)

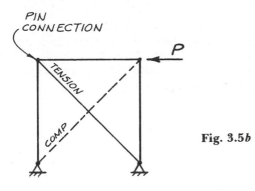

Fig. 3.5*b*

Lateral loads develop *axial* forces in the vertical truss. Slender compression members are typically ignored and the lateral load is taken by the tension brace.

Shearwall

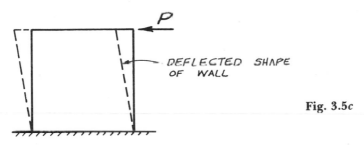

Fig. 3.5*c*

Segments of walls can be designed to function as *shear*-resisting elements in carrying lateral loads. The deflected shape shows shear deformation rather than bending.

Rigid frames, whether statically determinate or indeterminate, resist lateral loads by bending in the frame members. That is to say, the members have relatively small depths compared with their lengths and the stresses induced as the structure deforms under lateral loads are essentially flexural. Some axial forces are also developed.

Vertical trusses or braced frames are analyzed in a manner similar to horizontal trusses: connections are assumed to be pinned, and loads are assumed to be applied at the joints. Vertical trusses often take the form of cross or X-bracing. The vertical truss in Fig. 3.5*b* appears to be statically indeterminate. As such, both diagonal members must be designed to function simultaneously (one in tension and the other in compression). Perhaps a more common practice is to ignore the member in compression. This approach has the advantage of not requiring the design of a compression member with a long unbraced length. The resulting truss is statically determinate. The compres-

61

sion member, although ignored, cannot be omitted because it will in turn function as the tension member when the lateral load is applied in the opposite direction.

Steel rods are often used for X-bracing. Obviously such a slender compression member will buckle, and the rod in compression *must* be ignored. The concept of ignoring the member in compression is, however, not limited to rod X-bracing.

Shearwall structures make use of specially designed wall sections to resist lateral loads. A shearwall is essentially a vertical cantilever with the span of the cantilever equal to the height of the wall. The depth of these members (i.e., the length of the wall element parallel to the applied lateral load) is large in comparison with the depth of the structural members in the rigid-frame LFRS. For a member with such a large depth compared with its height, *shear* deformation replaces bending as the significant action (hence, the name shearwall).

It should be mentioned that the LFRSs in Example 3.3 are the *vertical resisting elements* of the system (i.e., the vertical components). Because buildings are three-dimensional structures, some horizontal system must also be provided to carry the lateral loads to the vertical elements. See Example 3.4. A variety of horizontal systems can be employed:

1. Horizontal wall framing
2. Vertical wall framing with horizontal trusses at the story levels
3. Vertical wall framing with horizontal diaphragms at the story levels

---

EXAMPLE 3.4   Horizontal Elements in the LFRS

Horizontal Wall Framing

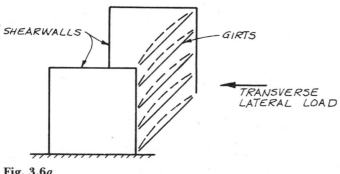

Fig. 3.6a

Horizontal wall members are known as *girts* and distribute the lateral load to the vertical LFRS. Dashed lines represent the deflected shape of the girts. See Fig. 3.6a. The lateral load to the shearwalls is distributed over the height of the wall.

Vertical Wall Framing and Horizontal Truss

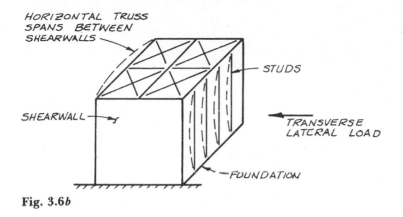

Fig. 3.6*b*

Vertical wall members are known as *studs*. The lateral load is carried by the studs to the roof level and the foundation. A horizontal truss in the plane of the roof distributes the lateral load to the transverse shearwalls. The diagonal members in the horizontal truss are often steel rods and are designed to function in tension only (like the vertical truss in Fig. 3.5*b*).

Vertical Wall Framing and Horizontal Diaphragm

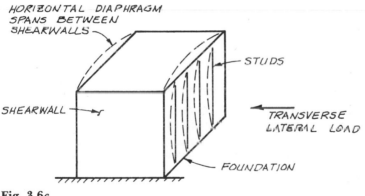

Fig. 3.6*c*

In Fig. 3.6*c* the LFRS is similar to the system in Fig. 3.6*b* except that the horizontal truss in the plane of the roof is replaced with a *horizontal diaphragm*. The use of vertical wall framing and horizontal diaphragms is the most common system in wood-frame buildings because the roof sheathing can be designed economically to function as both a vertical- and a lateral-load-carrying element. The horizontal diaphragm is designed as a beam spanning between the shearwalls. The design requirements for horizontal diaphragms and shearwalls are given in Chaps. 9 and 10.

The first two framing systems in Example 3.4 are relatively easy to visualize. The third system is also easy to visualize once the concept of a diaphragm is understood. A diaphragm can be considered to be a large, thin structural element that is loaded in its plane. In Fig. 3.6c the vertical wall members develop horizontal reactions at the roof level and at the foundation level (studs are assumed to span as simple beams between these two levels). The reaction of the studs at the roof level provides a load in the plane of the roof. The diaphragm acts as a large horizontal beam.

In wood buildings, or buildings with wood roof and floor systems and concrete or masonry walls, the roof or floor sheathing is designed and connected to the supporting framing members to function as a diaphragm. In buildings with concrete roof and floor slabs, the concrete slabs are designed to function as diaphragms.

The stiffness of a diaphragm refers to the amount of deflection that occurs in the horizontal diaphragm as a result of the in-plane lateral load (Fig. 3.6c shows the deflected shape). Wood diaphragms are not nearly as stiff as concrete slabs, and *wood diaphragms* are referred to as *flexible diaphragms. Concrete slabs* are known as *rigid diaphragms.*

In each of the sketches in Example 3.4, the transverse wind load is distributed horizontally to end shearwalls. The same horizontal systems shown in these sketches can be used to distribute lateral loads to the other basic vertical LFRSs (i.e., rigid frames or vertical trusses). Any combination of *horizontal* and *vertical* LFRSs can be incorporated into a given building to resist lateral loads.

The discussion of lateral loads in Example 3.4 was limited to loads in the transverse direction. In addition, a LFRS must be provided for loads in the longitudinal direction. This LFRS will consist of both horizontal and vertical components similar to those used to resist loads in the transverse direction.

Different types of *vertical* elements can be used to resist transverse lateral loads and longitudinal lateral loads in the same building. For example, a rigid frame can be used to resist lateral loads in the transverse direction, and a shearwall can be used to resist lateral loads in the other direction. The choice of LFRS in one direction does not necessarily limit the choice for the other direction.

In the case of the *horizontal* LFRS, it is unlikely that the horizontal system used in one direction would be different from the horizontal system used in the other direction. If the sheathing is designed to function as a horizontal diaphragm for lateral loads in one direction, it probably can be designed to function as a diaphragm for loads applied in the other direction. On the other hand, if the roof or floor sheathing is incapable of functioning as a diaphragm, a system of horizontal trusses spanning in both the transverse and longitudinal directions appears to be the likely solution.

The common types of LFRSs for conventional buildings have been summarized as a general introduction and overview. It should be emphasized that the large majority of wood-frame buildings, or buildings with wood roof and floor framing and concrete or masonry walls, use a combination of

1. Horizontal diaphragms
2. Shearwalls

to resist lateral loads. Because of its widespread use, only the design of this type of system is covered in this book.

3.4 Calculation of Lateral Loads in Box-Type Structures

The majority of wood structures use a box LFRS because the floor, roof, and wall sheathing, acting jointly with the supporting framing members, can be designed to function as a diaphragm. Thus, the sheathing material, which is normally required in any structure, can be used to function as the LFRS. Some additional stiffening or nailing may be required, but, functioning in this way, one material essentially serves two purposes (i.e., sheathing and lateral force resistance).

The purpose of the following examples is to illustrate how lateral loads are calculated and distributed in two different shearwall buildings. In the first example, a one-story building is analyzed, and in the second example a two-story building is examined.

The method used to calculate seismic forces in these two examples is based on the same criteria, but in the case of the one-story structure, a somewhat more direct solution is possible. In the one-story structure, the *distributed loads* to the roof diaphragm and shearwalls can be obtained directly. In the case of the two-story building, the *total base shear* must first be calculated and *then distributed* throughout the height of the structure. The procedure outlined in the second example is the one that must be used in all multistory structures. The more direct solution of the first example is limited to the case of a single-story building.

3.5 Design Problem: One-Story-Building Lateral Force Calculations

In this section a rectangular one-story building with a wood roof system and masonry walls is analyzed to determine both wind and seismic loads. See Fig. 3.7. The building chosen for this example has been purposely simplified so that the basic procedure is demonstrated without "complications."

The critical lateral load in the transverse direction at the roof level is to be determined. The lateral load is applied at the point where the roof diaphragm (sheathing) attaches to the wall. This will be the *reference point* in all lateral-load problems involving horizontal diaphragms. For this lateral load, the roof spans horizontally between the outside end shearwalls.

65

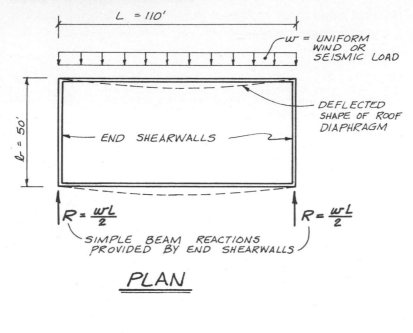

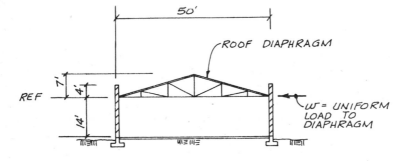

PLAN

TYP TRANSVERSE SECTION

Fig. 3.7 One-story building subjected to transverse lateral load.

Wind. The wind load is calculated by multiplying the design wind pressure by the projected vertical-height tributary to the roof diaphragm level. See Example 3.5. Because the wall is assumed to span vertically between the roof diaphragm and the foundation, the tributary height below the diaphragm is simply one-half of this wall height. Above the diaphragm, it is the horizontal projection to the top of the roof.

Seismic. For a one-story building, the seismic base shear expression ($V = ZIKCSW$) can be converted directly to a uniform load expression ($w = ZIKCSW_1$). W_1 indicates the weight of a one-foot-wide strip of dead load

EXAMPLE 3.5   Wind Load Calculation

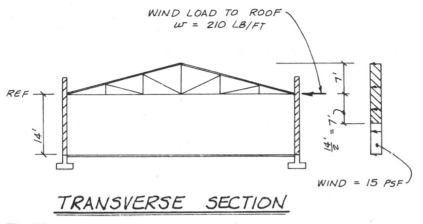

TRANSVERSE   SECTION

Fig. 3.8

Wind load from Code $= 15$ psf

$$\left.\begin{array}{l}\text{Trib. height to}\\ \text{roof diaphragm}\end{array}\right\} = \left\{\begin{array}{l}\text{trib. wall ht.}\\ \text{below ref. point}\end{array}\right\} + \left\{\begin{array}{l}\text{projected ht.}\\ \text{above ref.}\\ \text{point}\end{array}\right\}$$

$$\text{Trib. ht.} = \frac{14}{2} + 7 = 14 \text{ ft}$$

Wind load $= w =$ trib. ht. $\times$ wind pressure
$$= 14 \times 15$$
$$= 210 \text{ lb/ft}$$

tributary to the roof level. The reason that the uniform load w can be deter-mined directly is that there is only one term in the F_x expression for a one-story building (Sec. 2.12). Therefore, F_x *directly* equals the base shear.

For the seismic load in the transverse direction, the weight of the roof (plus some portion of the snow load, if any) and the two longitudinal walls (walls perpendicular to the applied load) need to be considered. The dead load of the transverse walls will be accounted for separately from w. In this manner, the lateral load is distributed directly "in proportion to the mass distribution" of the roof level. See Example 3.6.

In this example the procedure used to calculate the seismic coefficient is demonstrated. However, it should be pointed out that shearwall buildings usually have a low height-to-width ratio and are fairly rigid. As a result they tend to have low periods of vibration. For this reason, it is likely that low-rise buildings of this type will use the maximum value of C (Fig. 2.15) in the seismic coefficient. In addition, without a geotechnical study, the S coeffi-

67

cient will be taken as its maximum value. If the maximum values of C and S are used, the Code states that $C \times S$ need not exceed 0.14.

EXAMPLE 3.6 Seismic Load Calculation

Determine the seismic coefficient and load for the transverse direction. See Fig. 3.9.

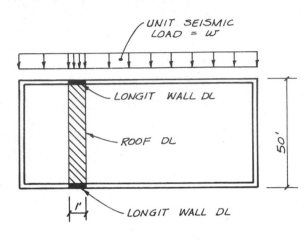

PLAN

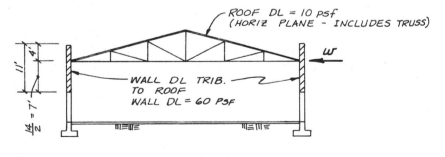

TRANSVERSE SECTION

Fig. 3.9

$Z = 1.0$ zone 4
$I = 1.0$ nonessential
$S = 1.5$ T_s is unknown
$K = 1.33$ box system

Fundamental period:

$$T = \frac{0.05h_n}{\sqrt{D}} = \frac{0.05(14)}{\sqrt{50}} = 0.099$$

Response value:

$$C = \frac{1}{15\sqrt{T}} = \frac{1}{15\sqrt{0.099}} = 0.212 > 0.12$$

∴ Use $C = 0.12$

$$CS = (0.12)(1.5) = 0.18 > 0.14$$

∴ Use $CS = 0.14$

$$ZIKCS = 1.0(1.0)(1.33)(0.14) = 0.186$$

For a one-story building,

$$w = ZIKCS(W_1) = 0.186\,W_1$$

where $W_1 = $ DL of 1-ft-wide strip trib. to roof level.
DL trib. to roof level for a 1-ft-strip:

Roof DL $= 10$ psf $\times 50$ ft $\quad = \quad 500$ lb/ft
Wall DL $= 60$ psf $\times 11$ ft $\times 2$ walls $= 1320$
$W_1 = \overline{1820 \text{ lb/ft}}$

$w = 0.186\,W_1 = 0.186(1820)$
$w = 339$ lb/ft $=$ uniform seismic load
$w = 210$ lb/ft $=$ uniform wind load < 339
∴ *Seismic governs*

As shown in the calculations, the use of the above values for a "nonessential" building in earthquake zone 4 yields a seismic coefficient of 0.186. This coefficient is often used directly for the typical building considered in this example. However, the designer should be aware of the basis for the seismic coefficient so that the proper coefficient can be determined if conditions different from those stated are encountered.

The 0.186 seismic coefficient can be interpreted as a loading criterion that requires approximately 19 percent of gravity (g force) to be applied horizontally at the roof level. The g force for an "essential facility" is 1.5 times larger, or approximately 28 percent of gravity.

One of the criteria used to design diaphragms and shearwalls is the *unit shear*. Although the actual design of diaphragms is covered in Chaps. 9 and 10, the calculation of unit shear will be illustrated in this example. The unit shear in the horizontal diaphragm is demonstrated first.

In the building being considered, only the exterior end walls are used as shearwalls. Therefore, the horizontal diaphragm must span as a simple beam between the two transverse walls. The reaction of the horizontal diaphragm on the transverse shearwall is $wL/2$. The shear diagram for a simple beam shows that the internal shear is equal to the external reaction. The maximum

69

total shear is converted to a *unit shear* by distributing it along the width of the diaphragm available for resisting the shear. See Fig. 3.10.

The loads and shears considered to this point have been those in the horizontal diaphragm. The next step in the process is to consider similar quantities in the shearwalls. In the calculation of the uniform load to the horizontal diaphragm, it will be recalled that only the DL of the roof and

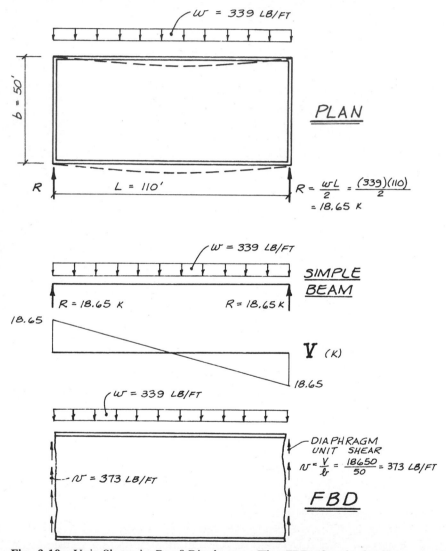

Fig. 3.10 Unit Shear in Roof Diaphragm. The FBD (free-body diagram) of the horizontal diaphragm is cut next to transverse shearwalls. For simplicity the calculations for unit shear are shown using the nominal length and width of the building (i.e., wall thickness ignored).

the longitudinal walls was included in the seismic load. The inertia force generated in the transverse walls was not included in the load to the roof diaphragm. The reason for this is that the *shearwalls carry directly their own seismic force* and they do not, therefore, contribute to the load in the horizontal diaphragm.

Several different approaches are used by designers to compute wall seismic forces. Two methods are illustrated here.

In the first approach, the unit shear in the shearwall is calculated at the *midheight* of the wall. See Example 3.7. This convention developed because the length of the shearwall, *b*, used to compute the unit shear is often a minimum at this location. For example, any openings in the wall (both doors and windows) are typically intersected by a horizontal line drawn at the mid-height of the wall. In addition, this approach is consistent with the *lumped-mass* model presented in Chap. 2.

The seismic force generated by the top half of the wall is given the symbol

---

EXAMPLE 3.7   **Unit Shear in Shearwall: Method 1**

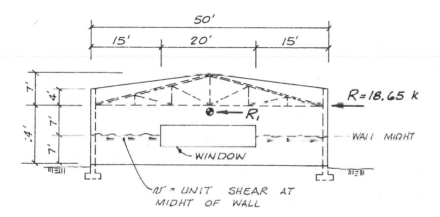

END ELEVATION

Fig. 3.11

Seismic Load

Determine seismic load generated by top half of wall DL.

$$\text{Wall area} = (11 \times 50) + \tfrac{1}{2}\,(3 \times 50) = 625 \text{ ft}^2$$
$$\text{Wall DL} = 625 \times 60 = 37.5 \text{ k} \qquad \text{neglect window reduction}$$
$$R_1 = (ZIKCS)W = 0.186\,W = 0.186 \times 37.5$$
$$R_1 = 6.98 \text{ k}$$

71

DESIGN OF WOOD STRUCTURES

Wall Shear

$$\left.\begin{array}{l}\text{Total shear at} \\ \text{midht. of wall}\end{array}\right\} = \left\{\begin{array}{l}\text{Sum of all forces on} \\ \text{shearwall above midht. (FBD)}\end{array}\right.$$

$$V = R + R_1 = 18.65 + 6.98$$

$$V = 25.63 \text{ k}$$

$$\text{Unit wall shear} = v = \frac{V}{b} = \frac{25630}{30}$$

$$\boxed{v = 854 \text{ lb/ft}} \text{ *}$$

---

*For design of masonry shearwalls, this load must be increased by a factor of 1.5 (UBC Chap. 24). For wall openings not symmetrically located, see Sec. 9.6.

R_1 and can be computed as the wall DL of the top portion of the wall times the seismic coefficient. The total shear at the midheight of the wall is the sum of all forces above this level. The unit shear can be computed once the total shear has been obtained.

In the second approach, the shear in the shearwall is calculated at the *base* of the wall. This practice uses a wall seismic force R_1 which is computed as the entire wall DL times the seismic coefficient. This second method will result in a larger total force to the shearwall.

The magnitude of the *unit wall shear* using this method may be larger or smaller than the shear obtained using the first method. The shears depend on the type and size of the wall openings (doors and windows). If a large door is the only opening in the wall, higher unit wall shears are obtained at the base of the wall than at the midheight. See Example 3.8.

If the second method of calculating unit wall shears is applied to the wall analyzed in Example 3.7, a *conservative* design value can be obtained by using the *force at the base* of the wall and the *length at the midheight* of the wall.

The question of which procedure should be used in a particular building design is a matter of judgment. The choice depends to some extent on the magnitude of the wall dead loads (light wall dead loads may have little effect on final results). Both procedures are found in the design of these types of buildings.

As mentioned earlier, the unit shears in the roof diaphragm and shearwalls constitute one of the main parameters in the design of these elements. There are additional design factors that must be considered and these will be covered in detail in subsequent chapters.

These examples have considered only transverse lateral loads, and a similar analysis can be used for the longitudinal direction. Roof diaphragm shears are usually critical when lateral loads in the transverse direction are considered, but both directions should be analyzed. Shearwalls may be critical in either the transverse or longitudinal directions depending on the size of the wall openings.

EXAMPLE 3.8   Unit Shear in Shearwall: Method 2

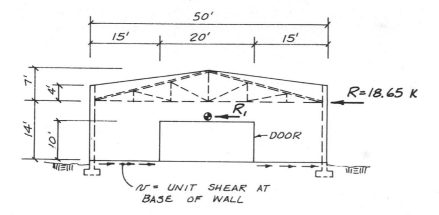

$\mathcal{V}$ = UNIT SHEAR AT BASE OF WALL

$\underline{END \quad ELEVATION}$

Fig. 3.12

Seismic Load

Determine seismic load generated by the total wall DL. Consider reduced DL because of wall opening.

$$\text{Wall area} = \tfrac{1}{2}(3 \times 50) + (18 \times 50) - (10 \times 20)$$
$$= 775 \text{ ft}^2$$
$$\text{Wall DL} = 775 \times 60 = 46.5 \text{ k}$$
$$R_1 = (ZIKCS)W = 0.186W = 0.186 \times 46.5$$
$$R_1 = 8.65 \text{ k}$$

Wall Shear

$$V = R + R_1 = 18.65 + 8.65$$
$$V = 27.3 \text{ k}$$
$$v = \frac{V}{b} = \frac{27{,}300}{30} = \boxed{910 \text{ lb/ft}} \; *$$

3.6   Design Problem: Two-Story-Building Lateral Force Calculations

A multistory building has a number of F_x seismic forces. The *direct* calculation of uniform diaphragm loads that was used in the one-story building example cannot be applied to a multistory building. The total seismic base shear must

* See footnote, Example 3.7.

73

first be calculated and then distributed in accordance with the Code expression for F_x. Wind load calculations are similar to those for the one-story building.

The one-story-building example was divided into a number of separate calculations and sketches for clarity. In the two-story example, the explanations of the one-story example are not repeated. A basic set of calculations is provided which illustrates the differences between single-story and multi-story buildings. See Example 3.9.

EXAMPLE 3.9 Two-Story Lateral Force Calculation

Calculate the critical lateral loads in the transverse direction for the building shown in Fig. 3.13a. Determine unit shears in the transverse direction in the roof and second-floor diaphragms. Also determine the unit shear in the transverse shearwalls at the midheight of the first- and second-story walls. Assume that openings in the masonry shearwalls are so small that they can be neglected.

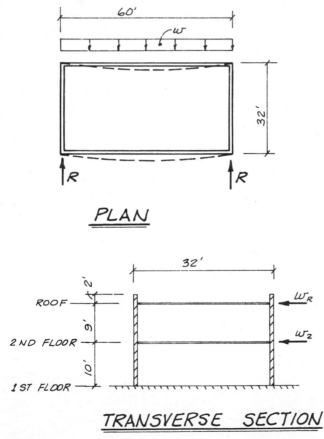

PLAN

TRANSVERSE SECTION

Fig. 3.13a

Loads

Wind Loads

Roof DL $= 20$ psf

Floor DL $= 15$ psf

Wall DL $= 60$ psf

Wind $= 15$ psf

Roof $w_r = 15\left(2 + \dfrac{9}{2}\right)$

$w_r = 97.5$ lb/ft

2d floor $w_2 = 15\left(\dfrac{9}{2} + \dfrac{10}{2}\right)$

$w_2 = 142.5$ lb/ft

Seismic coefficients:

$$Z = 1.0$$
$$I = 1.0$$
$$K = 1.33$$
$$S = 1.5$$

Seismic Loads

For the given seismic information, it is expected that the seismic coefficient will be 0.186. Show calculations to verify

$$\text{Period} = T = \frac{0.05 h_n}{\sqrt{D}} = \frac{0.05(19)}{\sqrt{32}} = 0.168 \text{ seconds} < 0.7$$

$$\therefore F_t = 0$$

$$\text{Response value} = C = \frac{1.0}{15\sqrt{T}} = \frac{1.0}{15\sqrt{0.168}}$$

$$= 0.163 > 0.12$$

$$\therefore \text{Use } C = 0.12$$

$$CS = 0.12 \times 1.5 = 0.18 > 0.14$$

$$\therefore \text{Use } CS = 0.14$$

$$\text{Seismic coefficient} = ZIKCS = (1.0)(1.0)(1.33)(0.14)$$
$$= 0.186 \quad \text{as expected}$$

Tributary Roof Dead Loads

Roof DL	$= 20(32)$	$= 640$ lb/ft
Wall DL (2 walls) $= 2(60 \text{ psf})\left(\dfrac{9}{2} + 2\right)$		$= \underline{780}$
DL per foot of roof (1-ft strip)		$= 1420$ lb/ft
Roof $+$ longit. wall DL $= W_r' = 1420(60)$		$= 85.2$ k
DL of 2 end walls $= 2(60 \text{ psf})(32)\left(\dfrac{9}{2} + 2\right)$		$= \underline{25.0}$
DL trib to roof (total)		$W_r = 110.2$ k

where W_r' represents the mass that generates seismic forces in the roof diaphragm, and W_r represents the total mass tributary to the roof level ($W_r' +$ transverse wall DL).

These quantities will be used to distribute the seismic load in proportion to the mass distribution of the structure. Similar quantities are now calculated for the second floor.

Tributary Second-Floor Dead Loads

$$\text{Floor DL} \qquad = 15(32) \qquad\qquad = \ 480 \text{ lb/ft}$$

$$\text{Wall DL (2 walls)} = 2(60 \text{ psf})\left(\frac{9}{2}+\frac{10}{2}\right) = \underline{1140}$$

$$\text{DL per foot of floor (1-ft strip)} \qquad = 1620 \text{ lb/ft}$$

$$\text{2d floor} + \text{longit. wall DL} = W'_2 = 1620(60) = \ \ 97.2 \text{ k}$$

$$\text{DL of 2 end walls} = 2(60 \text{ psf})(32)\left(\frac{9}{2}+\frac{10}{2}\right) = \ \underline{36.5}$$

$$\text{DL trib. to 2d floor (total)} \qquad\qquad W_2 = 133.7 \text{ k}$$

Calculations of seismic loads for multistory buildings are conveniently carried out in tabular form. The following table can easily be expanded for taller buildings or to include additional items such as overturning moments. The coefficient for calculating the story forces F_x is calculated below the table.

Story	h_x, ft	w_x, k	$w_x h_x$	F_x, k	V_x, k
R	19	110.2	2094	27.7	
					27.7
2	10	133.7	1337	17.7	
					45.4
1	0	—			
Σ		243.9	3431	45.4	

Base Shear

$$V = 0.186W = 0.186 \times 243.9 = 45.4 \text{ k}$$

Distribution of Base Shear Throughout Height

$$F_x = \frac{(V - F_t)\, w_x h_x}{\Sigma\, w_i h_i} \qquad F_t = 0$$

$$F_x = \frac{(45.4)\, w_x h_x}{3431} = (0.01323) w_x h_x$$

The expression for F_x is used to complete the table for story forces.

Story Shear

The story shear is defined as the shear between any two story levels. The story shear can be obtained by taking a free-body diagram (FBD) of the structure above the level in question (Fig. 3.13b).
Calculations for story shear are carried out in the above table.

Distribute Story Loads to Diaphragms

The previously calculated dead loads are used to determine the portion of the total story seismic load that is applied to the horizonal diaphragm.

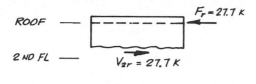
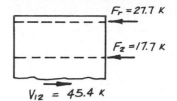

STORY FBD

BETWEEN 2ND FLOOR
AND ROOF

STORY FBD

BETWEEN 1ST AND
2ND FLOOR

Fig. 3.13b

Load to roof diaphragm:

$$F'_r = \left(\frac{W'_r}{W_r}\right) F_r = \left(\frac{85.2}{110.2}\right) 27.7 = 21.4 \text{ k}$$

$$w_r = \frac{F'_r}{L} = \frac{21400}{60 \text{ ft}} = 357 \text{ lb/ft} > 97.5 \text{ wind}$$

∴ Seismic governs

Load to second-floor diaphragm:

$$F'_2 = \left(\frac{W'_2}{W_2}\right) F_2 = \left(\frac{97.2}{133.7}\right) 17.7 = 12.9 \text{ k}$$

$$w_2 = \frac{F'_2}{L} = \frac{12900}{60 \text{ ft}} = 214 \text{ lb/ft} > 142 \text{ wind}$$

∴ Seismic governs

Diaphragm Unit Shears

$$\text{Roof } V_r = R_r = \frac{w_r L}{2} = \frac{357(60)}{2} = 10.7 \text{ k}$$

$$v_r = \frac{V_r}{b} = \frac{10700}{32} = \boxed{335 \text{ lb/ft}}$$

$$\text{Floor } V_2 = R_2 = \frac{w_2 L}{2} = \frac{214(60)}{2} = 6.43 \text{ k}$$

$$v_2 = \frac{V_2}{b} = \frac{6430}{32} = \boxed{201 \text{ lb/ft}}$$

Transverse Shearwalls—Unit Shears

The difference between the total story force and the total load to the horizontal diaphragm is the total seismic force generated by all shearwalls at a given level. Recall that the shearwalls are the walls parallel to the lateral load. Forces to individual walls can be determined by the proportion of the wall length to the total wall lengths.

In this example there are two equal-length shearwalls at each level. As in the one-story example, R_1 is the symbol used to represent the seismic force developed by the shearwall.

77

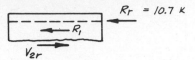

WALL FBD　　　　　　　**Fig. 3.13c**

2ND STORY SHEARWALL

$$R_1 = (F_r - F'_r)\left(\frac{32}{64}\right) = (27.7 - 21.4)\left(\frac{1}{2}\right)$$

$$= 3.15 \text{ k}$$

$$V_{2r} = R_r + R_1 = 10.7 + 3.15 = 13.85$$

For this building, with two equal-length shearwalls, this figure can alternatively be calculated as one-half the story shear from the tabular solution. For one wall,

$$V_{2r} = \tfrac{1}{2} V_{2r} = \tfrac{1}{2}(27.7) = 13.85 \text{ k}$$

$$\text{Unit wall shear} = v_{2r} = \frac{V_{2r}}{b} = \frac{13850}{32} = \boxed{433 \text{ lb/ft}} \text{ *}$$

Similar calculations are made for the first-story wall shears:

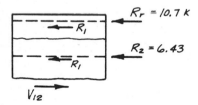

WALL FBD　　　　　　　**Fig. 3.13d**

1 ST STORY SHEARWALL

$$R_1 = (F_2 - F'_2)\left(\frac{32}{64}\right) = (17.7 - 12.9)\left(\frac{1}{2}\right)$$

$$= 2.40 \text{ k}$$

$$v_{12} = \Sigma R + \Sigma R_1$$

$$= 10.7 + 6.43 + 3.15 + 2.40 = 22.7 \text{ k}$$

Alternatively this quantity can be computed as one-half the story shear for a building with two equal shearwalls. For one wall,

$$V_{12} = \tfrac{1}{2} V_{12} = \tfrac{1}{2}(45.4) = 22.7 \text{ k}$$

$$\text{Unit wall shear} = v_{12} = \frac{V_{12}}{b} = \frac{22700}{32} = \boxed{709 \text{ lb/ft}} \text{ *}$$

---

* For design of masonry shearwalls, this load must be increased by a factor of 1.5 (UBC Chap. 24).

In this example the lower portion of the first-story wall DL was disregarded (as in the one-story example given in Fig. 3.11). The solution presented here can be extended to include the lower portion of the wall in a manner similar to that shown in Fig. 3.12.

The above analysis is for the lateral load in the transverse direction. A similar analysis can be used in the longitudinal direction.

3.7 Problems

3.1 The purpose of this problem is to compare the design values of shear and moment for a girder with different assumed load configurations (see Fig. 3.3).

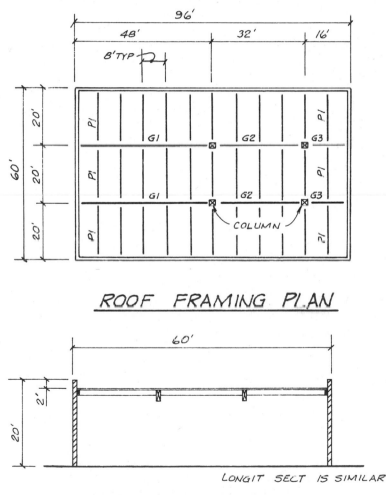

ROOF FRAMING PLAN

TRANSVERSE SECTION

LONGIT SECT IS SIMILAR

Fig. 3.A

Given: The roof framing plan in Fig. 3.A with girders G1, G2, and G3 support-ing purlin loads from P1. Roof DL = 13 psf. Roof LL is to be obtained from UBC Table 23C, Method 1.

Find: *a.* Draw the shear and moment diagrams for girder G1 (DL + LL), assum-ing
      1. A series of concentrated reaction loads from the purlin, Pl.
      2. A uniformly distributed load over the entire span (unit load times the trib. width).
  b. Rework part *a* for girder G2.
  c. Rework part *a* for girder G3.

3.2   Problem 3.2 is the same as Prob. 3.1 except that the Roof DL = 23 psf.

3.3   *Given:* UBC Chap. 23 load requirements.

        Find: The definition of
          a. Box system
          b. LFRS
          c. Shearwall
          d. Braced frame

3.4   *Given:* The plan and section of the building in Fig. 3.B. The wind load is 15 psf.

        Find: *a.* Uniform wind load on the roof diaphragm in the transverse direction. Draw the loading diagram.
          b. The wind load distribution on the roof diaphragm in the longitudinal direction. Draw the loading diagram.
          c. The total diaphragm shear and the unit diaphragm shear at line 1.
          d. The total diaphragm shear and the unit diaphragm shear at line 4.

3.5   *Given:* The plan and section of the building in Fig. 3.B. Roof DL = 15 psf on a horizontal plane, and wall DL = 12 psf. The seismic base shear coefficient has been calculated as 0.186.

        Find: *a.* Uniform seismic load on the roof diaphragm in the transverse direc-tion. Draw the loading diagram.
          b. The seismic load distribution on the roof diaphragm in the longitudi-nal direction. Draw the loading diagram noting the lower load at the overhang.
          c. The total diaphragm shear and the unit diaphragm shear at line 1.
          d. The total diaphragm shear and the unit diaphragm shear at line 4.

3.6   Problem 3.6 is the same as Prob. 3.4 except that the wind load is 20 psf.

3.7   *Given:* The plan and section of the building in Fig. 3.B. Roof DL = 10 psf, and wall DL = 8 psf. The seismic base shear coefficient has been calculated as 0.279.

        Find: *a.* Uniform seismic load on the roof diaphragm in the transverse direc-tion. Draw the loading diagram.
          b. The seismic load distribution on the roof diaphragm in the longitudi-nal direction. Draw the loading diagram noting the lower load at the overhang.
          c. The total diaphragm shear and the unit diaphragm shear at line 1.
          d. The total diaphragm shear and the unit diaphragm shear at line 4.

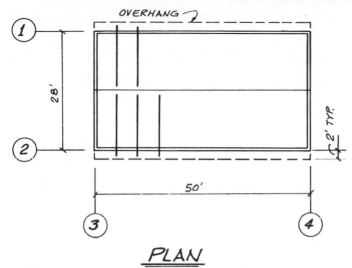

PLAN

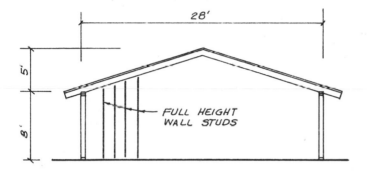

SECTION

Fig. 3.B

3.8 *Given:* The plan and section of the building in Fig. 3.A. The horizontal wind load is 20 psf.

  Find: *a.* The trib. wind load to the roof diaphragm in lb/ft. Draw the loading diagram.
  b. The total diaphragm shear and the unit diaphragm shear at the 60-ft transverse end walls.
  c. The total diaphragm shear and the unit diaphragm shear at the 96-ft longitudinal side walls.

3.9 Problem 3.9 is the same as Prob. 3.8 except that the wind load is 15 psf.

3.10 Problem 3.10 is the same as Prob. 3.8 except that the wind load is 25 psf.

3.11 *Given:* The plan and section of the building in Fig. 3.A. Roof DL = 10 psf, and the walls are 7½-in.-thick concrete. The building is located in seismic zone 4, and $I = 1.0$. T_s is unknown.

  Find: For the transverse direction

  a. The seismic coefficient.
  b. The uniform load to the roof diaphragm in lb/ft. Draw the loading diagram.
  c. The total diaphragm shear and the unit diaphragm shear at the transverse walls.
  d. The total shear and the unit shear at the midheight of the transverse shearwalls.

3.12 Problem 3.12 is the same as Prob. 3.11 except that the longitudinal direction is to be considered.

3.13 *Given:* The plan and section of the building in Fig. 3.A. Roof DL = 12 psf, and the walls are 6-in.-thick concrete. The building is located in seismic zone 3. The building is used for conventions where 600 people occupy one room. T_s is unknown.

  Find: For the transverse direction

  a. The seismic coefficient.
  b. The uniform load to the roof diaphragm in lb/ft. Draw the loading diagram.
  c. The total diaphragm shear and the unit diaphragm shear at the transverse walls.
  d. The total shear and the unit shear at the bottom of the transverse shearwalls.

3.14 Problem 3.14 is the same as Prob. 3.13 except that the longitudinal direction is to be considered.

3.15 *Given:* The elevation of the end shearwall of a building as shown in Fig. 3.C. The load from the roof diaphragm to the shearwall is 10 k. The wall DL = 20 psf, and the seismic coefficient is 0.186.

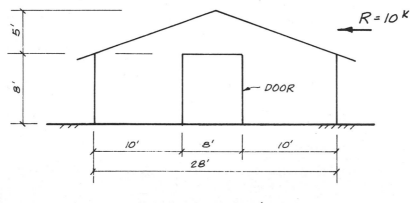

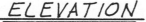

ELEVATION

Find: *a.* The total shear and the unit shear at the midheight of the wall.
       b. The total shear and the unit shear at the base of the wall.

3.16   *Given:*   The elevation of the side shearwall of a building as shown in Fig. 3.D. The load from the roof diaphragm to the shearwall is 50 k. The wall DL = 65 psf, and the seismic coefficient is 0.186.

    Find: *a.* The total shear and the unit shear at the midheight of the wall.
         b. The total shear and the unit shear at the base of the wall.

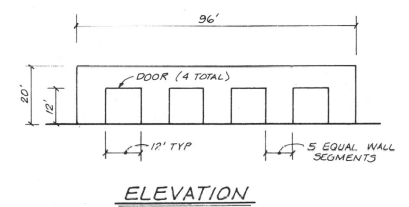

ELEVATION

Fig. 3.D

3.17   *Given:*   The elevation of the side shearwall of a building as shown in Fig. 3.D. The load from the roof diaphragm to the shearwall is 43 k. The wall DL = 75 psf, and the seismic coefficient is 0.233.

    Find: *a.* The total shear and the unit shear at the midheight of the wall.
         b. The total shear and the unit shear at the base of the wall.

3.18   *Given:*   The plan and section of the building in Fig. 3.E. Roof DL = 15 psf, floor DL = 20 psf, wall DL = 53 psf. Wind load = 15 psf. The seismic factors are $Z = 1.0$, $I = 1.25$, $K = 1.33$, and $S = 1.5$. Neglect any wall openings.

    Find: For the transverse direction
         a. The unit shear in the roof diaphragm.
         b. The unit shear in the floor diaphragm.
         c. The unit shear in the second-floor shearwall.
         d. The unit shear in the first-floor shearwall.

3.19   Problem 3.19 is the same as Prob. 3.18 except that the longitudinal direction is to be considered.

3.20   *Given:*   The plan and section of the building in Fig. 3.E. Roof DL = 10 psf, floor DL = 30 psf, wall DL = 16 psf. Wind load = 20 psf. The seismic factors are $Z = I = 1.0$, $K = 1.33$, and $S = 1.5$. Neglect any wall openings.

    Find: For the transverse direction
         a. The unit shear in the roof diaphragm.
         b. The unit shear in the floor diaphragm.

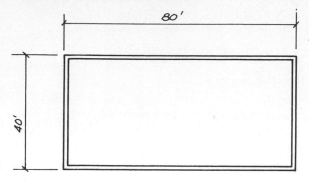

PLAN

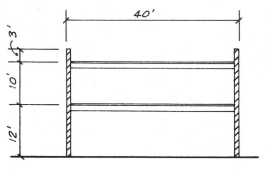

LONGIT SECT IS SIMILAR

TRANSVERSE SECTION

Fig. 3.E

    c. The unit shear in the second-floor shearwall.
    d. The unit shear in the first-floor shearwall.

3.21 Problem 3.21 is the same as Prob. 3.20 except that the longitudinal direction is to be considered.

CHAPTER FOUR

Properties of Wood and Lumber Grades

4.1 Introduction

The designer should have a basic understanding of the characteristics of wood, especially as they relate to the functioning of structural members. Many of these properties are treated at length in the publications listed in Appendix F. Chapter 4 will summarize the important items for the structural designer.

4.2 Classification of Species of Trees

There are a relatively large number of trees that are used to produce structural lumber. The NDS and the UBC have tabulated allowable structural design stresses for the following categories:

Aspen (bigtooth-quaking)*

Balsam fir

Black cottonwood*

California redwood

Coast Sitka spruce

Coast species

Douglas fir-larch

Douglas fir-larch (north)

---

* Hardwood.

Douglas fir-larch (south)

Eastern hemlock-tamarack

Eastern hemlock-tamarack (north)

Eastern spruce

Eastern white pine

Eastern white pine (north)

Eastern woods

Engelmann spruce-alpine fir (Engelmann spruce—lodgepole pine)

Hem-Fir

Hem-Fir (north)

Idaho white pine

Lodgepole pine

Mountain hemlock—Hem-Fir

Northern aspen*

Northern pine

Northern species

Northern white cedar

Ponderosa pine—Sugar pine (ponderosa pine—lodgepole pine)

Red pine

Sitka spruce

Southern pine

Spruce—Pine-Fir

Western cedars

Western cedars (north)

Western hemlock

Western white pine

White woods (Western woods)

The above list represents *commercial* combinations of species rather than a strict *botanical* classification. Certain botanical species are grown, harvested, manufactured, and marketed together. Commercial designations may contain several different botanical species which have similar performance characteristics. Tabulated allowable stresses for a given marketing group are based on the minimum values in the combination.

At first glance the number of species may seem overwhelming and the designer may question which species to use. The choice of species is essentially a matter of economics. For a given location, a relatively few number of species

---

* Hardwood.

will be available, and a check with local lumber distributors will narrow the choices considerably.

It should be noted that as the cost of some of the traditionally popular structural species becomes high because of shortages of supply, some other species may become economically competitive.

Trees are classified into two main categories: *hardwoods* and *softwoods*. These terms are not necessarily a description of the physical properties of the wood, but rather are classifications of trees. Hardwoods are broadleafed deciduous trees. Softwoods, on the other hand, have narrow, needlelike leaves, are evergreen, and are known as conifers. By far the majority of structural lumber comes from the softwood category.

Douglas fir and Southern pine are two species of trees that are widely used in structural applications. Although these species are classified as softwoods, they are relatively dense and have structural properties that exceed those of many hardwoods.

4.3   Cellular Makeup

As a biological material, wood represents a unique structural material because its supply can be renewed by growing new trees in forests which have been harvested. Proper forest management is necessary to ensure a continuing supply of lumber.

Wood is composed of elongated, round, or rectangular tubelike cells. These cells are much longer than they are wide and the length of the cells is essentially parallel with the length of the tree. The cell walls are made up of cellulose, and the cells are bound together by material known as lignin.

If the cross section of a log is examined, concentric rings will be seen. One ring represents the amount of wood material which is deposited on the outside of the tree during one growing season. One ring then is termed an *annular* or *annual ring*. See Fig. 4.1.

The annular rings develop because of differences in the wood cells that are formed in the early portion of the growing season compared with those that are formed toward the end of the growing season. Large, thin-walled cells are formed at the beginning of the growing season. These are known as *earlywood* or *springwood* cells. The cells deposited on the outside of the annual ring toward the end of the growing season are smaller and have thicker walls and are known as *latewood* or *summerwood* cells. It should be noted that annular rings occur only in trees that are located in climate zones which have distinct growing seasons. In tropical zones, trees produce wood cells which are essentially uniform throughout the entire year.

Because summerwood is denser than springwood, it is stronger (the more solid material per unit volume, the greater the strength of the wood). The annular rings, therefore, provide one of the *visual* means by which the strength of a piece of wood may be evaluated. The more summerwood in relation

87

to the amount of springwood (other factors being equal), the stronger the piece of lumber. This comparison is normally made by counting the number of growth rings per inch.

In addition to annular rings, two different colors of wood may be noticed in the cross section of the log. The darker center portion of the log is known as *heartwood*. The lighter portion of the wood near the exterior of the log is known as *sapwood*. The relative amount of heartwood compared with sapwood

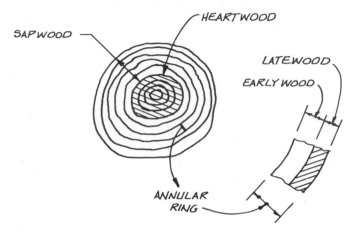

Fig. 4.1 Cross section of a log.

varies with the species of tree. Heartwood, because it occurs at the center of the tree, is obviously much older than sapwood, and, in fact, heartwood represents wood cells which are dead. These cells, however, provide strength and support to the tree. Sapwood, on the other hand, represents wood cells which are living. The strength of heartwood and sapwood is essentially the same. Heartwood is somewhat more decay-resistant than sapwood, but sapwood more readily accepts penetration by wood-preserving chemicals.

4.4 Moisture Content and Shrinkage

The *solid* portion of wood is made of a complex cellulose-lignin compound. The cellulose comprises the framework of the cell wall, and the lignin cements and binds the cells together.

In addition to the solid material, wood contains moisture. The moisture content is measured as the percentage of water to the dry weight of the wood

$$MC = \frac{\text{moist weight} - \text{dry weight}}{\text{dry weight}} \times 100 \text{ percent}$$

The moisture content in a living tree can be as high as 200 percent (i.e., in some species the weight of water contained in the tree can be 2 times the weight of the solid material in the tree). However, the moisture content of structural lumber in service is much less. The average moisture content that lumber assumes in service is known as the *equilibrium moisture content* (EMC). Depending on atmospheric conditions, the EMC of structural framing lumber in a covered structure (dry conditions) will range somewhere between 7 and 14 percent. In most cases, the MC at the time of construction will be higher than the EMC of a building (perhaps 2 times higher). See Example 4.1.

Moisture is held within wood in two ways. Water contained in the cell cavity is known as *free water*. Water contained within the cell walls is known as *bound water*. As wood dries, the first water to be driven off is the free water. The moisture content that corresponds to a complete loss of free water but 100 percent of the bound water is known as the *fiber saturation point* (FSP). No loss of bound water occurs as lumber dries above the fiber saturation point. In addition, no volume changes or changes in other structural

---

EXAMPLE 4.1   Bar Chart Showing Different MC Conditions

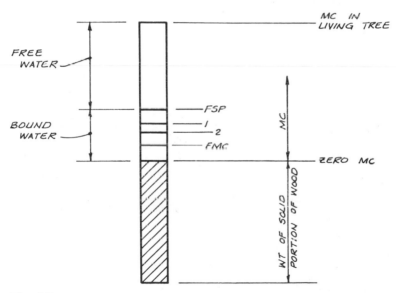

Fig. 4.2

Figure 4.2 shows the moisture content in lumber in comparison with its solid weight. The values indicate that the lumber was fabricated (point 1) at an MC below the FSP. Some additional drying occurred before the lumber was used in construction (point 2). The EMC is shown to be less than the MC at the time of construction. This is typical for most buildings.

properties are associated with changes in moisture content above the fiber saturation point.

However, with moisture content changes below the fiber saturation point, volume changes occur. If moisture is lost, wood shrinks; if moisture is gained, wood swells. Decreases in moisture content below the fiber saturation point are accompanied by increases in all strength properties. The accompanying shrinkage can tend to offset some of these increases, but the net effect of a loss in moisture content below the fiber saturation point will be a benefit. For most species of wood, the FSP corresponds to a moisture content of approximately 30 percent.

The drying of lumber in order to increase its structural properties is known as *seasoning*. As noted, the MC of lumber in a building typically decreases after construction until the EMC is reached. Although this drying in service can be called seasoning, the term seasoning often refers to a controlled drying process. Controlled drying can be performed by air drying or kiln drying (k.d.), and both increase the cost of lumber.

The volume changes that occur in wood as it looses moisture are greater parallel to the annular ring than normal to the annular ring. See Fig. 4.3*a* in Example 4.2. These nonuniform volume changes can result in seasoning defects known as *checks* which are radial cracks.

Three shrinkage values for the various species of wood have been compiled in Ref. 3 and other references included in Appendix F. The shrinkage values given are radial, tangential, and volumetric shrinkage. See Fig. 4.3*b*. The

---

EXAMPLE 4.2   Shrinkage of Lumber

Nonuniform Volume Changes

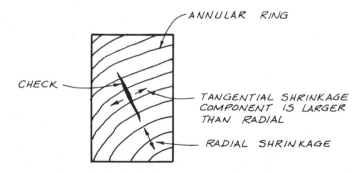

LUMBER   SECTION

Fig. 4.3*a*

Cracks that are essentially radial are the result of nonuniform volume changes and are known as *checks*.

Shrinkage Components

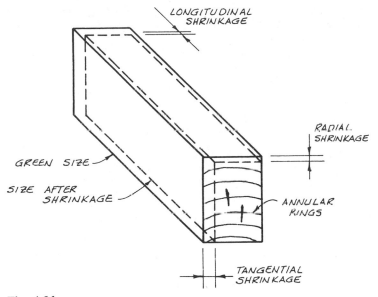

Fig. 4.3*b*

Tabulated values of radial, tangential, and volumetric shrinkage are percentages and therefore reflect *unit* changes.

For simplicity the annular rings are shown parallel to the short faces of the member. In practice the annular rings can have any orientation.

---

main concern is with the change in dimensions of the cross section of a piece of lumber. Hence the radial and tangential values are of primary interest. Very little shrinkage occurs parallel to the length of a piece of wood and longitudinal shrinkage values are not tabulated. However, the volumetric shrinkage is the sum of the radial, tangential, and longitudinal shrinkages. The longitudinal shrinkage, if desired, can be obtained from the three tabulated values.

As an example of shrinkage values, Douglas fir (coast) has a radial value of 4.8, a tangential value of 7.6, and a volumetric value of 12.4. These values represent the percent of shrinkage from a green condition to an oven-dry moisture content. The green condition can be taken as the fiber saturation point, or approximately 30 percent. For purposes of estimating shrinkage due to changes in moisture content, the shrinkage can be considered a linear function of moisture content ranging between 30 percent (no shrinkage) and zero (maximum shrinkage).

In detailing a wood structure, it may be desirable to actually calculate the amount of shrinkage that occurs in the cross section of a piece of wood as a member seasons in place. See Example 4.3.

In order to calculate the shrinkage, it is necessary to first estimate the EMC of the lumber in service. It was previously stated that the EMC in most buildings ranges between 7 and 14 percent. Reference 9 gives typical EMC values for several broad atmospheric zones. The *average* EMC for framing lumber in the "dry southwestern states" (eastern California. Nevada, southern Oregon, southwest Idaho, Utah, and western Arizona) is given as 9 percent. The MC in most covered structures in this area is expected to *range* between 7 and 12 percent. For the remainder of the United States, the *average* EMC is given as 12 percent with an expected *range* of 9 to 14 percent. These average values and ranges can be used to estimate the EMC for typical buildings. Special conditions must be analyzed individually.

EXAMPLE 4.3   Shrinkage Calculations

1. Determine the change in depth of the 2 × 12 in Fig. 4.4 as the moisture content changes from 19 percent to an EMC of 10 percent. Assume Douglas fir (coast).

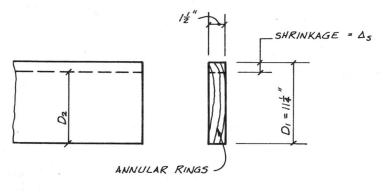

Fig. 4.4

Nominal dimensions are 2 × 12 in. Standard net dimensions are $1\frac{1}{2} \times 11\frac{1}{4}$ in. The formula for calculating shrinkage is

$$\Delta_s = \frac{(MC_1 - EMC)D_1}{\left(\dfrac{30 \times 100}{SV}\right) - 30 + MC_1}$$

where D_1 = initial depth
D_2 = final depth
MC_1 = initial MC
SV = shrinkage value

From the sketch, the required Δ_s is parallel (tangent) to the annular rings

Tangential SV = 7.6 percent   (from Ref. 3)

$$\Delta_s = \frac{(19-10)11.25}{\left(\dfrac{3000}{7.6}\right) - 30 + 19}$$

$$= 0.264 \approx \tfrac{1}{4} \text{ in. change in depth}$$

Depth at EMC = 10 percent:

$$D_2 = 11\tfrac{1}{4} - \tfrac{1}{4} \approx \boxed{11 \text{ in.}}$$

2. For the same size member, determine the change in depth if the annular rings are at right angles to those shown above.
In this case the radial shrinkage must be evaluated.

Radial SV = 4.8 percent   (from Ref. 3)

$$\Delta_s = \frac{(19-10)11.25}{\left(\dfrac{3000}{4.8}\right) - 30 + 19}$$

$$= 0.165 \approx \tfrac{3}{16} \text{ in.}$$

Depth at EMC = 10 percent:

$$D_2 = 11\tfrac{1}{4} - \tfrac{3}{16} \approx \boxed{11\tfrac{1}{16} \text{ in.}}$$

3. For a 2 × 12 with annular rings oriented at some intermediate angle, a final depth between 11 in. and $11\tfrac{1}{16}$ in. can be expected.

Example 4.3 illustrates shrinkage calculations. Calculations of this type are not required in all, or even most, structural designs. The designer, however, should be aware that wood is not a static material and that significant changes in dimensions can result because of atmospheric conditions. There are definitely cases where dimensional changes should be taken into account.

Possible movements of wood (shrinkage or expansion) should *always* be borne in mind when detailing wood structures, even though formal shrinkage calculations may not be performed. These considerations are especially important in connection design. The proper detailing of connections to avoid *built-in stresses* due to changes in moisture content is covered in Chap. 13.

4.5 Effect of Moisture Content on Lumber Sizes

The moisture content of a piece of lumber obviously affects the cross-sectional dimensions. The width and depth of a member are used to calculate the section properties used in structural design. These include area A, moment of inertia I, and section modulus S.

Fortunately for the designer, it is not necessary to compute section proper-

93

ties based on a consideration of the initial MC and EMC and the resulting shrinkage (or swelling) that occurs in the member. Lumber grading practices have established the *dry* size of lumber as the basis for structural calculations. This means that only one set of cross-sectional properties needs to be considered in design.

This is made possible by fabricating lumber to different cross-sectional dimensions based on the moisture content of the wood at the time of fabrication. Therefore, lumber which is fabricated from *green* wood will be somewhat larger at the time of fabrication. However, when this wood reaches a dry moisture content condition, the cross-sectional dimensions will closely coincide with those for lumber fabricated in the dry condition. This discussion has been based on the fabrication practices for *dimension* lumber (Sec. 4.8).

Larger-size wood members known as *timbers,* because of their large cross-sectional dimensions, are not produced in a dry condition since an excessive amount of time would be required to season these members. For this reason, cross-sectional dimensions that correspond to a green condition have been established as the basis for design calculations for these members. Allowable stresses have been adjusted to account for the higher moisture content of *timbers.*

Dry lumber is defined as lumber having a moisture of 19 percent or less. Lumber 20 percent and over in moisture content is defined as *unseasoned* or *green* lumber.

The *grade stamp* on a piece of lumber may contain one of three different moisture-content designations which will indicate the condition of the lumber at the time of fabrication. When unseasoned lumber is grade stamped, the term "S-GRN" (surfaced green) will appear. "S-DRY" (surfaced dry) indicates that the lumber was manufactured with a moisture content of 19 percent or less. *Dimension lumber* or *boards* seasoned to 15 percent or less in moisture content will be marked "MC 15." Examples of grade stamps are given in Sec. 4.12.

4.6   Pressure-Treated Lumber

Wood can be destroyed by several different instruments including

1.   Decay
2.   Termites
3.   Marine borers
4.   Fire

When required, chemicals can be impregnated by *pressure treatment* to prevent or effectively retard the destruction. Pressure treatments force the chemicals into the treated zone so that adequate amounts are retained to ensure performance. Surface treatments are generally unsatisfactory.

Decay is caused by fungi which feed on the cellulose or lignin of the wood. These fungi must have food, moisture (MC greater than approximately 20 percent), air, and favorable temperatures. All of these items are required in order for decay to occur (even so-called dry rot requires moisture).

If any of the requirements are not present, decay will not occur. Thus, untreated wood that is continuously dry (MC < 20 percent, as in most covered structures), or continuously wet (submerged in fresh water—no air), will not decay. Exposure to the weather (alternate wetting and drying) can set up the conditions necessary for decay to develop. Pressure treatment introduces chemicals that poison the food supply of the fungi.

Termites can be found in most areas of the United States, but they are more of a problem in the warmer-climate areas. Subterranean termites are the most common, but drywood and dampwood species also exist. Subterranean termites nest in the ground and enter wood which is near or in contact with damp ground. The cellulose forms the food supply for termites.

The UBC requires a minimum clearance of 18 in. between the bottom of unprotected floor joists (12 in. for girders) and grade. Good ventilation of crawl spaces and proper drainage also aid in preventing termite attack. Lumber which is near or in contact with the ground, and wall plates on concrete ground floor slabs and footings, must be pressure-treated to prevent termite attack. (Foundation-grade redwood has a natural resistance and can be used for wall plates.) The same pressure treatments provide protection against decay and termites.

Marine borers are found in salt waters, and they present a problem in the design of marine piles. Pressure-treated piles have an extensive record in resisting attack by marine borers.

A brief introduction to the fire-resistive requirements for buildings was given in Chap. 1. Where necessary to meet building code requirements, or where the designer decides that an extra measure of fire protection is desirable, *fire-retardant-treated wood* may be used. This type of treatment involves the use of chemicals in formulations that have good fire-retardant properties. Some of the types of chemicals that are used are preservatives and thus also provide decay and termite protection. Fire-retardant treatment, however, requires higher concentrations of chemicals in the treated wood than normal preservative treatments.

The three basic types of pressure preservatives are

1. Creosote and creosote solutions

2. Oilborne treatments (Pentachlorophenol and others dissolved in one of four hydrocarbon solvents)

3. Waterborne salts

There are a number of variations in each of these categories. The choice of the preservative treatment and the required retentions depend on the application. Detailed information on pressure treatments and their uses can be obtained from the American Wood Preservers Institute (AWPI), 1651 Old Meadow Road, McLean, Virginia 22102.

A good introduction to pressure treatments is given in Ref. 9. This reference covers fire-resistive requirements and fire hazards, as well as preservative and fire-retardant treatments.

4.7 Growth Characteristics of Wood

Some of the more important growth characteristics that affect the structural properties of wood are density, moisture content, knots, checks, shakes, splits, slope of grain, reaction wood, and decay. The effects of density, and how it can be measured visually by the annular rings, were described previously. Likewise, moisture content and its effects have been discussed at some length. The remaining natural growth characteristics also affect the strength of lumber, and limits are placed on the size and number of these structural defects which are permitted in a given stress grade. These items are briefly discussed here.

Knots constitute that portion of a branch or limb that has been incorporated into the main body of the tree. See Fig. 4.5. In lumber, knots are classified

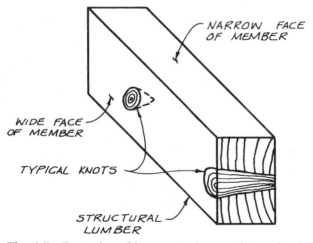

Fig. 4.5 Examples of knots. Lumber grading rules for the commercial species have different limits for knots occurring in the *wide* and *narrow* faces of the member.

as to form, size, quality, and occurrence. Knots decrease the mechanical properties of the wood because the knot displaces *clear wood* and because the slope of the grain is forced to deviate around the knot. In addition, stress concentrations occur because the knot interrupts wood fibers. Checking also may occur around the knot in the drying process. Knots have an effect on both tension and compression capacity, but the effect in the tension zone is greater. Lumber grading rules for different species of wood describe the size, type, and distribution (that is, location and number) of knots that are allowed in each stress grade.

Checks, shakes, and *splits* all constitute separations of wood fibers. See Fig. 4.6. Checks have been discussed earlier and are radial cracks that are caused

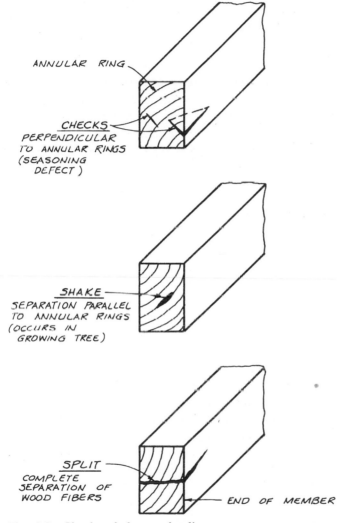

ANNULAR RING

CHECKS
PERPENDICULAR
TO ANNULAR RINGS
(SEASONING
DEFECT)

SHAKE
SEPARATION PARALLEL
TO ANNULAR RINGS
(OCCURS IN
GROWING TREE)

SPLIT
COMPLETE
SEPARATION OF
WOOD FIBERS — *END OF MEMBER*

Fig. 4.6 Checks, shakes, and splits.

by nonuniform volume changes as the moisture content of wood decreases (Sec. 4.4). Recall that more shrinkage occurs tangentially to the annular ring than radially. Checks therefore are seasoning defects. Shakes, on the other hand, are cracks which are usually parallel to the annular ring and develop in the standing tree. Splits represent complete separations of the wood fibers **97** throughout the thickness of a member. A split may result from a shake or seasoning or both. Splits are measured as the penetration of the split from the end of the member parallel to its length. Again, lumber grading rules provide limits on these types of defects.

The term *slope of grain* is used to describe the deviation of the wood fibers from a line that is parallel to the edge of a piece of lumber. Slope of grain is expressed as a ratio (e.g., 1 in 8, 1 in 15, etc.). See Fig. 4.7. In structural lumber, the slope of grain is measured over a sufficient length and area to

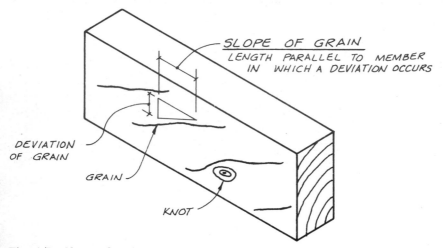

Fig. 4.7   Slope of grain.

be representative of the general slope of wood fibers. Local deviations, such as around knots, are disregarded in the general slope measurement. Slope of grain has a marked effect on the structural capacity of a wood member. Lumber grading rules provide limits on the slope of grain that can be tolerated in the various stress grades.

Reaction wood (known as compression wood in softwood species) is abnormal wood that forms on the underside of leaning and crooked trees. It is hard and brittle and its presence denotes an unbalanced structure in the wood. Compression wood is not permitted in readily identifiable and damaging form in stress grades of lumber. *Decay* is a disintegration of the wood caused by the action of fungi. Grading rules establish limits on the decay allowed in stress-grade lumber. Section 4.6 describes the methods of preserving lumber in service against decay attack.

4.8   Sizes of Structural Lumber and Size (Use) Categories

Structural calculations are based on the *standard net size* of a piece of lumber. The effects of moisture content on the size of lumber are discussed in Sec.

4.5. The designer may have to allow for shrinkage when detailing connections, but standard dimensions are accepted for stress calculations.

Most structural lumber is *dressed lumber*. In other words, the lumber is *surfaced* to the standard net size, which is less than the *nominal* (stated) size. See Example 4.4. Lumber is dressed on a planing machine for the purpose of obtaining smooth surfaces and uniform sizes. Typically lumber will be S4S (surfaced four sides), but other finishes can be obtained (e.g., S2S1E indicates surfaced two sides and one edge).

Dressed lumber is used in many structural applications, but large timbers are commonly *rough sawn* to dimensions that are close to the standard net sizes. The textured surface of rough-sawn lumber may be desired for architectural purposes and may be specially ordered in smaller sizes. The cross-sectional dimensions of rough-sawn lumber are approximately ⅛ in. larger than the standard dressed size. A less common method of obtaining a rough surface would be to specify *full-sawn* lumber. In this case, the actual size of the lumber should be the same as the nominal size. Cross-sectional properties for rough-sawn and full-sawn lumber are not included in the NDS because of their relative infrequent use.

EXAMPLE 4.4   Dressed, Rough-Sawn and Full-Sawn Lumber

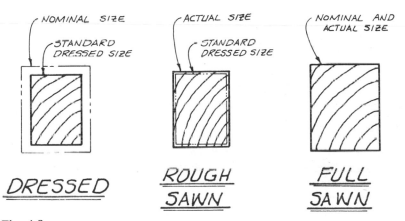

Fig. 4.8

Consider an 8 × 12 member (nomimal size = 8 in. × 12 in.)

1.   *Dressed lumber.* Standard net size = 7½ in. × 11½ in. Refer to NDS Appendix for dressed lumber sizes.

2.   *Rough-sawn lumber.* Approximate size = 7⅝ in × 11⅝ in. Rough size is approximately ⅛ in. larger than the dressed size.

3.   *Full-sawn lumber.* Minimum size 8 in. × 12 in. Full-sawn lumber is not generally available.

The NDS Appendix *Properties of Structural Lumber* gives cross-sectional dimensions and section properties for standard-dressed (S4S) lumber. Notice that the NDS table gives section properties calculated using the second dimension in the callout of a member size as the depth of the member. The first dimension is assumed to be the width of the member. For example, a 2 × 8 will have cross-sectional properties tabulated on the assumption that the member is being stressed about its strong axis. Conversely, an 8 × 2 will have properties tabulated for the member being stressed about the weak axis. See Example 4.5. Section properties for lumber S4S can conservatively be used for rough-sawn or full-sawn lumber. As an alternative, the section properties for rough-sawn or full-sawn lumber can be calculated by the designer.

EXAMPLE 4.5   Section Properties for Dressed Lumber

The cross-sectional properties of dressed lumber are tabulated in the NDS Appendix *Properties of Structural Lumber*. Show calculations to verify the tabulated values of a typical member.

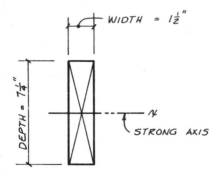

Fig. 4.9a

2 x 8 SECTION

Strong-Axis Section Properties—2 × 8

$$A = bh = 1\tfrac{1}{2} \times 7\tfrac{1}{4}$$
$$= 10.875 \text{ in.}^2$$
$$I = \frac{bh^3}{12} = \frac{1\tfrac{1}{2}(7\tfrac{1}{4})^3}{12}$$
$$= 47.635 \text{ in.}^4$$
$$S = \frac{bh^2}{6} = \frac{1\tfrac{1}{2}(7\tfrac{1}{4})^2}{6}$$
$$= 13.141 \text{ in.}^3$$

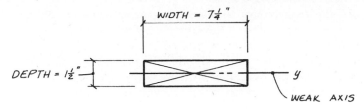

$$8 \times 2 \quad SECTION$$

Fig. 4.9b

Weak-Axis Section Properties—8 × 2

$$A = bh = 7\tfrac{1}{4} \times 1\tfrac{1}{2}$$
$$= 10.875 \text{ in.}^2$$
$$I = \frac{bh^3}{12} = \frac{7\tfrac{1}{4}(1\tfrac{1}{2})^3}{12}$$
$$= 2.039 \text{ in.}^4$$
$$S = \frac{bh^2}{6} = \frac{7\tfrac{1}{4}(1\tfrac{1}{2})^2}{6}$$
$$= 2.719 \text{ in.}^3$$

EXAMPLE 4.6   Size (Use) Categories

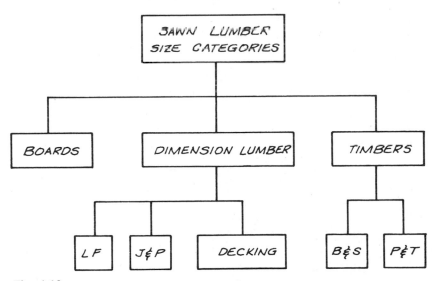

Fig. 4.10

DESIGN OF WOOD STRUCTURES

The three main categories of lumber (boards, dimension, and timbers) are divided into a large number of subcategories. The five divisions shown in Fig. 4.10 are those which constitute most structural lumber. In general, the other subcategories are used for nonstructural applications. Sizes in the five basic subcategories are summarized in the following table.

Symbol	Name	Nominal dimensions		Examples of sizes
		Thickness	Width	
LF	Light Framing*	2 to 4 in.	2 to 4 in.	2×2, 2×4, 4×4
J&P	Joist and Plank*	2 to 4 in.	5 in. and wider	2×6, 2×14, 4×10
	Decking*	2 to 4 in.	4 in. and wider	2×4, 2×8, 4×6
B&S	Beam and Stringer	5 in. and thicker	More than 2 in. greater than thickness	6×10, 6×14, 12×16
P&T	Post and Timber	5 in. and thicker	Not more than 2 in. greater than thickness	6×6, 6×8, 12×14

 * Some of the same sizes occur in Decking and LF, and Decking and J&P. Decking will be stressed about its minor axis, and lumber graded as LF or J&P will normally be stressed about its strong axis. In this book, lumber in these sizes will be assumed to be LF or J&P (instead of Decking) unless otherwise stated.

Now that the dimensions of structural lumber have been described, the *size* categories can be explained. The lumber grading rules which establish allowable stresses for use in structural design have developed over many years. In this development process, the *relative size* of a piece of wood was used as a guide in *anticipating* the application or "use" that a member would receive in the field. For example, members with rectangular cross sections make more efficient beams than members with square (or approximately square) cross sections. If the final application of a piece of wood were known, stress-grading rules could be established which account for the primary function (axial strength or bending strength) of the member. For this reason, lumber is divided into a number of *size (use)* categories. See Example 4.6. Notice that the emphasis is on size rather than use, and it is the size of a member that determines the category.

Size categories *must* be understood by the designer. The reason for this is that *different allowable stresses* apply to the *same stress grade* in the *different size categories*. For example, Select Structural (a stress grade) is available in LF, J&P, B&S, and P&T size categories. However, the allowable stresses are different for Select Structural in *each* of these size categories. See Example 4.7. The NDS Supplement emphasizes these differences by printing allowable stresses for the different size categories in colored bands.

EXAMPLE 4.7   Stress Grades

Typical stress grades within the various size categories (Douglas fir-larch; other species have similar stress grades).

1. *Light Framing* (LF)
    Dense Select Structural
    Select Structural
    Dense No. 1
    No. 1
    Dense No. 2
    No. 2
    No. 3
    Appearance
    Stud

2. *Joist and Plank* (J&P)
    Dense Select Structural
    Select Structural
    Dense No. 1
    No. 1
    Dense No. 2
    No. 2
    No. 3
    Appearance
    Stud

3. *Decking*
    Selected Decking
    Commercial Decking

4. *Beam and Stringer* (B&S)
    Dense Select Structural
    Select Structural
    Dense No. 1
    No. 1

5. *Post and Timber* (P&T)
    Dense Select Structural
    Select Structural
    Dense No. 1
    No. 1

The importance of clearly understanding the size categories can be seen from the above list of stress grades. The stress grade of *Select Structural*, for example, is available in each of the five size categories. Different allowable stresses are given for *Select Structural* in each of the five size categories.

As noted, the lumber grading rules developed considering the *anticipated use* of a wood member based on its size, but no such restriction exists for the *actual use* of the member by the designer. In other words, lumber that falls into the Beam and Stringer (B&S) size category was originally anticipated to be used as a bending member. As a rectangular member, a B&S bending about its strong axis is a more efficient beam (because of its larger section modulus) than a square (or essentially square) member such as a P&T. However, allowable stresses are tabulated for tension and compression as well as bending for all size categories. The designer can use a B&S in any of these applications.

Although size and use are related, it must be emphasized again that the *allowable stresses depend on the size* of a member rather than its *use*. Thus, a member in the Post and Timber (P&T) size category is always graded as a P&T even though it may eventually be used as a beam. Therefore, if a 6×8 is used as a beam, the allowable bending stress for a P&T applies. Similarly, if a 6×10 is used as a column, the compression value for a B&S must be used.

4.9   Stress Grades of Lumber

Lumber grading rules which establish the number, size, and distribution of strength-reducing characteristics provide allowable stresses for a number of stress grades. As noted in the previous section, the size (use) category affects the magnitude of the allowable stresses assigned to a particular stress grade. The stress grades that are available in the different size categories for one species of wood (Douglas fir-larch) are summarized in Example 4.7. Similar stress grades are used for other species of wood.

The following allowable stresses are tabulated for structural lumber:

Bending, F_b

Tension parallel to grain, F_t

Horizontal shear, F_v

Compression parallel to grain, F_c

Compression perpendicular to grain, $F_{c\perp}$

Modulus of elasticity, E

Tables of these allowable stresses are given in the NDS Supplement, Table 4A, *Design Values for Visually Graded Structural Lumber,* and in Chap. 25 of the UBC. All of these tabulated stresses include a factor of safety (except modulus of elasticity) and are, therefore, termed *allowable* stresses.

The development of allowable stresses for visually graded sawn lumber is covered by ASTM Standard D 245 and ASTM Standard D 2555. These are included in Ref. 29. Briefly, the process involves:

1.   A statistical analysis of a large number of ultimate clear-wood strength values for the various commercial species. With the exception of $F_{c\perp}$ and E, the 5 percent *exclusion value* serves as the starting point for the development of allowable stresses. Out of 100 clear wood specimens, 95 could be expected to fail at or above the 5 percent exclusion value and 5 could be expected to fail below this value.

2.   The 5 percent exclusion value is then adjusted for different *moisture contents*. Reductions in moisture content result in an increased stress.

3.   *Strength ratios* are used to adjust the clear-wood values to account for the strength-reducing defects permitted in a given stress grade.

4. The stresses from (3) are further reduced by a *general adjustment factor* which accounts for the duration of the test used to establish the initial clear-wood values, a manufacture and use adjustment, and several other factors.

The combined effect of these adjustments is to provide an average factor of safety on the order of 2.5. Because of the large number of variables in a wood member, the factor of safety for a given member may be considerably larger or smaller than the average. However, for 99 out of 100 pieces, the factor of safety will be greater than 1.25. Reference 8 gives a comprehensive description of the development of allowable stresses and the factors of safety. If the factors of safety provided by the tabulated allowable stresses are, in the judgment of the designer, insufficient for a particular design, the allowable stresses should be adjusted accordingly.

It was noted above that the modulus of elasticity does not contain a factor of safety. There are several reasons for this. One main use of the modulus of elasticity is in the calculation of the deflection of a beam. Because it may be desirable to know the actual amount of deflection, the value of E is a representative value. The factor of safety for deflection is taken into account in the establishment of allowable deflection limits. The modulus of elasticity is also used to calculate column strength. The factor of safety for column capacity is provided in the formulas for allowable column stress (Chap. 7) and the representative value of E, again, is used directly.

It should be noted that the *tabulated allowable stresses* simply represent a *starting point* in determining the *true allowable stresses* for a given set of design circumstances. Wood is a biological material with a number of growth characteristics which cause it to behave somewhat differently from other structural materials.

Generally speaking, wood structures are designed according to the principles of engineering mechanics and strength of materials. The calculation of actual stresses uses the same basic linear elastic theory that is applied to other structural materials (e.g., structural steel design). However, the unique aspects of wood must be reflected in design. There are exceptions, but usually this is accomplished by adjusting the *tabulated allowable stresses* to reflect the important factors that affect the strength of wood for a given set of design circumstances.

In most cases, adjusted allowable stresses can be obtained by *multiplying* tabulated allowable stresses by the appropriate adjusting factor(s). See Example 4.8. These factors should be used to adjust allowable stress and should not be used to modify design loads. Consistency in this approach is highly recommended.

Use of the appropriate factors for adjusting allowable stresses will be emphasized throughout this text. Some of the adjusting factors will cause the allowable stress to decrease, and others will cause the allowable stress to increase.

When factors that reduce strength are considered, a larger member size   **105**

EXAMPLE 4.8   Symbols and Adjustments of Stresses

Symbols for stresses in wood design have been standardized. These symbols are similar to those adopted by the American Institute of Steel Construction (AISC) for use in designing structural steel members. A list of symbols follows the table of contents.

Actual Stresses

Actual stresses are calculated from known loads and member sizes. These stresses are given the symbol of a lower case f and a subscript is used to indicate the type of stress involved. For example, the axial tension stress in a member is calculated as the force divided by the cross-sectional area. The notation is

$$f_t = \frac{P}{A}$$

Allowable Stresses

Members are checked by verifying that the actual stresses are less than the adjusted allowable stresses. Allowable stresses are given the symbol of an upper case F and a subscript is used to indicate the type of stress.

Tab. F_t = tabulated allowable tension stress
from the NDS Supplement.
Adj. F_t = adjusted allowable tension stress
for a given set of design conditions.
= (Tab. F_t) (product of adjusting factors)

Adjusting Factors

Some possible adjusting factors are

LDF = load duration factor
CUF = condition-of-use factor
C_F = size factor (bending stress only)
C_f = form factor (bending stress only)

Numerous other adjustments are possible. Those listed here are typical examples.

---

will be required to support a given load. On the other hand, when circumstances exist which produce increased strength, smaller, more economical members can result if these factors are taken into consideration. The point here is that a number of items can affect the strength of wood. These items *must* be considered in design when they result in a reduction of member capacity. Factors which increase the calculated strength of a member *may* be considered in the design.

This discussion emphasizes that a *conservative* approach (i.e., in the direction of greater safety) in structural design is the general rule. Factors which cause member sizes to increase *must* be considered. Factors which cause them to decrease *may* be considered or ignored. The question of whether the latter *can* be ignored has to do with economics. It may not be practical to ignore reductions in member sizes that result from a beneficial set of conditions.

Two of the factors in Example 4.8 apply to bending stresses only, and these are covered in Chap. 6 on beam design. The other two adjusting factors (condition-of-use and load duration factor) apply to most stress calculations and are treated in subsequent sections of this chapter.

Another somewhat general adjustment factor applies to lumber that has been pressure-treated (Sec. 4.6). Lumber that has a preservative treatment (i.e., lumber that is treated against decay, termites, or marine borers) has an allowable stress-multiplying factor of 1.0. In other words, no reduction of allowable stress is required for preservative treatments. On the other hand, the higher concentrations of chemicals in fire-retardant-treated lumber require a 10 percent reduction of allowable stresses (0.90 multiplying factor).

4.10   Condition-of-Use Factor (CUF)

The moisture content in wood can increase or decrease depending on local conditions. Strength properties are affected by the moisture content, and the tabulated allowable stresses in the NDS (and in the Code) apply to a specific set of moisture-content conditions. When different conditions exist, allowable stresses are to be adjusted to reflect the changes in strength.

Condition-of-Use Factor (CUF) is used in this book to refer collectively to any adjustment factor related to moisture content. It may be necessary to consider the moisture content at several different stages. These stages can include the moisture content

1.   At the time of surfacing (dressing of the lumber)
2.   In service (EMC)
3.   At the time of fabrication (this is especially important in the design of connections)

Tabulated allowable stresses for sawn lumber are given in the NDS Supplement, Table 4A, *Design Values for Visually Graded Structural Lumber.* These allowable stresses apply to lumber fabricated (surfaced) at a specified maximum moisture content and used under *dry* conditions such as in most covered structures. For most species of wood, tabulated allowable stresses apply to lumber *surfaced dry or surfaced green* and *used at a moisture content of 19 percent or less.* For certain species, such as Southern pine, there are separate tables for different moisture-content conditions.

When lumber is used in conditions which are different from those described, adjusting factors (CUFs) are provided in the *footnotes* to the NDS table. As an example, consider the adjustments to the allowable bending and compression parallel to grain for a 4 × 12 (not Southern pine) that is used at an EMC greater than 19 percent. The NDS Supplement, Table 4A, footnote 9, gives the following adjustments

$$\text{Adj. } F_b = (\text{CUF}) F_b = 0.86 F_b$$
$$\text{Adj. } F_c = (\text{CUF}) F_c = 0.70 F_c$$

CUFs for other allowable stresses and other moisture-content conditions can be obtained from the NDS.

4.11   Load Duration Factor (LDF)

Wood has a unique structural property. Generally speaking, wood can support higher stresses if these stresses are applied for a short period of time. This is particularly significant when it is realized that if an overload occurs, it is probably a temporary load.

The *load duration factor* (LDF) is used in this book to refer to adjustment factors to account for the length of time that a load is applied. The total accumulated length of time that a load is applied to a structure is referred to as the duration of loading.

In considering duration, it is the *full* design load that is of concern, and not the length of time that a fraction of the load is applied. For example, the duration associated with wind loading is taken as one day. Obviously, some wind or air movement is almost always present, but the maximum wind pressure during the lifetime of the structure is assumed to act for a total of one day.

Tabulated allowable stresses are based on a *normal* duration of load. This is taken as 10 years, and floor live loads are associated with this duration. Because tabulated allowable stresses apply directly to floor live loads, the LDF for this type of loading is 1.0. The LDF for other types of loads ranges from 2.0 to 0.9. See Example 4.9. It should be noted that the LDF applies to allowable stresses only, and it does not apply to modulus of elasticity E.

The curve in Example 4.9 shows the load duration factor plotted against total accumulated duration of load. The curve also indicates the durations that are commonly assigned to certain design loads.

It was previously stated that adjusting factors, including the LDF, should be applied as a multiplying factor for adjusting allowable stresses. Modifications of actual stresses or modifications of applied loads should not be used to account for load duration. A consistent approach in the application of the LDF to the allowable stress will avoid confusion.

The stresses that occur in a structure are usually not the result of a single applied load. On the other hand, they are normally caused by a combination of loads that act simultaneously. The question arises then as to which load duration factor should be applied when checking a stress caused by a combination of loads. It should be noted that the load duration factor applies to the *entire combination of loads* and not just to that portion of the stress caused by a load of a particular duration. The LDF to be used is the one associated with the *shortest* duration load in the combination.

For example, consider the possible load combinations on a floor beam that also carries a column load from the roof. What are the appropriate LDFs for the various load combinations? If stresses under DL alone are checked,

EXAMPLE 4.9 **Load Duration Factors**

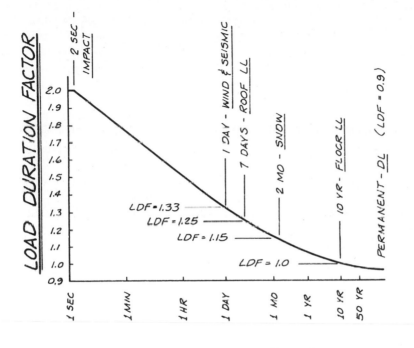

DURATION OF LOAD (TIME)

Fig. 4.11

Shortest duration load in combination	LDF
Dead load	0.9
Floor live load	1.0
Snow load	1.15
Roof live load	1.25
Wind or seismic	1.33
Impact	2.0

NOTES: 1. Check all Code-required load combinations.
2. LDF associated with the shortest duration load in the combination is used to adjust allowable stress.
3. The critical combination of loads is the one that requires the largest-size structural member.

the LDF $= 0.9$. If stresses under (DL + FLL) are checked, the shortest-duration load in the combination is FLL, and the LDF $= 1.0$. For (DL + FLL + snow), the LDF $= 1.15$. If no snow load occurs, the last combination becomes (DL + FLL + RLL), and the LDF $= 1.25$.

In this manner, it is possible for a smaller-magnitude load of long duration (with a small LDF) to actually be more critical than a larger load of shorter duration (with a large LDF). Whichever combination of loads, together with the appropriate duration factor, produces the largest member size is the one that must be used in the design of the structure. It may be necessary, therefore, to check several different combinations of loads in order to determine which combination governs the design. With some practice, the designer can often tell by inspection which combinations of loads need to be checked. In many cases, one or possibly two combinations only need to be checked. See Example 4.10.

EXAMPLE 4.10   Comparison of Load Combinations

Determine the design loads and the critical load combination for the beam in Fig. 4.12. The tributary width to the beam and the design unit loads are given.

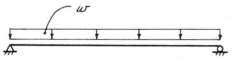

Fig. 4.12

$$\text{Trib. width} = 10 \text{ ft}$$
$$\text{Roof DL} = 20 \text{ psf}$$
$$\text{Roof LL} = 16 \text{ psf}$$

Part a
Load combination 1 (DL alone):

$$w_{DL} = 20 \times 10 = 200 \text{ lb/ft}$$
$$\text{LDF} = 0.90$$

Load combination 2 (DL + RLL):

$$w_{TL} = (20 + 16)10 = 360 \text{ lb/ft}$$
$$\text{LDF} = 1.25$$

The tabulated allowable stresses for the beam must be multiplied by 0.90 for load combination 1, and they must be multiplied by 1.25 for load combination 2. Theoretically both load combinations must be considered. However, with some practice, the designer will be able to tell from the relative magnitude of the loads which combination is critical. For example, 360 lb/ft is so large in comparison with 200 lb/ft that load combination 2 will be critical. Therefore, calculations for load combination 1 are not required. If it cannot be determined by inspection which loading is critical, calculations for both load cases should be performed.

In some cases calculations for two or more load cases *must* be performed. Often this occurs in members with combined axial and bending loads. These types of problems are considered in Chap. 7.

Part b

Show calculations which verify the critical load case for the beam in part *a* without complete stress calculations. Remove "duration" by dividing the design loads by the appropriate LDFs.*

Load combination 1:

$$\frac{w_{DL}}{LDF} = \frac{200}{0.90} = 222 \text{ lb/ft}†$$

Load combination 2:

$$\frac{w_{TL}}{LDF} = \frac{360}{1.25} = 288 \text{ lb/ft}†$$

$$222 < 288 \qquad \therefore \text{ load combination 2 governs}$$

When some designers first encounter the adjustment for duration of loading, they like to have a *system* for determining the critical loading condition. See Example 4.10, part *b*. The system presented is included as an introductory exercise only. It does not apply to all members, and it is, therefore, of limited value. The problem with the system has to do with column design. Not all column ranges (short, intermediate, and long—Chap. 7) use the LDF. The system is useful, however, in introducing the general subject of load duration.

Essentially the system involves removing the question of duration from the problem so that several load combinations can be compared directly. If the sum of the loads in a given combination is divided by the LDF for the combination, duration is removed from the load. If this is done for each required load combination, the resulting loads can be compared. The largest modified load represents the critical combination.

There is a second objection to the system just described. It runs counter to the recommendation that adjustment factors be applied to the allowable stress and not to the design loads. Thus, if this analysis is used, the calculations should be done separately (perhaps on scrap paper). Once the critical combination is known, the actual design loads (not modified) can then be used in formal calculations, and the LDF can be applied to the allowable stress in the usual manner.

---

* This method cannot be applied to all types of members.

† These modified loads are to be used to determine the critical load combination only. Actual design loads (e.g., $w = 360$ lb/ft) should be used in calculations. The LDF $= 1.25$ should be applied to allowable stresses.

111

4.12 Lumber Grade Stamps

Lumber which is to be used in structural applications must be stress graded. The majority of softwood lumber is stress graded by visual inspection. A small percentage of lumber is machine-stress rated by subjecting each piece of wood to a nondestructive test. Each piece of wood as it is visually inspected is stamped with a grade mark. See Fig. 4.13.

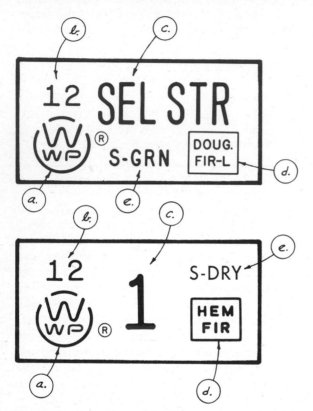

Fig. 4.13 Examples of lumber grade stamps. (a) Lumber grading association. (b) Mill number. (c) Lumber grade. (d) Commercial lumber species. (e) Moisture content at the time of surfacing. (*Courtesy of WWPA.*)

The design values (allowable stresses) contained in the NDS Supplement are obtained from the grading rules published by the following agencies:

1. National Lumber Grades Authority (a Canadian agency)
2. Northeastern Lumber Manufacturers Association
3. Northern Hardwood and Pine Manufacturers Association
4. Redwood Inspection Service

5. Southern Pine Inspection Bureau

6. West Coast Lumber Inspection Bureau

7. Western Wood Products Association

The designer should be aware that more than one set of grading rules can be used to grade some commercial species. For example, Douglas fir-larch can be graded under the Western Wood Products Association (WWPA) rules or under the West Coast Lumber Inspection Bureau (WCLIB) rules. There are some differences in allowable stresses between the two sets of rules.

Both the WWPA and WCLIB rules have "full length" grading in all sizes (LF, J&P, and P&T) *except* B&S. For the Beam and Stringer sizes the WWPA has full-length grading, and a higher allowable tension stress can be justified. The WCLIB rules apply the more restrictive rules in the middle third of the member only (the location of high bending stresses in a simple beam), and less restrictive rules are used for the outer thirds. F_t is smaller for the latter rules because of the greater defects in the outer thirds. Other differences in allowable stresses are a result of the rounding procedures that were applied when the grading agencies developed the allowable stress tables.

Because the designer has no control over which set of grading rules will be used, the lower allowable stress should be used when conflicting values are tabulated. The higher allowable stresses are justified in a review of the design of an existing member that has the appropriate grade stamp present.

It was previously noted that lumber grading rules establish limits on the size and number of growth characteristics that are permitted in the various grades. If lumber that has previously been graded is resawn (for example, several smaller pieces of lumber produced from one large member), the initial grade stamp no longer is valid and the lumber must be regraded. The reason for this requirement is that the acceptable size of a defect in the original large member may not be permitted in the same grade for the smaller resawn members.

4.13 Problems

4.1 Describe the following types of trees:
   a. Softwoods
   b. Hardwoods
   What types of trees are used for most structural lumber?

4.2 Sketch the cross section of a log. Label and define the following items:
   a. Annular ring
   b. The two types of wood cells
   c. Heartwood and sapwood

4.3 Define the following terms:
   a. Moisture content

113

      b. Fiber saturation point
      c. Equilibrium moisture content

4.4   Give the moisture content ranges for:
      a. Dry lumber
      b. Green lumber

4.5   What is the average EMC for an enclosed building in southern California?

4.6   Determine the dressed size, area, moment of inertia, and section modulus for the following members. Tables may be used.
      a. 2×4 and 4×2
      b. 2×8 and 8×2
      c. 4×10 and 10×4
      d. 6×16 and 16×6

4.7   Give the size limitations for the following categories of lumber:
      a. Light Framing
      b. Joist and Plank
      c. Beam and Stringer
      d. Post and Timber

4.8   Give the size (use) categories for the following members:
      a. 10×12
      b. 14×14
      c. 4×8
      d. 4×4
      e. 2×12
      f. 6×12
      g. 8×12
      h. 8×10

4.9   *Given:*   No. 1 Douglas fir-larch

      Find: The tabulated allowable single-member bending stress F_b, tension parallel to grain (F_t), and modulus of elasticity E for the following members:
            a. 2×4
            b. 2×6
            c. 4×6
            d. 6×6
            e. 6×10
            f. 8×10

4.10   Problem 4.10 is the same as Prob. 4.9 except that the stress grade is No. 1 Hem-Fir.

4.11   Problem 4.11 is the same as Prob. 4.9 except that the stress grade is Select Structural Spruce-Pine-Fir.

4.12   Problem 4.12 is the same as Prob. 4.9 except that the stress grade is No. 3 White woods (Western woods).

4.13   Give the load duration factor (LDF) associated with the following loads:
      a. Snow
      b. Wind

  c. Floor live load
  d. Roof live load
  e. Dead load

4.14 A column in a building is subjected to several different loadings.

  Find: For each of the following subproblems determine the critical load combi-
    nation. Assume that the column is in the *short* (defined in Chap. 7) column
    range, and, therefore, the LDF applies.
    a. DL = 5 k; roof LL = 3 k.
    b. Roof DL = 3 k; roof LL = 5 k; floor DL = 6 k; floor LL = 10 k.
    c. Roof DL = 3 k; roof LL = 5 k; floor DL = 6 k; floor LL = 10 k;
      wind load = 10 k (resulting from overturning loads on the lateral-
      force-resisting system).

4.15 A column in a building is subjected to several different loadings.

  Find: For each of the following subproblems determine the critical load combi-
    nation. Assume that the column is in the *short* (defined in Chap. 7) column
    range, and, therefore, the LDF applies.
    a. Roof DL = 10 k; roof LL = 2 k.
    b. Roof DL = 10 k; roof LL = 2 k; floor DL = 8 k; floor LL = 10 k.
    c. Roof DL = 10 k; roof LL = 2 k; floor DL = 8 k; floor LL = 10 k;
      wind load = 6 k (axial column force resulting from overturning in
      the LFRS).

4.16 A column in a structure is supporting a water tank. The axial load from the
  tank plus water is P_w, and the axial load resulting from the lateral overturning
  load is P_l. Because the contents of the tank are present much of the time,
  this load is considered a permanent load.

  Find: The critical load combination for each of the following loadings. Assume
    that the column is in the *short* (defined in Chap. 7) column range, and,
    therefore, the LDF applies.
    a. $P_w = 60$ k; $P_l = 10.5$ k.
    b. $P_w = 60$ k; $P_l = 40.3$ k.
    c. $P_w = 60$ k; $P_l = 29.1$ k.

4.17 A roof beam supports a tributary load from roof rafters.

  Find: The critical combination of uniform loads for the following loading con-
    ditions.
    a. Roof DL = 200 lb/ft.
    b. Roof DL = 200 lb/ft; roof LL = 150 lb/ft.
    c. Roof DL = 150 lb/ft; snow load = 400 lb/ft.

4.18 A roof beam supports a tributary load from roof rafters.

  Find: The critical combination of uniform loads for the following loading con-
    ditions.
    a. Roof DL = 175 lb/ft; roof LL = 350 lb/ft.
    b. Roof DL = 150 lb/ft; snow load = 500 lb/ft.
    c. Roof DL = 350 lb/ft; roof LL = 175 lb/ft.

4.19 A floor beam supports a tributary load from floor joists.

  Find: The critical combination of uniform loads for the following loading con-
    ditions.

    a. Floor DL = 600 lb/ft; floor LL = 1000 lb/ft.
    b. Floor DL = 800 lb/ft; floor LL = 2000 lb/ft.
    c. Floor DL = 600 lb/ft; floor LL = 800 lb/ft.

4.20 A 2 × 4 Select Structural Hem-Fir member is fabricated at a maximum moisture content of 15 percent (MC15). It is used in a building where the EMC = 15 percent.

  Find: *a.* The CUFs (if any) for this condition of use.
          b. The adjusted allowable stresses for this condition.

4.21 Problem 4.21 is the same as Prob. 4.20 except that the member is a 4 × 10.

4.22 A 4 × 12 Select Structural Douglas fir (south) member is exposed to the weather such that the moisture content of the wood will exceed 19 percent for extended periods of time.

  Find: *a.* The CUFs for this condition of use.
          b. The adjusted allowable stresses for this condition.

4.23 Problem 4.23 is the same as Prob. 4.22 except that the member is a 4 × 4.

4.24 Problem 4.24 is the same as Prob. 4.22 except that the member is a 6 × 14.

4.25 Problem 4.25 is the same as Prob. 4.22 except that the member is a 6 × 8.

CHAPTER FIVE

Structural Glued-Laminated Timber

5.1 Introduction

Sawn lumber is fabricated in a large number of sizes and grades (Chap. 4) and is used for a wide variety of structural members. However, the cross-sectional dimensions and lengths of these members are limited by the size of the trees available to produce this type of lumber.

When the span becomes long or when the loads become large, the use of sawn lumber may become impractical. In these circumstances (and possibly for architectural reasons) structural glued-laminated timber (glulam) can be used.

Glulam members are fabricated from relatively thin laminations of wood. These laminations can be glued together and spliced in such a way to produce wood members of practically any size and length. Lengths of glulam members are limited by handling systems and length restrictions imposed by highway transportation systems rather than by the size of the tree.

Chapter 5 provides an introduction to glulam lumber and its design characteristics. The similarities and differences of glulam and sawn wood members are also noted.

5.2 Sizes of Glulam Members

The pieces of wood that are used to form laminations in a glulam member are typically either ¾ in. or 1½ in. in thickness. These dimensions are the actual thicknesses of the laminations. The total depth of the member, then, will be a multiple of the lamination thickness.

Glulam members may be straight or they may be bent in order to provide a curved member. See Fig. 5.1. If a member is sharply curved, the smaller-thickness (¾-in.) laminations should be used in the fabrication because smaller built-in stresses will result. However, a straight member, or a member with only a small degree of curvature, will be less expensive if the thicker (1½-in.) laminations are used. Cost is heavily influenced by the number of glue lines in a member.

Only the design of straight and slightly curved rectangular members is

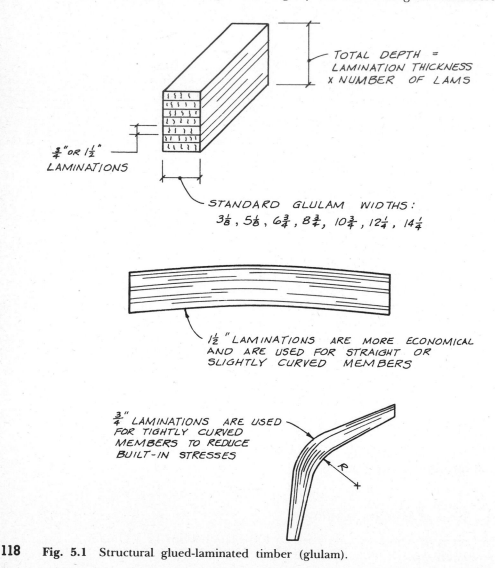

TOTAL DEPTH = LAMINATION THICKNESS × NUMBER OF LAMS

¾" OR 1½" LAMINATIONS

STANDARD GLULAM WIDTHS:
$3\frac{1}{8}$, $5\frac{1}{8}$, $6\frac{3}{4}$, $8\frac{3}{4}$, $10\frac{3}{4}$, $12\frac{1}{4}$, $14\frac{1}{4}$

1½" LAMINATIONS ARE MORE ECONOMICAL AND ARE USED FOR STRAIGHT OR SLIGHTLY CURVED MEMBERS

¾" LAMINATIONS ARE USED FOR TIGHTLY CURVED MEMBERS TO REDUCE BUILT-IN STRESSES

R

118   **Fig. 5.1**   Structural glued-laminated timber (glulam).

included in this text. The design of special tapered members and curved members (including arches) is covered in Ref. 3.

The sizes of glulam members are called out on plans by giving their actual dimensions (unlike sawn lumber which uses "nominal" sizes). The standard widths of glulam members are given in Fig. 5.1, and cross-sectional properties and unit weights are given in Appendix C. References 3 and 28 also have tables of glulam section properties.

5.3 Fabrication of Glulam Members

Most structural glulam members are produced using Douglas fir, Southern pine, and, more recently, Hem-Fir. Quality control standards ensure the production of a quality product. In fact, the structural properties of glulam members in most cases exceed the structural properties of sawn lumber.

The reason that the structural properties for glulam are so high is that the material included in the member can be selected from relatively high-

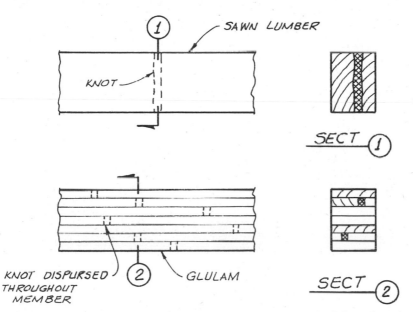

Fig. 5.2 Dispersion of growth defects in glulam. Growth defects found in sawn lumber can be eliminated or (as shown in this sketch) dispersed throughout the member to reduce the effect at a given cross section.

quality laminating stock. The growth defects that limit the structural capacity of a large solid wood member can simply be excluded in the fabrication of a glulam member.

In addition, laminating optimizes material by dispersing the strength-reducing defects in the laminating material throughout the member. For example,   **119** consider the laminations that are produced from a sawn member with a knot that completely penetrates the member at one section. See Fig. 5.2. If this member is used to produce laminating stock which is later reassembled in a glulam member, it is unlikely that the knot defect will be reassembled in all the laminations at exactly the same location in the glulam member. Therefore, the reduction in cross-sectional properties at any section consists only of a portion of the original knot. The remainder of the knot is distributed to other locations in the member.

Besides dispersing defects, the fabrication of glulam members makes efficient use of available structural materials in another way. High-quality laminations are located in that portion of the cross section which is more highly stressed. For example, in a beam, wood of superior quality is located at the extreme fibers. This coincides with the location of maximum bending stresses. See Example 5.1. Research has demonstrated that the outer laminations in the tension zone are the most critical laminations in a beam. For this reason, additional grading requirements are used for the outer tension laminations.

EXAMPLE 5.1 **Distribution of Laminations in Glulam Beams**

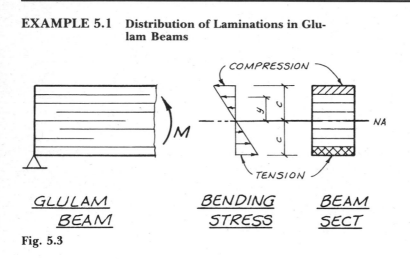

Fig. 5.3

Bending stress calculation:

$$\text{Arbitrary point} \qquad f_b = \frac{My}{I}$$

$$\text{Maximum stress} \qquad f_b = \frac{Mc}{I}$$

In glulam beams, high-quality laminations are located in areas of high stress (i.e., near the top and bottom of the beam). Lower-quality wood is placed near the neutral axis where the stresses are lower. The outer tension laminations are critical and require the highest-grade stock.

It should be pointed out that glulam bending members can be loaded with the transverse load applied either parallel or perpendicular to the wide face of the laminations. See Example 5.2.

EXAMPLE 5.2   Bending of Glulams

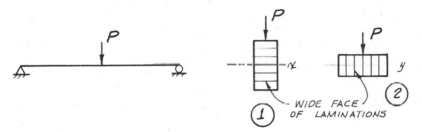

Fig. 5.4

Bending can occur about either the x or y axis of a glulam. In section 1 the load is perpendicular to the wide face of the laminations, and bending occurs about the major axis of the member. This is the more common situation. In section 2 the load is parallel to the wide face of the laminations, and bending occurs about the weak or minor axis of the member.

When the load is perpendicular to the wide face, the depth of the beam simply depends on the number of laminations that are used. Because efficient beams typically have depths much greater than their widths, most glulam beams will have the load applied perpendicular to the wide face of the laminations. Example 5.1 assumes this type of loading, and, unless otherwise noted, this loading condition should be assumed for other beams.

Laminations are selected and dried to a moisture content of 16 percent or less before gluing. Differences in moisture content between adjacent laminations should not exceed 5 percent in order to minimize internal stresses and checking.

Two types of glue are permitted in the fabrication of glulam members: (1) dry-use adhesives (casein glue) and (2) wet-use adhesives (phenol-resorcinol-base or resorcinol-base adhesives). Both types of glue are capable of producing joints which have horizontal shear capabilities in excess of the capacity of the wood itself.

Although both types of glue are permitted, currently the wet-use adhesives are used almost exclusively. This became common practice when room-temperature-setting glues were developed for exterior use. Wet-use adhesives, as the name implies, can withstand severe conditions of exposure.

121

The laminations in a glulam member run parallel to the length of the member. The efficient use of materials and the long length of many glulam members requires that effective end splices be developed in a given lamination. Several different configurations of lamination *end joint* splices are possible. See Fig. 5.5.

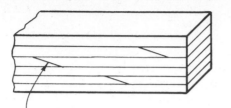

END JOINT SPLICES SHOULD BE WELL SCATTERED ESPECIALLY IN MEMBERS WITH HIGH TENSION STRESSES

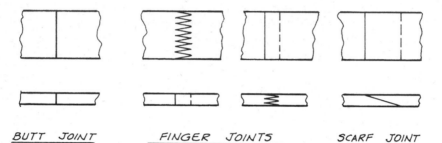

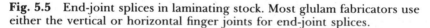

BUTT JOINT   FINGER JOINTS   SCARF JOINT

Fig. 5.5 End-joint splices in laminating stock. Most glulam fabricators use either the vertical or horizontal finger joints for end-joint splices.

Butt joints are poor splices and are considered ineffective in transmitting both tension and compression. Finger joints can produce high-strength joints when the fingers have relatively flat slopes. The fingers have very small blunt tips to permit glue squeeze-out. Finger joints also make efficient use of laminating stock because the lengths of the fingers are usually short in comparison with the lengths of scarfed joints. With scarfed joints, the flatter the slope of the joint, the greater the strength of the connection. Scarf slopes of 1 in 5 or flatter for compression and 1 in 10 or flatter for tension are recommended (Ref. 5).

If the width of the laminating stock is insufficient to produce the required width of glulam, more than one piece of stock can be used for a lamination. The *edge joints* in a lamination can be glued, or they can be staggered in adjacent laminations and a reduced shear stress can be used in design.

Although one should be aware of the basic fabrication procedures and concepts outlined in this section, the building designer does not have to be concerned about designing the individual laminations, splices, and so on. The designer can simply specify a particular species and grade of glulam and that fabrication, shipping, and erection are to conform to the appropriate American Institute of Timber Construction (AITC) Standards. Conformance to these standards should result in members that are capable of functioning satisfactorily under allowable design stresses.

5.4 Grades of Glulam Members

Grades of glulam members are given as combinations of laminations. The two main types are *bending combinations* and *axial combinations.*

As previously noted, beams which are bent in the normal manner (i.e., with the applied load perpendicular to the wide face of the laminations), have high-quality laminations at the extreme fibers. The designations used for the bending combinations are the allowable bending stress for the grade in hundreds of psi, followed by the letter F. For example, 24F indicates a bending combination with a tabulated allowable bending stress of 2400 psi for normal duration. For Douglas fir and larch, bending combinations range from 16F through 26F in increments of two (note that the 26F combination may not be readily available). Other species have similar combinations.

Members which are principally axial-load-carrying members have combinations of laminations numbered 1 through 5. Because axial-load members are uniformly stressed throughout the cross section, the distribution of lamination grades is uniform across the member section, compared with the distribution used for beams.

Glulam combinations are, in one respect, similar to the "use" categories of sawn lumber. The bending combinations anticipate that the member will be used as a beam, and the axial combinations assume that the member will be loaded axially.

Bending combinations are fabricated with higher-quality laminating stock at the outer fibers, and consequently they make efficient beams. This fact, however, does not mean that a bending combination cannot be loaded axially. Likewise, an axial combination can be designed for a bending moment. The combinations, then, have to do with efficiency, but they do not limit the use of a member. The ultimate use is determined by stress calculations.

Allowable stresses for glulam design can be obtained from several AITC publications (Refs. 3 and 28), UBC Chap. 25, and the NSD Supplement, Tables 5A and 5B, *Design Values for Structural Glued Laminated Softwood Timbers.* Tabulated allowable stresses for glulam include

Bending, F_b

Tension parallel to grain, F_t

Horizontal shear, F_v

Compression parallel to grain, F_c

Compression perpendicular to grain, $F_{c\perp}$

Modulus of elasticity E

As might be expected, two allowable bending stresses are given. One applies when the transverse load is parallel to the wide face of the laminations; the other is used when the load is perpendicular to the wide face of the laminations (Fig. 5.4). In addition, two allowable values for compression perpendicular to the grain are given. One value applies to bearing on the tension face, and the other applies to bearing on the compression face. The allowable bearing may be larger for the tension face because of the higher-quality laminations used in the tension zone in certain bending combinations.

It is interesting to compare several tabulated allowable stresses for glulam and sawn lumber. The maximum allowable bending stress for the highest-quality Douglas fir glulam is 2600 psi, compared with 1900 psi for a Douglas fir sawn member (Beam and Stringer). For the same grades, the tabulated allowable values for horizontal shear are 165 psi for the glulam and 85 psi for the sawn member.

It should be noted that although the laminations in a given combination may have different properties (strength and stiffness), the tabulated allowable stresses and the cross-sectional properties are used in practice on the assumption that the member has uniform properties throughout. Therefore, the design procedures for glulam are much the same as those used for sawn lumber.

In recent years the allowable stresses for axial tension in glulam members have been reduced. Although axial tension members occur less frequently than other types of members, the designer should be aware that numerous tables exist in older codes and handbooks which list allowable stresses that are no longer current. An attempt by the designer should be made to keep abreast of design changes as they are introduced.

Along these same lines, the designer should be aware of the current development of a new laminating specification by AITC. *AITC 117—79 Standard Specifications for Structural Glued Laminated Timber of Softwood Species* will allow manufacturers to select laminating combinations which fit their varying raw material supplies.

This new laminating specification will allow the mixing of different species of wood (e.g., Douglas fir and Hem-Fir) in the same member and make better utilization of forest resources. Some modification or expansion of the glulam grade combinations will probably occur under the new laminating specification. Because these are not yet available, the older glulam combinations are used throughout this book.

5.5 Adjustment Factors for Glulam Allowable Stresses

Glulam timbers are a wood product. Some structural design values may be larger for glulam than for sawn lumber, but this is essentially a result of the selective placement of laminations and the dispersion of imperfections.

Because of the nature of the material itself, glulams respond under load in much the same manner as sawn lumber members. To account for this behavior, allowable stresses for glulam members are modified in the same way allowable stresses are modified for sawn lumber design. In fact, some of the modification factors for glulam and sawn lumber are the same.

Most glulam members can support higher stresses for short periods of time. The same *load duration factors* (LDFs) are used to account for this in both glulam and sawn lumber. Adjustments for load duration are treated at length in Sec. 4.11.

Tabulated allowable stresses are for "dry conditions of use." For glulam this corresponds to a maximum moisture content of 15 percent in service (as in most covered structures). The concept of a *condition-of-use factor* (CUF) was introduced in Sec. 4.10. Its purpose is to modify allowable stresses when the moisture content in service is different than that for the tabulated values. When the MC of glulam in service will be 16 percent or greater, the allowable stresses should be multiplied by the CUFs for "wet conditions of use." These numerical values are different from those used for sawn lumber, but the concept is the same. CUFs for glulam design are given in the glulam allowable stress tables.

The LDF and CUF are perhaps the more common general adjustments for glulam design. Other factors may be required, but these normally apply to specific types of stresses. These other factors will be covered in the appropriate chapter (e.g., adjustment of allowable bending stress based on the depth of the member is covered in Chap. 6 on beam design).

5.6   Problems

5.1   What are the standard thicknesses of laminations used in the fabrication of glulam members? Under what conditions would these thicknesses be used?

5.2   What is the most common type of end-joint splice used in the fabrication of glulam members? Sketch the splice.

5.3   Describe the distribution of the laminating stock that would be used in the fabrication of a glulam beam. Where is the highest-quality lamination located?

5.4   *Given:* A 6¾ × 25.5 Combination 20F Douglas fir glulam.

    Find: *a.* The cross-sectional properties (A, S, I, and C_F) about the major or strong (x) axis of the member. Sketch the cross section.
             b. The tabulated allowable stresses (F_b, F_v, F_c, $F_{c\perp}$, F_t, and E) associated with the properties in *a.*
             c. The cross-sectional properties about the minor or weak (y) axis of the member.
             d. The tabulated allowable stresses associated with the properties in *c.*

5.5   Problem 5.5 is the same as Prob. 5.4 except that the member is a 5⅛ × 27 Combination 22F Southern pine glulam.

125

5.6 Problem 5.6 is the same as Prob. 5.4 except that the member is a 6¾ × 9 Combination 4 Douglas fir glulam.

5.7 Problem 5.7 is the same as Prob. 5.4 except that the member is a 8¾ × 10.5 Combination 3 California redwood glulam.

5.8 What are the condition-of-use factors (CUFs) for wet-service conditions for glulam members?

CHAPTER SIX

Beam Design

6.1 Introduction

The design of rectangular sawn wood beams and straight or slightly curved rectangular glulam beams is covered in this chapter. Because glulam members may be somewhat more complicated than sawn members, special design procedures may apply to glulam beams only. The special considerations that apply to glulam design only are noted. Where no distinction is made, it may be assumed that essentially the same procedures apply both to sawn lumber and glulam design.

The design of wood beams follows the same basic overall procedure used in the design of beams of other structural materials. The factors that need to be considered are

1. Bending (including lateral stability)
2. Shear
3. Deflection
4. Bearing

The first three items can govern the size of a wood member. The fourth item must be considered in the design of the supports. In many beams, the bending stress is the critical design item. For this reason, a trial size is often obtained from bending stress calculations. The remaining items are then simply checked with the trial size. If the trial size proves inadequate in any of the checks, the design is revised.

127

6.2 Bending

The discussion here assumes that the beam is braced to provide lateral stability and that bending stresses are parallel to the longitudinal axis of the member. Wood is very weak in tension applied across the grain, and bending stresses that are not parallel with the longitudinal axis of the member are usually not permitted. See Example 6.1. Bending may occur about either the strong or weak axis of the member, but, unless otherwise noted, bending about the strong axis is assumed.

EXAMPLE 6.1   Bending in Wood Members

Longitudinal Bending Stresses

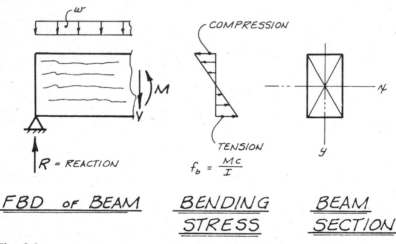

Fig. 6.1*a*

The FBD (Fig. 6.1*a*) shows a typical beam cut at an arbitrary point. The internal forces *V* and *M* are required for equilibrium. The bending-stress diagram indicates that the stresses developed by the moment are longitudinal stresses, and they are, therefore, parallel to the grain in the wood beam.

Bending is shown about the strong (*x*) axis of the member. Bending can also occur about the weak (*y*) axis, but bending about the strong axis will be assumed unless otherwise indicated.

Cross-Grain Bending—Not Allowed

Section 1 (Fig. 6.1*b*) shows a concrete wall connected to a wood horizontal diaphragm. The lateral load is shown to be transferred from the wall through the wood ledger by means of anchor bolts and nailing.

Section 2 indicates that the ledger cantilevers from the anchor bolt to the diaphragm level. Section 3 is an FBD showing the internal forces at the anchor bolt and the bending stresses that are developed in the ledger. The bending stresses in the ledger are across the grain (as opposed to being parallel to the grain). Wood is very weak in cross-grain bending and tension. This connection is introduced at this point to define the cross-grain bending problem. Tabulated allowable bending stresses apply to longitudinal bending stresses only.

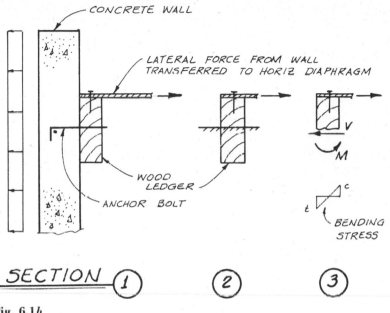

Fig. 6.1b

Because of failures in connections of this type, cross-grain bending and cross-grain tension are not permitted by the Code for the anchorage of seismic forces. No allowable stresses are assigned to cross-grain bending or tension.

It should be noted that the use of a wood ledger in a building with concrete or masonry walls is still a very common connection. However, additional anchorage hardware is required to prevent the ledger from being stressed across the grain. Anchorage for this type of connection is covered in detail in Chap. 14.

The check for bending stresses in a wood beam uses the formula

$$f_b = \frac{Mc}{I} = \frac{M}{S} \leqslant \text{adj.}\ F_b$$

This formula says that the actual (calculated) bending stress must be less than or equal to the adjusted allowable bending stress. The formula used to calculate the actual bending stress was developed for an "ideal" material. Such a material is defined as a solid, homogeneous, isotropic (having the same properties in all directions) material. In addition, plane sections before bending are assumed to remain plane during bending, and stress is assumed to be linearly proportional to strain.

129

Wood is made up of hollow cells, and because of the various growth characteristics (annular rings, knots, moisture content, slope of grain) it does not completely satisfy the criteria for which the formula was theoretically developed.

However, adequate results are obtained in design by applying the ordinary bending formula and adjusting the allowable stresses. The adjustments reflect important design factors which cause wood to perform differently from an "ideal" material. Some of the general stress adjustments were discussed in Chaps. 4 and 5. Adjustments, in most cases, take the form of factors to be multiplied by the tabulated allowable stresses. For example, the allowable stress for a beam of rectangular cross section is

$$\text{Adj. } F_b = \text{tab. } F_b \times \text{LDF} \times \text{CUF} \times C_F \times \cdots$$

where adj. F_b = adjusted allowable bending stress
Tab. F_b = tabulated allowable bending stress
LDF = load duration factor
CUF = condition-of-use factor
C_F = size factor
$\times \cdots$ = any other adjustment factors that may apply

The load duration factor and the condition-of-use factor were covered in detail in Chaps. 4 and 5. Recall that the LDF is used to account for the total length of time that the design load is applied. Tabulated allowable stresses are for "normal" duration. The CUF is basically a modification factor which accounts for moisture-content conditions (such as exposure to the weather) which are different from those that apply to tabulated allowable stresses.

Other less common adjustment factors (indicated by $\times \cdots$ in the adjustment calculation) may also be required. For example, if fire-retardant-treated sawn lumber is used, a 10 percent reduction (0.90 multiplying factor) must be applied to *all* allowable stresses. Another adjustment factor may be used to increase allowable stresses when 2-in.- to 4-in.-thick lumber is used flatwise (i.e., when it is bent about its minor axis). A number of these general adjusting factors are given in the footnotes to the allowable stress tables. The designer should carefully evaluate each design situation to determine if conditions are different from those for which allowable stresses are tabulated.

Size Factor C_F.   Tabulated allowable bending stresses apply to relatively shallow beams of rectangular cross section. Shallow beams are defined as those with a depth of 12 in. or less. For shallow beams $C_F = 1.0$.

As the depth increases over this limit, the moment capacity of a beam becomes less than that predicted by ordinary bending theory. The size factor C_F is used to adjust (reduce) the tabulated allowable bending stress for beams greater than 12 in. deep. See Example 6.2.

The size factor applies to both sawn lumber and glulam beams. Because the size of the tree limits the depth of sawn beams, large reductions in allow-

EXAMPLE 6.2   Size Factor

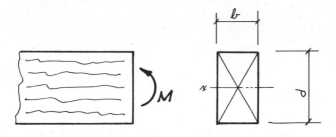

SECTION

Fig. 6.2a

Ordinary bending theory says that the allowable moment capacity of a beam is equal to the allowable bending stress times the section modulus:

$$\text{Allow. } M = F_b S$$

This expression is valid for wood beams with depths of 12 in. or less ($d \leq 12$ in.). When $d > 12$ in., the size factor is used to account for the reduced moment capacity:

$$\text{Allow. } M = (C_F F_b)S$$

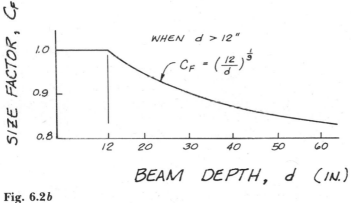

Fig. 6.2b

In a check of a given beam, the actual bending stress is compared with the adjusted allowable stress

$$f_b = \frac{M}{S} \leq C_F \times F_b$$

NOTE: *The size factor applies to bending stress calculations only.*

able bending stress are normally associated with glulam beams rather than sawn members.

The size factor given by the expression in Example 6.2 may be calculated, or it can be read directly from the table in Appendix C or similar tables in Refs. 3 and 28. It should be noted that the size factor defined by this expression applies to beams which meet the following criteria:

1. The member is simply supported.
2. The load is uniformly distributed.
3. The span-to-depth ratio (L/d) of the beam is equal to 21. L and d must be in consistent units.

The size factor defined in Example 6.2 is adequate for sawn lumber without further modification. It is also adequate for many glulam beam designs, but a refined value for C_F can be determined, and may be required in some circumstances. The procedure for modifying C_F for conditions other than those stated are given in the NDS Sec. 5.3.4, *Size Factor for Structural Glued Laminated Beams*, in UBC Chap. 25, and in Refs. 3 and 28.

Form Factor C_f. Tabulated allowable stresses, and those allowable stresses that have been adjusted by the size factor C_F, apply to beams of rectangular cross section. For other types of cross sections, the form factor is another factor used to adjust allowable stresses. Form factors are given in both the NDS and the Code.

A common example of a wood member with a nonrectangular cross section is a pole. The form factor for a circular cross section is $C_f = 1.18$, and in this case the adjusting factor provides an increase in allowable stresses. The allowable bending stresses for poles (UBC Chap. 25) and piles (NDS Part VI, *Round Timber Piles*) have the 1.18 form factor already included. (The *effect* of the 1.18 form factor can be obtained in a different manner: the moment capacity of a round bending member is the same as a beam with a square cross section of equal area.)

The form factor is cumulative with the size factor. In other words, if a beam with a nonrectangular cross section has a depth greater than 12 in., the allowable bending stress is

$$\text{Adj. } F_b = C_F C_f (\text{tab. } F_b)$$

The *form factor* and the *size factor* are to be used in the adjustment of *bending stresses only* and do not apply to other allowable stresses. Both factors are discussed in considerable detail in Ref. 7.

In the next part of this section the tabulated allowable bending stresses for wood beams will be reviewed.

Repetitive Members. In the NDS Supplement and UBC Chap. 25, two values of allowable bending stress for *sawn lumber* are tabulated. One value applies to single-member usage, and the other applies to a repetitive-member condition. The repetitive-member allowable bending stresses are approximately 15 percent greater than the allowable bending stresses for single-member uses. Repetitive members are defined as three or more parallel beams spaced not more than 24 in. o.c. and joined together by floor, roof, or other load-distributing elements. The larger allowable bending stresses for repetitive members reflect the ability of these members to share loads with adjacent parallel members. Members which do not qualify as repetitive members are to be designed according to the single member allowable bending stresses.

When a concentrated load is supported by a deck which distributes the load to parallel beams, the entire concentrated load need not be assumed to be supported by one beam. NDS Appendix D provides a method for the *lateral distribution of a concentrated load* to adjacent parallel beams.

It will be recalled from Sec. 5.4 that two allowable bending stresses also apply to glulam beams. However, these depend on whether bending occurs about the major or minor axis of the member, and no value for repetitive members is given. This is because glulam beams are usually large members that would not be spaced close enough to qualify as a repetitive member.

Notching of Beams.   The final part of this section deals with the effects of notches on the bending strength of beams.

The notching of structural members to accommodate piping or ductwork should be avoided whenever possible. A note on the building plans should prohibit the cutting of any structural member unless it is specifically detailed on the structural plans.

The critical location of a notch is in the tension zone of a beam. Besides reducing the depth available for resisting the load, stress concentrations are also developed. Stress concentrations are especially large for the typical square-cut notch.

The NDS says that the maximum depth of a notch shall not exceed one sixth of the depth of the member and that the notch shall not be located in the middle third of the member. Although not stated, it is apparent that this latter criterion applies to simply supported beams because of the high bending stresses in this area. In addition, the NDS *prohibits the notching* of the tension side of beams when the nominal width of the member is 4 in. or greater. For glulam beams the NDS refers the designer to Ref. 3, which essentially *recommends that notching* in the tension face in the area of high bending stress *be avoided.*

These negative statements should serve as a warning about the potential hazard that can be created by stress concentrations due to reentrant corners. Failures have occurred in beams with notches located some distance from the point of maximum bending, and at a load considerably less than the design load.

In an attempt to provide some guidance, the UBC states that the bending stress at a notch is to be calculated by assuming that there is an identical notch in the other edge of the beam. Thus, the section modulus would be evaluated using the beam depth minus two times the actual depth of the notch. This approach recognizes that the neutral axis of the beam does not shift abruptly at a notch. However, one should realize that, even with this severe penalty, *this method may not be conservative.* The problem is best handled by avoiding notches in the first place. In the case of an existing notch, some strengthening of the member at the notch may be advisable. The effects of shear stresses at the notched end of a beam are considered in Sec. 6.4.

6.3   Lateral Stability

When a member functions as a beam, a portion of the cross section is stressed in compression and the remaining portion is stressed in tension. If the compression zone of the beam is not braced to prevent lateral movement, the member may buckle at a bending stress that is less than the adjusted allowable

---

EXAMPLE 6.3   Lateral Buckling of Beams

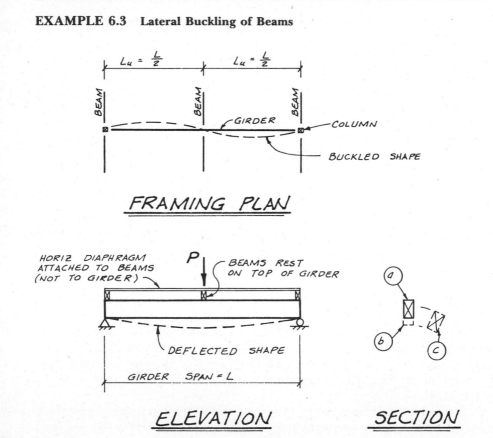

1. The beams that frame into the girder provide lateral support to the compression (top) side of the girder at a spacing of $L_u = L/2$. The distance between points of lateral support is known as the *unbraced length* L_u of a beam.

2. It is important to realize that the *span* and *unbraced length* of a beam are two different items. They may be equal, but they may also be quite different. The span is used to calculate stresses and deflections. The unbraced length, together with the cross-sectional dimensions, is used to analyze the stability of a beam.

3. The section view shows several possible conditions:
    a. The unloaded position of the girder.
    b. This represents the deflected position of the girder under a vertical load with no instability. Vertical deflection occurs if the beam remains stable.
    c. If the unbraced length is excessive, the compression side of the beam may buckle laterally in a manner similar to a column. This buckling takes place between points of lateral support. This buckled position is also shown in the plan view.

4. When the top of the beam is always in compression (positive moment everywhere) and when floor or roof sheathing is effectively connected directly to the beam, the unbraced length approaches zero. In this case lateral buckling is prevented, and the strength of beam depends on the bending strength of the material and not on stability factors.

---

stress defined in Sec. 6.2. (The allowable stress in Sec. 6.2 assumed lateral buckling was prevented by the presence of adequate bracing.)

The bending compressive stress can be thought of as creating an equivalent column buckling problem in the compressive half of the cross section. Buckling in the plane of loading is prevented by the presence of the stable tension portion of the cross section. Therefore, if buckling of the compression side occurs, movement will take place *laterally* between points of lateral support. See Example 6.3.

In many practical situations, the question of lateral instability is simply eliminated by providing lateral support to the compression side of the beam at close intervals. In fact, an *effective connection* (proper nailing) of a roof or floor *diaphragm* to the compression side of a beam causes the unbraced length to approach zero, and lateral instability is prevented by *full* (continuous) *lateral support*.

In the case of laterally unbraced *steel beams* (W shapes), the problem is amplified because cross-sectional dimensions are such that relatively *slender* elements are stressed in compression. Slender elements have large width-to-thickness (*b/t*) ratios and these elements are particularly susceptible to buckling.

In the case of rectangular *wood beams,* the dimensions of the cross section are such that the width- (depth-) to-thickness ratios (*d/b*) are relatively small. Common framing conditions and cross-sectional dimensions cause large reductions in allowable bending stresses to be the exception rather than the rule. Procedures are available, however, for taking lateral stability into account, and these are outlined in the remainder of this section.

Two methods of handling the lateral stability of beams are currently in use. One method is based on *rules of thumb* that developed over the years. These rules are applied to the design of sawn lumber beams. In this approach the required type of lateral support is specified on the basis of the depth-to-thickness ratio d/b of the member.

EXAMPLE 6.4   Lateral Support of Beams—Approximate Method

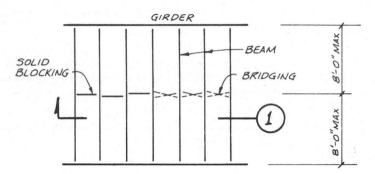

FRAMING PLAN

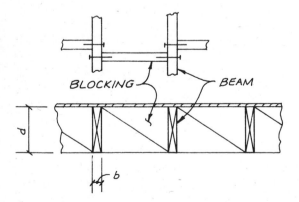

SOLID BLOCKING   BRIDGING

SECTION ①

When $d/b = 6$, lateral support can be provided by full-depth solid blocking or by bridging spaced at 8 ft-0 in. maximum. *Solid blocking* must be the same depth as the beams. Adjacent blocks may be staggered to facilitate construction (i.e., end nailing through beam). *Bridging* is cross-bracing made from wood (typically 1×3 or 1×4) or light-gauge steel (available prefabricated from manufacturers of hardware for wood construction).

These rules are outlined in the NDS Sec. 4.4.1, *Lateral Stability of Bending Members*, and in UBC Chap. 25. As an example, the rules state that if $d/b = 6$, bridging or full-depth solid blocking is required at intervals of 8 ft-0 in. maximum. These can be omitted *if* the compression edge is adequately supported laterally by sheathing *and* the beam is supported at bearing points so that rotation is prevented. See Example 6.4.

These requirements for lateral support are approximate because only the proportions of the cross section are considered. The second, more accurate method of accounting for lateral stability uses the *slenderness factor* C_s. See Example 6.5. The slenderness factor considers the unbraced length (distance between points of lateral support to the compression side of the beam) in addition to the dimensions of the cross section. This method was developed for large, important *glulam beams*, but it applies equally well to sawn beams.

EXAMPLE 6.5 Slenderness Factor for Unbraced Beams

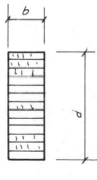

Fig. 6.5

BEAM
SECTION

The *slenderness factor* for beams measures the tendency of the member to buckle laterally between points of lateral support to the *compression side* of the beam. Dimensions are in inches.

$$C_s = \sqrt{\frac{L_e d}{b^2}}$$

where b = beam width

d = beam depth

L_u = unbraced length of the beam (distance between points of lateral support as in Fig. 6.3)

L_e = effective unbraced length

= $1.61L_u$ for a single-span beam with a concentrated load at center

= $1.92L_u$ for a single-span beam with a uniformly distributed load

= $1.84L_u$ for a single-span beam with equal end moments

= $1.69L_u$ for a cantilever beam with a concentrated load at the unsupported end

= $1.06\ L_u$ for a cantilever beam with a uniformly distributed load

= $1.92\ L_u$ is a conservative value for single span or cantilever beam with any load

These factors are available from a number of sources including Refs. 1, 2, and 3.

In calculating C_s the *effective unbraced length* is used like the effective length of a column (Chap. 7). For a beam, the effective length L_e depends both on the type of span (end conditions) and the type of load. The effective length is obtained by multiplying the actual unbraced length L_u by an effective length factor.

Where required, the slenderness factor is used to calculate a reduced allowable bending stress. (A similar practice is used in the design of laterally unbraced steel beams.) The allowable bending stress can be plotted against the slenderness factor. See Example 6.6. This corresponds to a "column curve" in which the allowable axial compressive stress in a column is plotted vs. the slenderness ratio.

Three different segments of the allowable bending curve are defined by the following ranges of C_s:

$$0 < C_s \leqslant 10 \qquad \text{short unbraced beam}$$
$$10 < C_s \leqslant C_k \qquad \text{intermediate unbraced beam}$$
$$C_k < C_s \leqslant 50 \qquad \text{long unbraced beam}$$

In the *short* unbraced range, no reduction of the tabulated allowable bending stress is required. The *long* unbraced range uses an elastic "Euler-type" buckling formula that depends on the stiffness of the beam. For the *intermediate* range a transition formula is used. Similar ranges and design expressions are used in column design (Sec. 7.4).

The use of the load duration factor in these stability expressions deserves some special attention. The expressions for F_b' are shown in the NDS and the UBC without the LDF. The designer must then determine when and where the LDF is to be used. For guidance, the LDF is included at the accepted locations in the expressions in Example 6.6.

It should be noted that the application of the LDF as shown in Example 6.6 results in a discontinuity of allowable stresses at C_k. This occurs because the full LDF is applied to F_b' in the intermediate range, and it is not applied in the long range. For beam design this discontinuity is simply accepted and should not be cause for concern. In column design the LDF is applied in a somewhat different manner. This alternate approach eliminates the discontinuity of allowable stress at the transition to the long range for columns.

The tabulated allowable F_b is used in the stability expressions for bending. The size effect (Sec. 6.2) is considered in a separate, independent check on bending stress. The lower allowable bending stress considering *lateral stability* or *size factor* is the one to be used in the design of the member.

EXAMPLE 6.6 **Allowable Bending Stress Considering Lateral Stability**

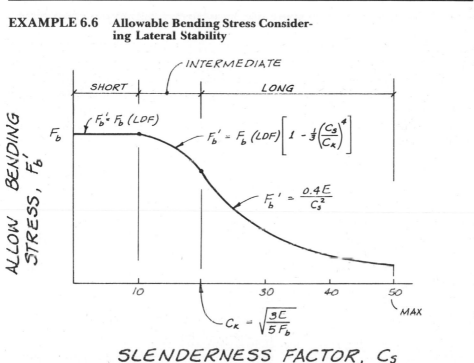

SLENDERNESS FACTOR, C_S

Fig. 6.6

F'_b = Allowable bending stress considering the effects of lateral stability
F_b = Tabulated allowable bending stress

The LDF is not used to calculate F'_b in the long range, or in the calculation of the limit C_k. In the calculation of F'_b for all three ranges, the coefficient to account for size effect (C_F) is not applied. Therefore, the design of a beam must finally be based on the smaller allowable bending stress considering

1. Lateral stability (F'_b)
2. Size factor (Adj. F_b—Sec. 6.2).

The application of the lateral stability expressions is demonstrated in an example in Sec. 6.12 and 6.13.

6.4 Shear

The shear stress in a beam is often referred to as "horizontal" shear. The shear stress at any point in a cross section can be computed by the formula

$$f_v = \frac{VQ}{Ib}$$

It may be helpful to compare the shear stress distribution given by this formula for a typical steel beam and a typical wood beam. See Example 6.7. Theoretically the formula applies to the calculation of shear stresses in both types of members. However, in design practice the shear stress in a steel W shape is approximated by a nominal (average web) shear calculation.

The average shear stress calculation gives reasonable results in typical steel beams, but it does not apply to rectangular wood beams. The maximum shear in a rectangular beam is 1.5 times the average shear stress. This difference is significant and cannot be disregarded.

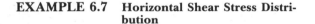

EXAMPLE 6.7   Horizontal Shear Stress Distribution

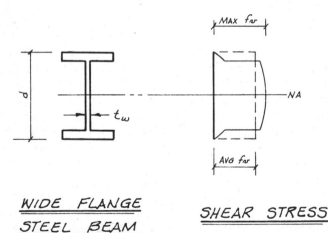

WIDE FLANGE STEEL BEAM

SHEAR STRESS

Fig. 6.7a

For a steel W shape, a nominal check on horizontal shear is made by dividing the total shear by the cross-sectional area of the web

$$\text{Avg. } f_v = \frac{V}{A_{\text{web}}} = \frac{V}{dt_w} \approx \text{max. } f_v$$

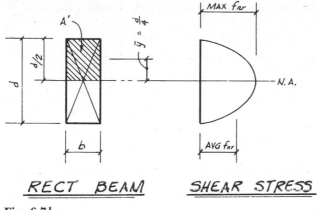

RECT BEAM   SHEAR STRESS

Fig. 6.7b

For rectangular beams the theoretical *maximum* horizontal shear *must* be used. The following development shows that the maximum shear is 1.5 times the average.

$$\text{Avg. } f_v = \frac{V}{A}$$

$$\text{Max. } f_v = \frac{VQ}{Ib} = \frac{VA'\bar{y}}{Ib} = \frac{V\left(b \times \dfrac{d}{2}\right)\left(\dfrac{d}{4}\right)}{\left(\dfrac{bd^3}{12}\right) \times b}$$

$$= \frac{3V}{2bd} = 1.5\frac{V}{A} = 1.5(\text{avg. } f_v)$$

A convenient formula for the calculation of horizontal shear stresses in a rectangular beam is thus developed. For wood cross sections of other configurations, the distribution of shear stresses will be different, and it will be necessary to use the basic horizontal shear formula, or some other appropriate check depending on the type of member involved. The check on shear for a rectangular wood beam is

$$f_v = \frac{1.5V}{A} \leq \text{adj. } F_v$$

The tabulated allowable stress again provides only a starting point in the determination of the true allowable stress for a given set of circumstances. The load duration factor (LDF), the condition-of-use factor (CUF), and any other adjustment factor that applies, must be taken into account in determining the adjusted allowable stress.

In addition to the typical multiplying factors for adjusting the allowable shear stress, there are a number of other adjustments that *may* be used in the horizontal shear check.

141

One common and easy-to-use adjustment applies to both sawn and glulam beams. This adjustment consists of neglecting all loads (both distributed and concentrated) within a distance *d* from the supports of the beam. See Example 6.8. For a single moving load, it can be located at distance *d* from the support rather than directly at the support.

EXAMPLE 6.8   Reduction of Loads for Horizontal
Shear Calculations

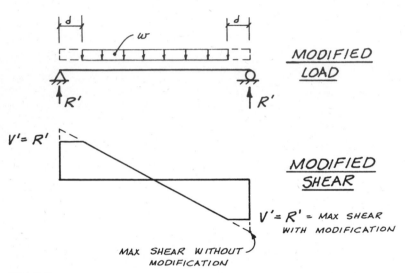

Fig. 6.8

---

1.   This shear adjustment consists of omitting the load within a distance *d* (the depth of the beam) from the support. This procedure applies to concentrated loads as well as uniformly distributed loads.

2.   The modified loads are for *horizontal shear stress* calculations in wood (sawn and glulam) beams *only*. The full design loads must be used for other design criteria.

3.   The concept of omitting loads within *d* from the support is based on an assumption that all loads are applied to the *top* of the beam. In this way the omitted loads are transmitted to the support by diagonal compression. A similar type of adjustment for shear is also used in concrete design.

Since a higher actual shear stress will be calculated without this adjustment, it is conservative not to apply the adjustment. In beams in which shear will likely not be critical, shear calculations often are shown based on the total shear at the support rather than the reduced shear.

142

It is convenient in calculations to adopt a notation which indicates whether the adjustment is being used. V alone represents a shear which is not adjusted, and V' indicates a shear which has been calculated using the adjustment. Correspondingly, f_v is the shear stress calculated using V, and f'_v is the shear based on V'. The modified load diagram is to be used for horizontal shear stress calculations only. Reactions and moments must be calculated using the full design loads.

Other more complicated adjustments for shear are included in the NDS. As noted earlier, bending stresses often govern the size of a beam, but secondary items, such as shear, *can* control the size under certain circumstances. However, the need to consider the more complicated horizontal shear adjustments is relatively infrequent, and they are not covered here.

It will be helpful if the designer can learn to recognize the type of beam in which shear is critical. As a general guide, shear is critical on relatively *short, heavily loaded spans.* With some experience, the designer will be able to identify by inspection what probably constitutes a "short, heavily loaded" beam. In such a case the design would start by picking a trial beam size

---

EXAMPLE 6.9   Shear in Notched Beams

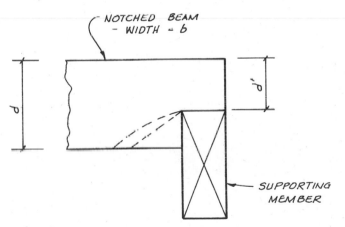

Fig. 6.9

For square-cut notches, the shear stress is increased by a stress concentration factor d/d':

$$f_v = \left(\frac{1.5 V}{bd'}\right)\left(\frac{d}{d'}\right)$$

Tapered notches can be used to relieve stress concentrations (dashed lines).

which satisfies the horizontal shear formula. Other items, such as bending and deflection, would then be checked.

If a beam is *notched* at a free end, the shear at the notch must be checked. To do this the theoretical formula for horizontal shear is applied with the actual depth at the notch, d', used in place of the total beam depth d. See Example 6.9. For square-cut notches, the actual stress must be increased by a stress concentration factor which is taken as the ratio of the total beam depth to the net depth at the notch (d/d').

Notches of other configurations which tend to relieve stress concentrations will have lower stress concentration factors. The magnitude of a reduced stress concentration factor is left to the designer to evaluate.

Additional provisions for horizontal shear at bolted connections in beams are covered in Sec. 12.3.

6.5 Deflection

The deflection limitations of UBC Chap. 23 were discussed in Chap. 2. Actual deflections for a trial beam are calculated for known span length, support conditions, and applied loads. Deflections may be calculated from a special deflection analysis or from standard beam formulas. The actual calculated deflection should be less than or equal to the allowable deflections given in Chap. 2.

EXAMPLE 6.10   Beam Deflection Criteria and Camber

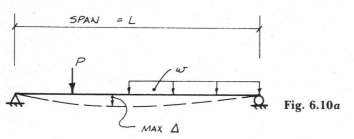

Fig. 6.10a

Actual deflection

The maximum deflection is a function of the loads, type of span, moment of inertia, and modulus of elasticity:

$$\text{Max. } \Delta = f\left(\frac{P, w, L}{I, E}\right)$$

Deflection criteria

$$\text{Max. } \Delta \leq \text{Allow. } \Delta$$

If this criterion is not satisfied, a new trial beam size is selected using the moment of inertia and the allowable deflection as the basis.

144

Camber

Camber is initial curvature built into a member which is opposite to the deflection under gravity loads.

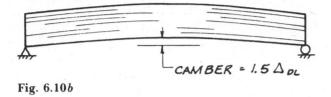

Fig. 6.10b

The material property that is used in the calculation of actual deflections is the modulus of elasticity. Tabulated values of modulus of elasticity E for wood are used without adjustment for duration of load. Other adjustments, such as those required for condition of use (CUF), may be required, depending upon service conditions.

In design of glulam beams and wood trusses, it is common practice to call for a certain amount of *camber* to be built into the member. Camber is defined as an initial curvature or deflection which is built into the member when it is fabricated. In glulam design, the typical camber is $1.5\Delta_{DL}$. This amount of camber is intended to account for (balance) long-term deflection, including creep. Additional camber may be provided to obtain adequate roof slope to prevent ponding (Chap. 2).

6.6   Design Summary

One of the three design criteria discussed in the previous sections (bending, shear, and deflection) will determine the required size of a wood beam. In addition, consideration must be given to the type of lateral support that will be provided in order to prevent lateral instability. If necessary the bending stress analysis will be expanded to take the question of lateral stability into account. With some practice, the structural designer may be able to tell by inspection which of the criteria will be critical.

The sequence of the calculations used to design a beam has been described in the above sections. It will be repeated here in summary.

For many beams the bending stress is the critical design item. Therefore, a trial beam size is often developed from the bending stress formula

$$\text{Req'd } S = \frac{M}{\text{adj. } F_b}$$

A trial member is chosen which provides a furnished section modulus S which is greater than the required value. Because the size factor C_F is not definitely known until the size of the beam has been chosen, it may be helpful

145

to summarize the actual vs. allowable bending stresses after a size has been established:

$$f_b = \frac{M}{S} < \text{adj. } F_b$$

After a trial size has been established, the remaining items (shear and deflection) should be checked. For a rectangular beam, the shear is checked by the expression

$$f_v = \frac{1.5V}{A} < \text{adj. } F_v$$

If this check proves unsatisfactory, the size of the trial beam is revised to provide a sufficient area A so that the shear is adequate.

The deflection is checked by calculating the actual deflection using the moment of inertia for the trial beam. The actual deflection is then compared with the allowable deflection:

$$\Delta \leqslant \text{allow. } \Delta$$

If this check proves unsatisfactory, the size of the trial beam is revised to provide a sufficient moment of inertia I so that the deflection criteria is satisfied.

It is possible to develop a trial member size by starting with something other than bending. For example, for a beam with a short, heavily loaded span, it is reasonable to establish a trial size using the shear calculation

$$\text{Req'd } A = \frac{1.5V}{\text{adj. } F_v}$$

The trial member should provide an area A which is greater than the required area.

If the structural properties of wood are compared with the properties of other materials, it is noted that the modulus of elasticity for wood is relatively low. For this reason, in fairly long span members, deflection can be the critical item. Obviously, if this case is recognized, or if more restrictive deflection criteria are being used in design, the trial member size should be based on satisfying deflection limits. Then the remaining criteria of bending and shear can be checked.

In lieu of other criteria, the most economical beam for a *given grade* of lumber is the one that satisfies all stress deflection criteria with the minimum cross-sectional area. Sawn lumber is purchased by the board foot (a board foot is a volume of wood based on nominal dimensions that corresponds to a 1×12 piece of wood 1 ft long). The number of board feet for a given member is obviously directly proportional to the cross-sectional area of a member.

A number of factors besides minimum cross-sectional area can affect the final choice of a member size. There are detailing considerations in which a member size must be chosen which fits in the structure and accommodates

other members and their connections. Secondly, a member size may be selected that is uniform with the size of members used elsewhere in the structure.
This may be convenient from a structural detailing point of view, and it
also can simplify material ordering and construction. Finally, availability of
lumber sizes must also be considered.

6.7 Bearing Stresses

Bearing stresses perpendicular to the grain in a wood beam occur at beam
supports or where loads from other members frame into the beam. See Example 6.11. The actual bearing stress is calculated by dividing the reaction by
the contact area between the two members. This actual stress must be less
than or equal to the adjusted allowable bearing stress:

$$f_{c\perp} = \frac{P}{A} \le \text{adj. } F_{c\perp}$$

EXAMPLE 6.11   Bearing Perpendicular to Grain

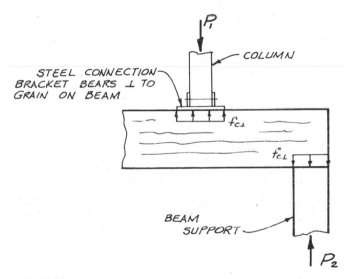

Fig. 6.11a

Bearing stress calculation:

$$f_{c\perp} = \frac{P}{A} \le \text{adj. } F_{c\perp}$$

where $f_{c\perp}$ = bearing stress perpendicular to grain
$\quad\quad P$ = applied load or reaction
$\quad\quad A$ = contact area
Adj. $F_{c\perp}$ = adjusted allowable bearing stress perpendicular to grain

147

Adjustment Based on Bearing Length
When the bearing length is less than 6 in. *and* when the distance from the end of
the beam to the contact area is more than 3 in., the allowable bearing stress may
be increased (multiplied) by the bearing length factor (BLF).

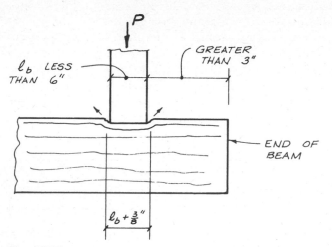

Fig. 6.11*b*

In effect the BLF increases the effective bearing length by ⅜ in. This accounts for
the additional wood fibers that resist the applied load after the beam becomes slightly
indented.
Bearing length factor:

$$\text{BLF} = \frac{l_b + 0.375}{l_b}$$

The tabulated allowable stress must be adjusted, depending upon actual design conditions:

$$\text{Adj. } F_{c\perp} = \text{tab. } F_{c\perp} \times \text{LDF} \times \text{CUF} \times \cdots$$

As shown, the load duration factor and the condition-of-use factor are
two factors that are used to adjust tabulated allowable bearing stresses. Other
factors may also apply.

There is one modifying factor that applies to compression perpendicular
to the grain only. It is based on the bearing length of an applied load or
reaction. Bearing length factor (BLF) is the term used in this book to account
for an effective increase in the bearing length. The bearing length l_b is the
contact length parallel to the length of the beam. The BLF may be used to
account for additional wood fibers beyond the length l_b that develop vertical
resisting components.

If the conditions shown in Example 6.11 are met, $F_{c\perp}$ may be multiplied by the BLF. Values for BLF can be obtained from the NDS Sect. 3.11.2, *Bearing Perpendicular to Grain,* and from UBC Chap. 25, or they may be calculated as illustrated in Example 6.11. It should be noted that compression perpendicular to the grain is not a matter of life safety, but rather relates to the amount of deformation that is acceptable in a given structure.

The allowable compressive stresses parallel to the grain (F_c) and perpendicular to the grain $(F_{c\perp})$ differ substantially. It is possible for bearing stresses to occur which are at some angle other than 0° or 90° with respect to the direction of the grain. In this case, an allowable bearing stress somewhere between F_c and $F_{c\perp}$ is determined from the Hankinson formula. See Example 6.12.

EXAMPLE 6.12   Bearing at an Angle to Grain

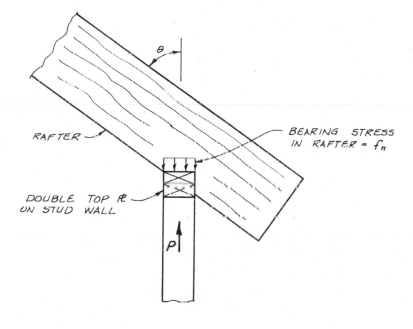

Fig. 6.12

For bearing at some angle to grain other than 0° or 90°:

$$f_n = \frac{P}{A} \leq \text{adj. } F_n$$

where f_n = bearing stress at angle θ
P = applied load or reaction
A = contact area
Adj. F_n = adjusted allowable bearing stress at angle θ to the grain

Hankinson Formula

The allowable bearing stress at angle θ to the grain is given by the Hankinson formula

$$F_n = \frac{F_c F_{c\perp}}{F_c \sin^2 \theta + F_{c\perp} \cos^2 \theta}$$

This formula can be solved mathematically or the graphical solution in the NDS Appendix may be used. The value of F_n is adjusted for duration of load (LDF). If condition-of-use factors (CUFs) are required, they can be applied to F_c and $F_{c\perp}$ prior to solving for F_n.

NOTE: The above connection is given to illustrate bearing at an angle θ. For the condition shown, bearing stresses will be governed by compression perpendicular to the grain $(F_{c\perp})$ in the double wall plate, rather than F_n in the rafter. If the top plate stresses are excessive, a bearing plate between the rafter and the top plate can be used to relieve bearing stresses on the top plate. Bearing in the rafter would then become critical.

6.8 Design Problem: Sawn Beam

In Example 6.13 a typical sawn beam in a flat roof is designed. The trial size is based on bending stress calculations. The repetitive member allowable bending stress applies because the beams are spaced less than 24 in. o.c., and they are connected together by the roof sheathing. This higher allowable stress reflects the ability of this type of system to share high local loads with adjacent parallel beams. Allowable stresses are adjusted for duration of load, but no other stress adjustments are required.

After bending, shear and deflection are checked. The shear stress is not critical, but the second deflection check indicates that deflection under DL + LL is slightly over the arbitrary allowable. The decision whether to accept this deflection is a matter of judgment. In this case it was decided to accept the deflection, and the trial beam size was chosen for the final design.

EXAMPLE 6.13   Sawn Beam Design

Design the roof beam shown below to support the given loads. Beams are spaced 16 in. o.c. (1.33 ft), and sufficient roof slope is provided to prevent ponding. The ceiling is gypsum board. Sheathing prevents lateral buckling. Use No. 1 Douglas fir-larch (DF-L). Roof DL = 15 psf, and roof LL = 20 psf. Allowable stresses and section properties are to be taken from the NDS.

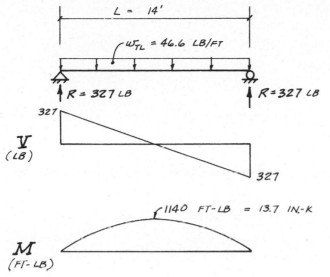

Fig. 6.13

Uniform loads.

$$RDL = 15 \times 1.33 = 20 \quad lb/ft$$
$$RLL = 20 \times 1.33 = \underline{26.6}$$
$$\text{Total load} = w_{TL} = 46.6 \, lb/ft$$

The required load combinations are DL alone (LDF = 0.9) and DL + RLL (LDF = 1.25). By a comparison of the loads and LDFs (Example 4.10), it is determined that the critical load combination is DL + RLL (i.e., total load).
Determine a trial size based on bending and then check other criteria.

Bending

Assume J&P size category. These roof beams meet the requirements for repetitive members:

$$F_b = 1750 \, psi$$

Assume

1. Beam depth will be less than 12 in., $C_F = 1.0$
2. EMC ≤ 19 percent, CUF = 1.0

$$\text{Adj. } F_b = (\text{tab. } F_b) \times LDF$$
$$= 1750 \times 1.25 = 2187 \, psi$$
$$\text{Req'd } S = \frac{M}{F_b} = \frac{13,700}{2187} = 6.26 \, in.^3$$
$$Try \ 2 \times 6. \qquad S = 7.56 > 6.26$$

and $C_F = 1.0$ because $d = 5.5 < 12$. ∴ Bending *OK*

Shear

Because it is judged that the shear stress for this beam is likely not to be critical, the maximum shear from the shear diagram will be used without modification.

$$f_v = \frac{1.5V}{A} = \frac{1.5(327)}{8.25} = 59.5 \text{ psi}$$

Adj. $F_v = (\text{tab. } F_v) \times \text{LDF} = 95 \times 1.25 = 119 \text{ psi} > 59.5$

∴ Shear OK

Deflection

The Code does not specify deflection criteria for roof beams that do not support plastered ceilings. The calculations below use the recommended deflection criteria from Ref. 3 (Sec. 2.7).

$$\Delta_{LL} = \frac{5w_{LL}L^4}{384EI} = \frac{5(26.6)(14)^4(12 \text{ in./ft})^3}{384(1,800,000)20.8} = 0.62 \text{ in.}$$

Allow. $\Delta_{LL} = \dfrac{L}{240} = \dfrac{14 \times 12}{240} = 0.70 > 0.62 \qquad OK$

Deflection under total load can be calculated by proportion

$$\Delta_{TL} = \Delta_{LL}\left(\frac{w_{TL}}{w_{LL}}\right) = 0.62\left(\frac{46.6}{26.6}\right) = 1.08 \text{ in.}$$

Allow. $\Delta_{TL} = \dfrac{L}{180} = \dfrac{14 \times 12}{180} = 0.93 < 1.08$

In the second deflection check, the actual deflection is slightly over the allowable. The decision whether to accept or reject the trial beam is a matter of judgment:

1. The deflection calculation for this beam was performed as a *guide* only, and it is not a Code *requirement* for this building.

2. The possible detrimental effects of this deflection must be weighed against the economics of using a larger beam throughout the structure.

After a consideration of the facts concerning this particular building, assume that it is decided to accept the trial size.

> Use 2 × 6 No. 1 DF-L
> MC ≤ 19 percent

Bearing

Evaluation of bearing stresses requires knowledge of the support conditions. Without such information, the minimum bearing length can simply be determined:

Adj. $F_{c\perp} = (\text{tab. } F_{c\perp}) \times \text{LDF} = 385 \times 1.25 = 481 \text{ psi}$

$$\text{Req'd } A = \frac{R}{F_{c\perp}} = \frac{327}{481} = 0.68 \text{ in.}^2$$

$$\text{Req'd } l_b = \frac{A}{b} = \frac{0.68}{1.5} = 0.45 \text{ in.}$$

All practical support conditions provide bearing lengths in excess of this minimum value.

6.9 Design Problem: Rough-Sawn Beam

In this example a large rough-sawn beam with a fairly short span is checked. The cross-sectional properties for dressed lumber (S4S) are smaller than those for rough-sawn lumber, and it would be conservative to use S4S section properties for this problem. However, the larger section properties obtained using the rough-sawn dimensions are used in this example.

Because the load (and the corresponding beam size) is relatively large in comparison to the span length, it is likely that shear will be the critical item. For this reason the shear capacity is checked first. In this problem the basic shear adjustment of neglecting any loads within a distance d from the support is used. See Example 6.14.

Bending and deflection are seen to be not critical. In fact, the stress grade for this beam could be reduced from Select Structural to No.1 and the given size would still be acceptable. The reason for this is that the allowable shear stress is the same for all stress grades in the Beam and Stringer size category. The lowest stress grade (No.1 DF) has an allowable bending stress that is greater than the actual:

$$\text{Adj. } F_b = (\text{tab. } F_b) \times LDF \times C_F \times \cdots$$
$$- 1300 \times 1.0 \times 0.96$$
$$- 1250 \text{ psi}$$
$$f_b = 890 < 1250 \qquad OK$$

EXAMPLE 6.14 Rough-Sawn Beam

Determine if the rough sawn 6 × 18 DF Select Structural beam (Fig. 6.14) is adequate to support the given loads. The load is combined DL + FLL. Lateral buckling is prevented, and the EMC is less than 19 percent. Allowable stresses are to be taken from the NDS.

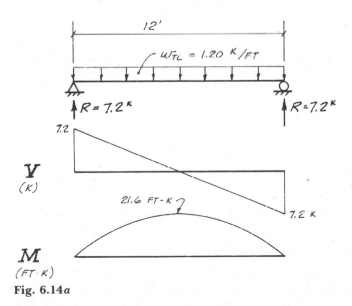

Fig. 6.14a

DESIGN OF WOOD STRUCTURES

Section Properties

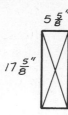

$5\frac{5}{8}''$

$17\frac{5}{8}''$

$A = bd = (5\frac{5}{8})(17\frac{5}{8}) = 99.14 \text{ in.}^2$

$S = \dfrac{bd^2}{6} = \dfrac{(5\frac{5}{8})(17\frac{5}{8})^2}{6}$

$S = 291.2 \text{ in.}^3$

$I = \dfrac{bd^3}{12} = \dfrac{(5\frac{5}{8})(17\frac{5}{8})^3}{12}$

$I = 2566 \text{ in.}^4$

6x 18 ROUGH **Fig. 6.14b**
(B ＆ S)

Shear

Start with the modified shear (the modified load diagram applies to shear calculations *only*).

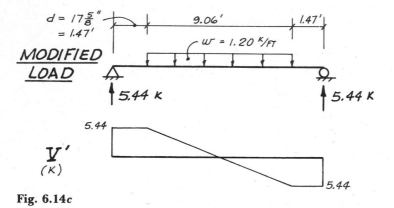

Fig. 6.14c

$$f_v = \frac{1.5V}{A} = \frac{1.5(5440)}{99.14} = 82.3 \text{ psi}$$

Adj. $F_v = (\text{tab. } F_v) \times \text{LDF} \times \cdots = 85 \times 1.0$

Adj. $F_v = 85 \text{ psi} > 82.3$ *OK*

Bending

$$M = 21.6 \times 12 = 259 \text{ in.-k}$$

$$f_b = \frac{M}{S} = \frac{259,000}{291.2} = 890 \text{ psi}$$

Adj. $F_b = (\text{tab. } F_b) \times \text{LDF} \times C_F \times \cdots$

$= 1600 \times 1.0 \times 0.96 = 1536 \text{ psi} > 890$ *OK*

Deflection

Because the percentages of DL and FLL were not given, only the total load deflection is calculated.

154

$$\Delta_{TL} = \frac{5w_{TL}\,L^4}{384\,EI} = \frac{5(1200)(12)^4(12 \text{ in./ft})^3}{384(1,600,000)(2566)} = 0.14 \text{ in.}$$

$$\text{Allow. } \Delta_{TL} = \frac{L}{240} = \frac{12(12)}{240} = 0.60 > 0.14 \qquad OK$$

Bearing

$$\text{Req'd } A_b = \frac{R}{F_{c\perp} \times \text{LDF}} = \frac{7200}{385 \times 1.0} = 18.7 \text{ in.}^2$$

$$\text{Req'd } l_b = \frac{A}{b} = \frac{18.7}{5.625} = 3.32 \text{ in.} \qquad Say \; 3\tfrac{1}{2} \text{ in.}$$

$$\boxed{\begin{array}{c} 6 \times 18 \text{ rough-sawn} \\ \text{Sel. Str. DF beam is } OK \end{array}}$$

NOTE: A lower stress grade could be used for this beam.

The deflection check would remain unchanged because the modulus of elasticity is the same for Select Structural and No. 1. A more economical beam would be obtained with the lower stress grade.

The importance of knowing the correct *size category* can be seen by comparing the allowable stresses for No. 1 DF-L B&S with those in the example in Sec. 6.8 for No. 1 DF-L J&P. For a given grade, the allowable stresses depend on the size (use) categories.

6.10 Design Problem: Sawn-Beam Analysis

The two previous examples have involved beams in the J&P and B&S size categories. Example 6.15 is provided to give additional practice in determining allowable stresses. The member is in the J&P size category, but the load duration factor and condition-of-use factors are different. Condition-of-use factors (CUFs) are obtained from the footnotes to the allowable stress tables.

6.11 Design Problem: Glulam Beam with Full Lateral Support

The examples in Secs. 6.11, 6.12, and 6.13 all deal with the design of the same glulam beam, but different conditions of lateral support for the beam are considered.

The first example deals with the design of a beam that has *full lateral support* to the compression side of the member. With this condition lateral stability is simply not a problem. See Example 6.16.

EXAMPLE 6.15   Sawn-Beam Analysis

Determine if the given beam (Fig. 6.15) is adequate if the (DL + snow load) is 250 lb/ft. Lumber is No.1 Hem-Fir, and lateral stability is not a factor. The beam is used in a factory where the EMC will exceed 19 percent.* Beams are 4 ft-0 in. o.c. Minimum roof slope for drainage is provided. Consider shear and bending stresses only. Deflection is not a problem. Take allowable stresses and section properties from the NDS.

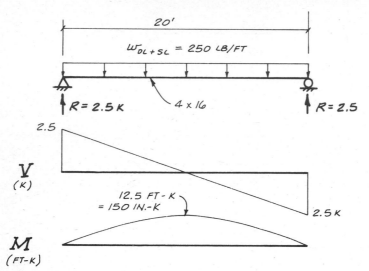

Fig. 6.15

Bending

$$f_b = \frac{M}{S} = \frac{150{,}000}{135.66} = 1105 \text{ psi}$$

$$\text{Adj. } F_b = (\text{tab. } F_b) \times C_F \times \text{LDF} \times \text{CUF}$$
$$= 1200 \times 0.97 \times 1.15 \times 0.86$$
$$= 1150 \text{ psi} > 1105 \text{ psi} \qquad OK$$

Shear

$$f_v = \frac{1.5V}{A} = \frac{1.5(2500)}{53.375} = 70.3 \text{ psi}$$

$$\text{Adj. } F_v = (\text{tab. } F_v) \times \text{LDF} \times \text{CUF}$$
$$= 75 \times 1.15 \times 0.97$$
$$= 83.7 \text{ psi} > 70.3 \text{ psi}† \qquad OK$$

> 4 × 16 No.1
> Hem-Fir beam   OK

---

* The need for pressure-treated lumber to prevent decay should be considered (Sec. 4.6).

† If f_v had exceeded F_v, the design shear could have been reduced in accordance with Sec. 6.4.

EXAMPLE 6.16   Glulam Beam—Full Lateral Support

Determine the required size of 22F DF glulam for the simple-span beam shown in Fig. 6.16. Assume a dry condition of use. Roof DL = 200 lb/ft and snow load = 800 lb/ft. By inspection the critical load combination is

$$DL + SL = TL = 200 + 800 = 1000 \text{ lb/ft}$$

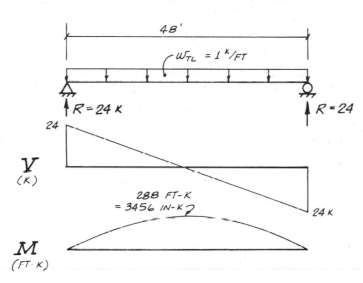

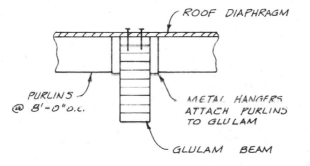

BEAM SECTION

Fig. 6.16

Bending

The allowable bending stress in wood beams is governed by one of two criteria:

1. Size (depth) effect
2. Lateral stability

157

DESIGN OF WOOD STRUCTURES

The sketch of the beam cross section shows that the compression side of the beam (positive moment places the top side in compression) is restrained from lateral movement by an effective connection to the roof diaphragm. The unbraced length is zero, and the beam automatically falls into the short unbraced range. Lateral buckling is thus prevented. In this case only the bending stress based on the size factor needs to be considered.

Assume $C_F = 0.90$,

$$\text{Trial } F_b = (\text{tab. } F_b) \times C_F \times \text{LDF}$$
$$= 2200 \times 0.90 \times 1.15 = 2277 \text{ psi}$$

$$\text{Req'd } S = \frac{M}{F_b} = \frac{3,456,000}{2277} = 1518 \text{ in.}^3$$

Try 6¾ × 37.5 (twenty-five 1½-in. lams):

From Appendix C,

$$A = 253.1 \text{ in.}^2$$
$$S = 1582 \text{ in.}^3$$
$$I = 29,663 \text{ in.}^4$$
$$C_F = 0.88 \text{ (tabulated)}$$

$$\frac{L}{d} = \frac{48 \times 12}{37.5} = 15.4 < 21$$

∴ C_F can be increased, but is not required. Use tabulated C_F.

Stress summary:

$$f_b = \frac{M}{S} = \frac{3,456,000}{1582} = 2185 \text{ psi}$$

$$\text{Adj. } F_b = (\text{tab. } F_b) \times C_F \times \text{LDF}$$
$$= 2200 \times 0.88 \times 1.15$$
$$= 2226 > 2185 \qquad OK$$

Shear

Ignore shear adjustment (conservative).

$$f_v = \frac{1.5V}{A} = \frac{1.5(24,000)}{253.1} = 142 \text{ psi}$$

$$\text{Adj. } F_v = (\text{tab. } F_v) \times \text{LDF} = 165 \times 1.15$$
$$= 190 > 142 \qquad OK$$

Deflection

$$\Delta_{TL} = \frac{5w_{TL} L^4}{384EI} = \frac{5(1000)(48)^4(12 \text{ in./ft})^3}{384(1,800,000)(29,663)} = 2.24 \text{ in.}$$

$$\frac{\Delta}{L} = \frac{2.24}{48 \times 12} = \frac{1}{257} < \text{allow. } \Delta_{TL} \text{ of } \frac{1}{180} \qquad OK$$

By proportion

$$\Delta_{SL} = \left(\frac{800}{1000}\right) \Delta_{TL} = 0.8(2.24) = 1.79 \text{ in.}$$

$$\frac{\Delta}{L} = \frac{1.79}{48 \times 12} = \frac{1}{321} < \frac{1}{240} \qquad OK$$

$$\text{Camber} = 1.5 \, \Delta_{DL} = 1.5\left(\frac{200}{1000}\right)(2.24) = 0.67 \text{ in.}$$

Bearing

The support conditions are unknown so the required bearing length will simply be determined. Use minimum $F_{c\perp}$.

$$\text{Req'd } A = \frac{R}{F_{c\perp} \times \text{LDF}} = \frac{24{,}000}{385 \times 1.15} = 54.2 \text{ in.}^2$$

$$\text{Req'd } l_b = \frac{52.4}{6.75} = 8.03 \text{ in.} \qquad \textit{Say } l_b = 8 \text{ in. min.}$$

> Use 6¾ × 37.5 (twenty-five 1½-in. lams)
> 22F DF glulam—camber 0.67 in.

In Sec. 6.12 the example is reworked with lateral supports at 8 ft-0 in. o.c. This spacing is obtained when the purlins rest on top of the glulam. With this arrangement the sheathing is separated from the beam, and the distance between points of lateral support becomes the spacing of the purlins.

In Sec. 6.13, the beam is analyzed for an unbraced length of 48 ft-0 in. In other words, only the ends of the beam are stayed against translation and rotation. This condition would exist if no diaphragm action developed in the sheathing (i.e., the sheathing for some reason is not capable of functioning as a diaphragm), or if no sheathing or effective bracing is present along the beam. Fortunately, this situation is not common in ordinary building design.

Allowable stresses can be obtained from the NDS, UBC Chap. 25, or Refs. 3 and 28. Section properties can be read from Appendix C.

6.12 Design Problem: Glulam Beam with Lateral Support at 8 ft-0 in.

In order to design a beam with an unbraced compression zone, it is necessary to check both lateral stability and size effect. To check lateral stability, a trial beam size is required so that the slenderness factor C_s can be computed. This is similar to column design, where a trial size is required before the slenderness ratio and the strength of the column can be evaluated.

All criteria except unbraced length are the same for this example and the previous problem. Therefore, initial trial beam size is taken from Example 6.16. The 6¾ × 37.5 trial represents minimum beam size based on the size effect criterion. Because all other factors are the same, only the lateral stability criteria are considered in this example. See Example 6.17. The calculations indicate that lateral stability is less critical than the size effect for this particular beam. The trial size, then, is adequate.

159

EXAMPLE 6.17   Glulam Beam—Lateral Support at 8 ft-0 in.

Rework Example 6.16 using the modified lateral support condition shown in the beam section view in Fig. 6.17. All other criteria are the same. See Fig. 6.16 for the load, shear, and moment diagrams.

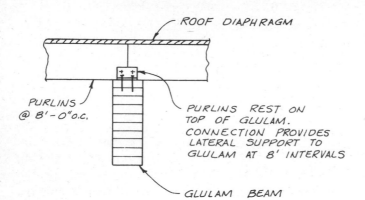

ROOF DIAPHRAGM

PURLINS @ 8'-0"o.c.

PURLINS REST ON TOP OF GLULAM. CONNECTION PROVIDES LATERAL SUPPORT TO GLULAM AT 8' INTERVALS

GLULAM BEAM

BEAM SECTION

Fig. 6.17

Bending
Size effect—see Example 6.16.
Lateral stability:

$$\text{Try } 6\tfrac{3}{4} \times 37.5 \quad \text{minimum section based on size effect}$$
$$\text{Unbraced length} = L_u = 8 \text{ ft} = 96 \text{ in.}$$

Effective unbraced length:

$$L_e = 1.92 L_u = 1.92 \times 96 = 184 \text{ in.}$$

Slenderness factor:

$$C_s = \sqrt{\frac{L_e d}{b^2}} = \sqrt{\frac{184 \times 37.5}{(6.75)^2}}$$
$$= 12.3 > 10$$

Lateral support conditions exceed the short unbraced length limit (Fig. 6.6).
  Determine the limit C_k to see whether the beam has an intermediate or long unbraced length. The LDF is not used to calculate C_k.

$$C_k = \sqrt{\frac{3E}{5F_b}} = \sqrt{\frac{3(1,800,000)}{5(2200)}} = 22.2 > 12.3$$

$$10 < C_s < C_k \qquad \therefore \text{ intermediate}$$

160   The LDF is used to calculate F_b' in the intermediate range.

$$F'_b = F_b \text{ (LDF)} \left[1.0 - \frac{1}{3} \left(\frac{C_s}{C_k} \right)^4 \right]$$

$$= (2200 \times 1.15) \left[1.0 - \frac{1}{3} \left(\frac{12.3}{22.2} \right)^4 \right] = 2450 \text{ psi}$$

From Example 6.16—size effect calculations:

$$F_b = 2200 \times 1.15 \times 0.88 = 2226 \text{ psi}$$
$$2226 < 2450 \qquad \therefore \text{ size effect governs}$$

∴ The trial size is adequate for bending. The other design criteria was shown to be adequate in the previous example.

> Use 6¾ × 37.5 22F DF glulam

6.13 Design Problem: Glulam Beam with Lateral Support at 48 ft-0 in.

The purpose of this brief example is to illustrate the use of the formula for lateral stability calculations for long unbraced lengths. See Example 6.18. As with the previous example, the initial trial size is taken from Example 6.16. A trial size is required in order to calculate the slenderness factor.

This example illustrates why it is desirable to have at least some intermediate lateral bracing. The very long unbraced length causes the trial size to be considerably overstressed, and a new trial beam size is required.

The example is not carried beyond the point of checking the initial trial beam. The purpose of the example is satisfied without completing the design. A larger trial size would be evaluated in a similar manner.

EXAMPLE 6.18 Glulam Beam—Lateral Support at 48 ft-0 in.

Rework the beam design problem in Examples 6.16 and 6.17 with lateral supports at the ends of the span only. See Figure 6.16 for the load, shear, and moment diagrams.

Bending
Size effect—see Example 6.16.
Lateral stability:

> *Try* 6¾ × 37.5 minimum section based on size effect
> Unbraced length = L_u = 48 ft = 576 in.

For such a long unbraced length, it is likely that stability will control. Effective unbraced length:

$$L_e = 1.92 L_u = 1.92 \times 576 = 1106 \text{ in.}$$

Slenderness factor:

$$C_s = \sqrt{\frac{L_e d}{b^2}} = \sqrt{\frac{1106 \times 37.5}{(6.75)^2}}$$

$$C_s = 30.2$$

From Example 6.17

$$C_k = 22.2 < 30.2$$

$$\therefore \text{ Long unbraced length (Fig. 6.6)}$$

The LDF is not used to calculate F_b' in the long range.

$$F_b' = \frac{0.4E}{C_s^2} = \frac{0.4(1,800,000)}{(30.2)^2} = 791 \text{ psi}$$

$$791 < 2226 \text{ psi for size effects}$$

$\therefore$ Stability governs.

$$f_b = \frac{M}{S} = \frac{3,456,000}{1582} = 2185 > 791 \qquad NG$$

$\therefore$ A revised trial size is required.

6.14 Cantilever Beam Systems

Cantilever beam systems are often used in glulam construction. The reason for this is that a smaller-size beam can be used with a cantilever system compared with a series of simply supported beams. The cantilever length L_c used in the cantilever system is important. See Example 6.19. By properly choosing the cantilever length, an optimum beam size can be obtained.

For the design of roofs, UBC Chap. 23 requires that *if* the RLL is less than 20 psf, the case of DL on all spans plus RLL on alternate spans (unbalanced RLL) must be considered in addition to full (DL + RLL) on all spans. For floor beams, the case of (DL + full FLL) and (DL + unbalanced FLL) must be considered for the design of beams regardless of the magnitude of the FLL. See Example 6.20.

The case of unbalanced live loads can complicate the design of cantilever beam systems. This is particularly true if deflections are considered in addition to bending and shear. The cantilever lengths provided in Ref. 3 consider *constant uniform loads over all spans.* When unbalanced live loads are considered, the optimum cantilever span length L_c will be different from those established for the same uniform load on all spans.

In a cantilever system the compression side of the member is not always on the top of the beam. This will require a *lateral stability* analysis of bending stresses even though the top of the girder is connected to the horizontal diaphragm. See Example 6.21.

EXAMPLE 6.19   Cantilever Beam Systems

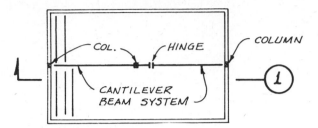

FRAMING PLAN

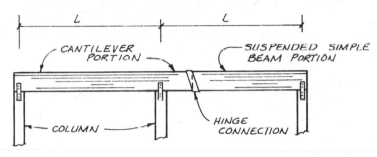

SECTION ①

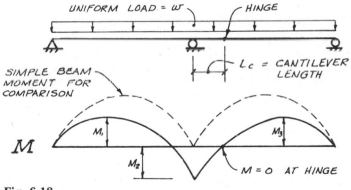

Fig. 6.18

The *bending strength* of a cantilever beam system can be optimized by choosing the cantilever length L_c so that the local maximum moments M_1, M_2, and M_3 will all be equal. For the two-equal-span cantilever system shown above, with a constant uniform load on both spans, the cantilever length

$$L_c = 0.172L$$

163

gives equal local maximum moments

$$M_1 = M_2 = M_3 = 0.086wL^2$$

This maximum moment is considerably less than the maximum moment for a uniformly loaded simple beam

$$M = \frac{wL^2}{8} = 0.125wL^2$$

Recommended cantilever lengths for a number of cantilever beam systems with *equal uniform loads on all spans* are given in the table in Ref. 3 titled *Cantilever Beam Coefficients*.

EXAMPLE 6.20   Load Cases for Cantilever Beam Systems

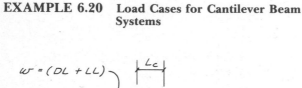

LOAD CASE 1

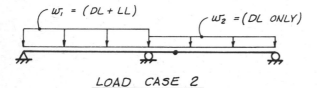

LOAD CASE 2

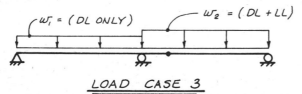

LOAD CASE 3

Fig. 6.19

Load Case 1   **(DL + LL on all spans)**

This load constitutes the maximum total load and can produce the critical design moment, shear, and deflection.

Load Case 2   (**DL + unbalanced LL on left span**)

When unbalanced LL is required, this load will produce the critical positive moment in the left span.

Load Case 3   (**DL + unbalanced LL on right span**)

This load case is normally not critical but will produce the minimum reaction at the left support. For a large LL and a long cantilever length, it is possible to develop an uplift reaction at this support. In this case the connection must be capable of anchoring the beam.

EXAMPLE 6.21   **Lateral Stability of Cantilver Systems**

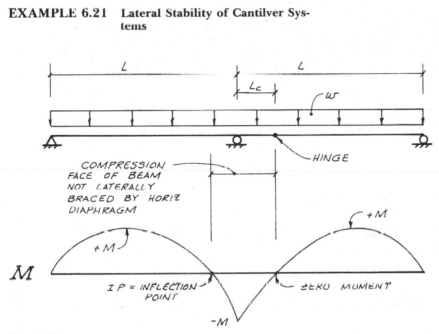

Fig. 6.20*a*

Moment diagram sign convention:

> Positive moment = compression on the top of the beam
> Negative moment = compression on the bottom of the beam

In areas of negative moment, the horizontal diaphragm is connected to the tension side of the beam, and this does not provide lateral support to the compression side of the beam. If the lower face of the beam is braced (see Fig. 6.20*b*) at the interior column, the unbraced length L_u for evaluating lateral stability is the cantilever length L_c or the distance from the column to the IP. For the given beam these lengths are equal (Fig. 6.20*a*).

165

If restraint at the column is not provided, L_u is the full length of the lower side of the beam that is in compression. Several types of knee braces can be used to brace the bottom side of the beam. A *prefabricated metal knee brace* and a *lumber knee brace* are shown below. The distance between knee braces, or the distance between a brace and a point of zero moment, is the unbraced length.

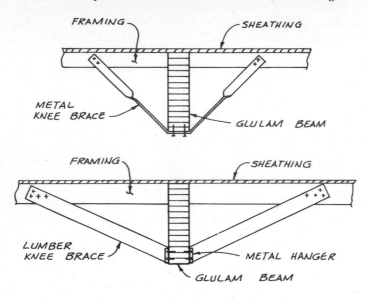

TYP BRACES FOR NEGATIVE MOMENT

Fig. 6.20*b*

6.15 Design Problem: Cantilever Beam System

In this example a cantilever beam system with two equal 50-ft spans is designed. See Example 6.22. The initial step is to determine the cantilever length L_c.

The girder is designed for a reduced roof live load, and this requires that both full and unbalanced loading be considered. For this loading L_c is taken as $0.2L$. Two different beam sizes are developed because the three local maximum moments are not equal for the two required load cases. The larger beam is required for the cantilever beam member AD, and the smaller size is for the suspended beam member DF.

For the cantilever member AD it is necessary to check lateral stability for the portion of the member where negative moment occurs (compression on the bottom of the beam).

The final part of the example considers the camber provisions for the girder. In cambering members, most glulam manufacturers are able to set jigs at 4-ft intervals. However, the designer in most cases does not have to specify the camber settings at these close intervals. Typically the required camber would be specified at the midspans, at the internal hinge points, and perhaps at the point of inflection. The manufacturer, then, would establish the camber at various points along the span using a parabolic or circular curve. Camber tolerance is roughly $\pm \frac{1}{4}$ in.

EXAMPLE 6.22 Cantilever Beam System

Design a cantilever roof beam system using UBC roof design loads. Determine the optimum location of the hinge. Use 22F Douglas fir glulam. Tributary width to the girder is 20 ft. Roof DL = 14 psf, including an estimated 2 psf (40 lb/ft) for the weight of the girder. There is no snow load, and the beam does not support a plastered ceiling. CUF = 1.0. Allowable stresses can be obtained from the NDS, UBC Chap. 25, and Refs. 3 and 28. Section properties can be found in Appendix C.

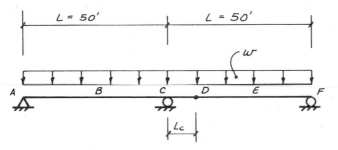

Fig. 6.21a

Loads

As noted in Sec. 6.14, the Code requires that unbalanced roof LL be considered unless the live load is 20 psf or greater. Thus, one possible design approach is to use a 20-psf RLL on all spans.

$$w_{DL} = 14 \times 20 = 280 \text{ lb/ft}$$
$$w_{LL} = 20 \times 20 = \underline{400}$$
$$w_{TL} = 680 \text{ lb/ft}$$

An alternate design can be made using a reduced roof LL, considering (DL + LL) on all spans or (DL + unbalanced LL), whichever is critical.

$$\text{Trib. } A > 600 \text{ ft}^2 \qquad \therefore \text{ RLL} = 12 \text{ psf}$$

$$w_{DL} = 14 \times 20 = 280 \text{ lb/ft}$$
$$w_{LL} = 12 \times 20 = \underline{240}$$
$$w_{TL} = 520 \text{ lb/ft}$$

A lighter section will be obtained with the reduced live load.
    In Example 6.19 a cantilever length of

$$L_c = 0.172L$$

DESIGN OF WOOD STRUCTURES

was recommended for a two-span cantilever system with a constant uniform load on both spans. When unbalanced live load is also considered, a different cantilever length will give approximately equal positive and negative moments for the cantilever segment. This length is

$$L_c = 0.2L = 0.2(50) = 10 \text{ ft}$$

With this the shear and moment diagrams for the two critical loading conditions can be drawn.

Load Case 1 (DL + RLL) on all spans

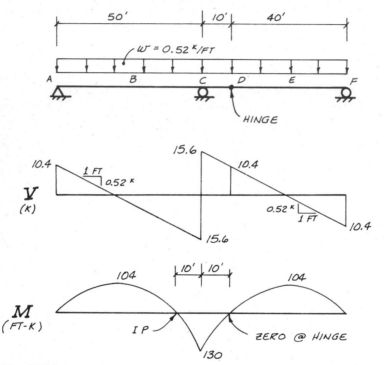

Fig. 6.21*b*

NOTE: For comparison the moment for a simple beam is

$$M = \frac{wL^2}{8} = \frac{(0.52)(50)^2}{8} = 162 \text{ ft-k}$$

Member AD

BENDING:

Max. moments from Fig. 6.21*b* and Fig. 6.21*c*:
Max. + $M \approx$ Max. $- M = 130$ ft-k $= 1560$ in.-k
Trial $F_b = ($tab. $F_b) \times$ LDF $\times C_F$

Assume $C_F = 0.90$:

Trial $F_b = 2200(1.25)(0.90) = 2475$ psi

$$\text{Req'd } S = \frac{M}{F_b} = \frac{1,560,000}{2475} = 630 \text{ in.}^3$$

Load Case 2 **(DL + unbalanced RLL) on left span**

NOTE: Load case 3 of Example 6.20 does not govern the size of the beam, and it is not considered in this example.

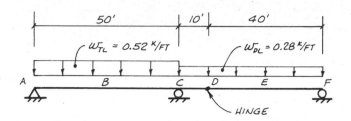

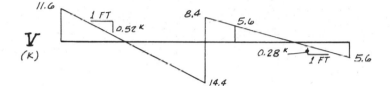

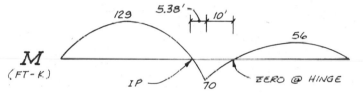

Fig. 6.21c

Try 5⅛ × 27
From Appendix C,

$$A = 138.4 \text{ in.}^2$$
$$S = 622.7 \text{ in.}^3$$
$$I = 8406 \text{ in.}^4$$
$$C_F = 0.91$$

Although the furnished S of 622.7 appears to be less than the required value, the actual value of the size factor is larger than the assumed value. Show stress summary:

$$f_b = \frac{M}{S} = \frac{1,560,000}{622.7} = 2505 \text{ psi}$$

Adj. $F_b = (\text{tab. } F_b) \times \text{LDF} \times C_F$

$$= 2200 \times 1.25 \times 0.91 = 2503 \approx 2505 \qquad OK$$

LATERAL STABILITY:

The top of the girder will be attached to the roof diaphragm. Therefore, full lateral support is provided to the member in areas of positive moment.

Negative moment occurs at the interior column support. Provide a brace to the compression side (bottom) of the beam at the column. The unbraced length of the **169**

DESIGN OF WOOD STRUCTURES

beam is the distance from the interior column to the point of zero moment (to the hinge or inflection point—whichever is larger).

Load Case 1 is critical for lateral stability considerations because this loading produces the maximum negative moment. In addition, the distance from the column to the inflection point is greater for this load case.

$$\text{Unbraced length} = L_u = 10 \text{ ft} = 120 \text{ in.}$$

$$\text{Effective unbraced length} = L_e = 1.69 L_u \quad \text{for a cantilever with a concentrated load at the free end}$$

$$L_e = 1.06 L_u \quad \text{for a cantilever with a uniformly distributed load}$$

For this beam an intermediate L_e could be determined on the basis of percentage of the total moment at the column that is produced by the concentrated load and the distributed load. For simplicity use the factor for the concentrated load

$$L_e = 1.69 L_u = 1.69 \times 120 = 203 \text{ in.}$$

$5\frac{1}{8} \times 27$:

$$C_s = \sqrt{\frac{L_e d}{b^2}} = \sqrt{\frac{203 \times 27}{(5.125)^2}} = 14.4$$

$$C_k = \sqrt{\frac{3E}{5F_b}} = \sqrt{\frac{3(1,800,000)}{5(2200)}}$$

$$= 22.2 > 14.4$$

$$10 < 14.4 < 22.2 \qquad \therefore \text{ intermediate unbraced range}$$

$$F_b' = F_b \text{ (LDF)} \left[1.0 - \frac{1}{3}\left(\frac{C_s}{C_k}\right)^4 \right]$$

$$= 2200(1.25) \left[1.0 - \frac{1}{3}\left(\frac{14.4}{22.2}\right)^4 \right] = 2585$$

$$F_b' = 2585 \text{ psi} \qquad \text{lateral stability}$$

$$F_b = 2503 \text{ psi} \qquad \text{size effect (governs)}$$

No modification of the design is required for stability.

SHEAR:

$$\text{Max. } V = 15.6 \text{ k} \qquad \text{neglect reduction (conservative)}$$

$$f_v = \frac{1.5V}{A} = \frac{1.5(15,600)}{138.4} = 169 \text{ psi}$$

$$F_v = (\text{tab. } F_v) \times \text{LDF} = 165 \times 1.25$$

$$= 206 > 169 \qquad OK$$

Member DF

BENDING:

$$\text{Max. } M = 104 \text{ ft-k} = 1248 \text{ in-k}$$

$$\text{Trial } F_b = (\text{tab. } F_b) \times \text{LDF} \times C_F$$

Assume $C_F = 0.92$:

$$\text{Trial } F_b = 2200 \times 1.25 \times 0.92 = 2530 \text{ psi}$$

$$\text{Req'd } S = \frac{M}{F_b} = \frac{1,248,000}{2530} = 493 \text{ in.}^3$$

Try 5⅛ × 24.
From Appendix C,

$$A = 123 \text{ in.}^2$$
$$S = 492 \text{ in.}^3$$
$$I = 5904 \text{ in.}^4$$
$$C_F = 0.93$$

Stress summary:

$$f_b = \frac{M}{S} = \frac{1,248,000}{492} = 2537 \text{ psi}$$

Adj. $F_b = 2200 \times 1.25 \times 0.93 = 2558$ psi   *OK*

SHEAR:

Max. $V = 10.4$ k

$$f_v = \frac{1.5V}{A} = \frac{1.5(10,400)}{123} = 127$$

Adj. $F_v = $ (tab. F_v) × LDF

$$= 165 \times 1.25 - 206 \text{ psi} > 127 \quad OK$$

Deflection and Camber

A comprehensive deflection analysis for a cantilever beam system can prove to be a cumbersome calculation. This is especially true if unbalanced loads are involved. Some designers choose to perform a complete deflection analysis, and others do not. This is really at the discretion of the designer because the Code does not provide deflection criteria for roofs (except for those that support a plastered ceiling).

In order to simplify this example, only the dead load deflection calculation will be illustrated. This is required in order to determine the camber for the beam (camber = 1.5 Δ_{DL}).

Various methods of calculating deflection can be used. Here the dead-load deflection will be calculated by the superposition of handbook deflection formulas (Refs. 3 and 12). The deflection will be evaluated at three points:

1. At the center of span *AC* (point *B*)
2. At the hinge (point *D*)
3. At the midspan of the suspended beam (point *E*)

NOTE: The camber provisions included in this example are for long-term deflection. A roof slope of ¼ in./ft (in addition to long-term DL deflection considerations) is required to provide drainage.

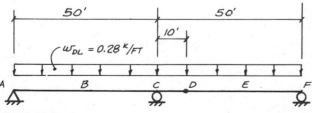

Fig. 6.21d

DESIGN OF WOOD STRUCTURES

CAMBER AT B:

Deflection due to uniform load on member AD:

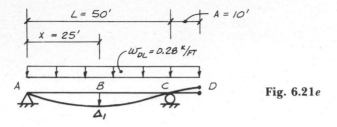

Fig. 6.21e

$$\Delta_1 = \frac{wX}{24\,EIL}(L^4 - 2L^2X^2 + LX^3 - 2A^2L^2 + 2A^2X^2)$$

$$= \frac{(0.28)(25)(12\text{ in./ft})^3}{24(1800)(8406)(50)}[(50)^4 - 2(50)^2(25)^2 + 50(25)^3 - 2(10)^2(50)^2 + 2(10)^2(25)^2]$$

$\Delta_1 = 2.35$ in. down

Deflection due to load on DF:

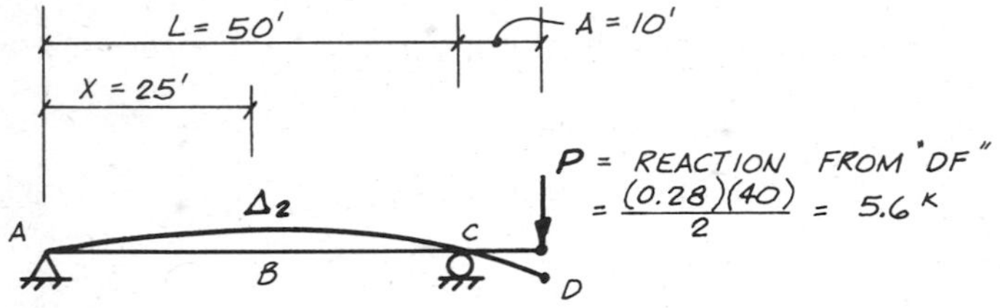

L = 50'

X = 25'

A = 10'

Δ_2

A   B   C   D

P = REACTION FROM "DF"

$= \dfrac{(0.28)(40)}{2} = 5.6$ K

Fig. 6.21f

$$\Delta_2 = \frac{PAX}{6\,EIL}(L^2 - X^2)$$

$$= \frac{(5.6)(10)(25)(12)^3}{6(1800)(8406)(50)}[(50)^2 - (25)^2]$$

$\Delta_2 = 1.0$ in. up

$\Delta_{DL} = \Delta_1 + \Delta_2 = 2.35 - 1.0 = 1.35$ in. down

Camber at $B = 1.5\ \Delta_{DL} = 1.5(1.35) = 2$ in. up

CAMBER AT HINGE D:

Deflection due to uniform load on AD:

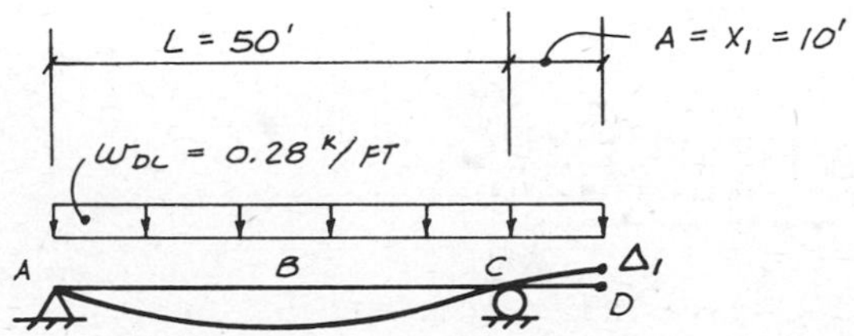

L = 50'

A = X₁ = 10'

W_{DL} = 0.28 K/FT

A   B   C Δ_1 D

Fig. 6.21g

172

$$\Delta_1 = \frac{wX_1}{24\ EI}(4A^2L - L^3 + 6A^2X_1 - 4AX_1^2 + X_1^3)$$

$$= \frac{(0.28)(10)(12\ \text{in./ft})^3}{24(1800)(8406)}[(4)(10)^2(50) - (50)^3 + 6(10)^2(10) - 4(10)(10)^2 + (10)^3]$$

$\Delta_1 = 1.36$ in. up

Deflection due to load on DF:

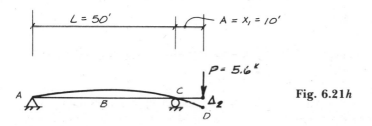

Fig. 6.21h

$$\Delta_2 = \frac{PX_1}{6EI}(2AL + 3AX_1 - X_1^2)$$

$$-\frac{(5.6)(10)(12)^3}{6(1800)(8406)}[(2)(10)(50) + 3(10)(10) - (10)^2]$$

$\Delta_2 = 1.28$ in. down

$\Delta_{DL} = \Delta_1 + \Delta_2 = -1.36 + 1.28 = 0.08$ in.—very small

Specify *no camber* at hinge D.

CAMBER AT E:

The left support of member DF (i.e., the hinge) has been found to have a very small deflection. The dead-load deflection calculation for point E is a simple beam-deflection calculation

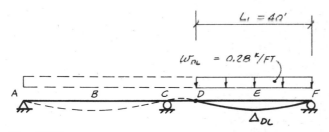

Fig. 6.21i

$$\Delta_{DL} = \frac{5wL_1^4}{384\ EI} = \frac{5(0.28)(40)^4(12)^3}{384(1800)(5904)}$$

$$= 1.52\ \text{in.}$$

Camber at $E = 1.5\ \Delta_{DL} = 1.5 \times 1.52 = 2\frac{1}{4}$ in. up

173

> Use $5\frac{1}{8} \times 27$ for member AD
> $5\frac{1}{8} \times 24$ for member $DF - 22F$ DF glulam
>
> Camber: 2 in. up at point B
> Zero camber at hinge D
> $2\frac{1}{4}$ in. up at point E

6.16   Lumber Roof and Floor Sheathing

Sheathing for roofs or floors may be lumber decking or plywood. Plywood is described in considerable detail in Chap. 8, and decking will be covered briefly in this section.

Timber decking is available as *solid decking* or *laminated decking*. Solid decking is made from kiln-dried lumber and is available in Select or Commercial

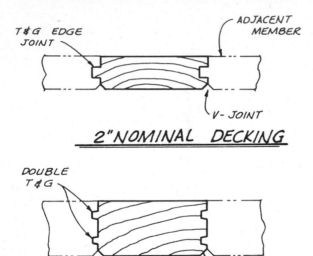

2" NOMINAL DECKING

3" & 4" NOMINAL DECKING

Fig. 6.22   Solid lumber decking. Decking can be obtained with various surface patterns if the bottom side of the decking is architecturally exposed. These sketches show a V-joint pattern.

grades in a number of commercial wood species. Common sizes are 2 × 6, 3 × 6, and 4 × 6 (nominal sizes). Various types of edge configurations are available, but tongue-and-groove (T&G) edges are probably the most common. See Fig. 6.22. A single tongue-and-groove is used on 2-in.-nominal decking, and a double tongue-and-groove is used on the larger thicknesses.

Glued laminated decking is fabricated from three or more individual kiln-dried laminations. Laminated decking also has T&G edge patterns.

Decking essentially functions as a series of parallel beams that span between the roof or floor framing. Bending stresses or deflection criteria usually govern the allowable loads on decking. Spans range from 3 ft up to 20 ft and more depending on the load, span type, grade, and thickness of decking. The layup of decking affects the load capacity because it determines the type of beam span of the decking. See Example 6.23.

EXAMPLE 6.23   Layup of Decking

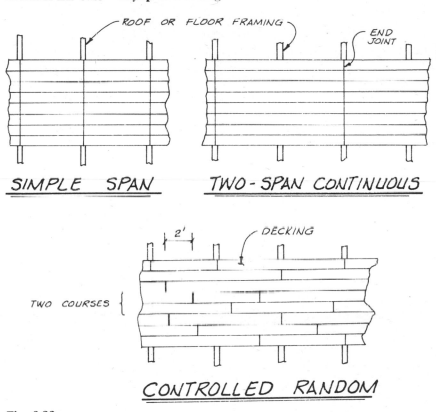

Fig. 6.23

Layup refers to the arrangement of end joints in decking. Five different layups are defined in Ref. 3, and three of these are shown in Fig. 6.23. Controlled random layup is economical and simply requires that end joints in adjacent members be well staggered (2 ft minimum). In addition, end joints that occur on the same transverse iine must be separated by at least two courses. For other requirements see Ref. 3.

Reference 3 gives the bending and deflection coefficients for the various types of layup. These can be used to calculate the required thickness of decking. Generally speaking, however, the designer can often simply refer to allowable span and load tables for decking requirements. UBC Table 25T gives the allowable span for 2-in. T&G decking. Reference 28 includes allowable load tables for controlled random layup for both solid and laminated decking. Reference 3 has expanded tables to cover the other types of layup. Information is also available from manufacturers and suppliers.

6.17 Problems

Some of the following beam problems have allowable stresses given. When they are not given, allowable stresses are to be taken from the NDS.

6.1 The beam in Fig. 6.A is a 4 × 8 wood member. $E = 1,600,000$ psi.

 Find: The maximum bending stress, shear stress, and deflection.

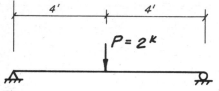

Fig. 6.A

6.2 The beam in Fig. 6.B is a 4 × 8 wood member. $E = 1,600,000$ psi.

 Find: The maximum bending stress, shear stress, and deflection.

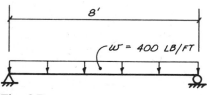

Fig. 6.B

6.3 The beam in Fig. 6.C is a 4 × 8 wood member. $E = 1,600,000$ psi.

 Find: The maximum bending stress, shear stress, and deflection.

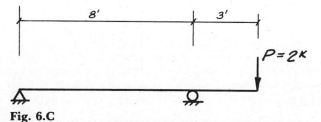

176 **Fig. 6.C**

6.4 A simply supported horizontal (¼ in./ft slope) roof beam has a 12-ft span and the following uniform loads: roof DL = 40 lb/ft; roof LL = 80 lb/ft. The beam has full lateral support, and the ceiling supported by the beam is not plaster. Tabulated allowable stresses are F_b = 1400 psi, F_v = 75 psi, E = 1,500,000 psi. CUF = 1.0.

Find: Minimum required beam size.

6.5 A simply supported floor beam has a span of 16 ft and the following uniform loads: floor DL = 50 lb/ft, floor LL = 200 lb/ft. Lateral buckling is prevented by continuous support. Tabulated allowable stresses are F_b = 1150 psi, F_v = 75 psi, E = 1,400,000 psi. CUF = 1.0.

Find: Minimum required beam size.

6.6 A simply supported horizontal (¼ in./ft slope) roof beam has a 12-ft span and the following uniform loads: roof DL = 500 lb/ft, roof LL = 550 lb/ft. No lateral buckling, and no plastered ceiling. Tabulated allowable stresses are F_b = 1600 psi, F_v = 85 psi, E = 1,600,000 psi. CUF = 1.0.

Find: Minimum required beam size.

6.7 Problem 6.7 is the same as Prob. 6.6 except the loads are roof DL = 350 lb/ft, roof LL = 400 lb/ft.

6.8 The roof rafters in Fig. 6.D are 24 in. o.c. Roof DL is 12 psf along the roof, and roof LL is in accordance with UBC Table 23C, Method 1. Calculate design shear and moment using the horizontal plane method of Example 2.5. Tabulated allowable stresses are F_b = 1500 psi (repetitive member), F_v = 95. Lateral stability is not a problem. Disregard deflection. CUF = 1.0.

Find: Minimum required rafter size.

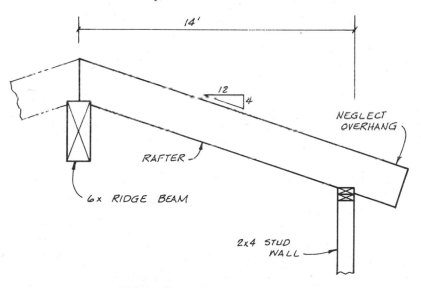

SECTION

Fig. 6.D

6.9 Problem 6.9 is the same as Prob. 6.8 except that the rafters are spaced 6 ft-0 in. o.c.

6.10 The roof rafters in Fig. 6.D are 24 in. o.c. The roof DL is 15 psf along the roof, and the design snow load is 50 psf. Calculate the design shear and moment using the horizontal plane method of Example 2.5. Disregard deflection. Lateral stability is not a problem. Lumber is No. 1 Douglas fir-larch. CUF = 1.0.

Find: The minimum rafter size.

6.11 Problem 6.11 is the same as Prob. 6.10 except that the roof slope is $\frac{6}{12}$.

6.12 The beam in Fig. 6.E is supported laterally at the ends only. The member is a 6 × 14 DF-L Select Structural. Load is a combination of DL plus floor LL. CUF = 1.10.

Find: *a.* The allowable bending stress considering size effect.
        b. The allowable bending stress considering the effects of lateral stability.

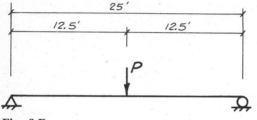

Fig. 6.E

6.13 Problem 6.13 is the same as Prob. 6.12 except that the member is a 5⅛ × 18 Southern pine glulam (20F).

6.14 *Given:* The rafter connection in Figure. 6.F. The load is a combination of roof DL plus snow load. Lumber is No. 1 Mountain hemlock. CUF = 1.0.

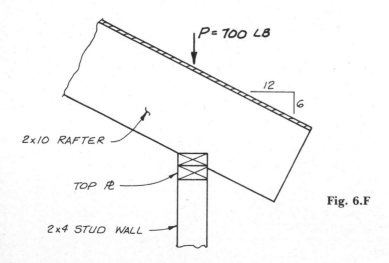

Fig. 6.F

Find: *a.* The actual bearing stress in the rafter and in the top plate of the wall.

      b. The allowable bearing stress in the rafter and in the top plate.

6.15 *Given:* The rafter connection in Fig. 6.F with the slope changed to $1\frac{2}{12}$. The load is a combination of DL plus roof LL. Lumber is No. 2 Southern pine (surfaced dry). CUF = 1.0.

Find: *a.* The actual bearing stress in the rafter and in the top plate of the wall.

      b. The allowable bearing stress in the rafter and in the top plate.

6.16 *Given:* The roof framing plan of the industrual building in Figure. 6.G. No ceiling. The total dead loads to the members are

$$\text{Subpurlin (2} \times \text{4 at 24 in. o.c.)} = 7.0 \text{ psf}$$
$$\text{Purlins (4} \times \text{14 at 8 ft-0 in. o.c.)} = 8.5$$
$$\text{Girder} = 10.0$$

Roof live loads are to be in accordance with UBC Table 23.C, Method 1. CUF = 1.0.

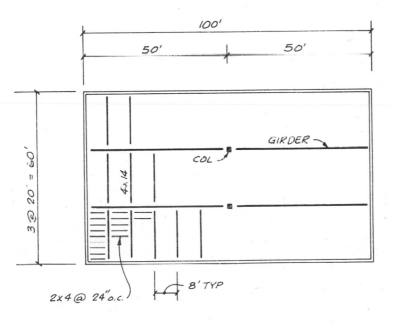

ROOF FRAMING PLAN

Fig. 6.G

Find: *a.* Check the subpurlins using No. 1 DF-L. Are the AITC recommended deflection criteria satisfied?

      b. Check the purlins using No. 1 DF-L. Are the AITC deflection limits met?

     c. Design the girder using 24F Douglas fir glulam. Determine the minimum size considering both strength and stiffness. Stability is not a problem.

6.17 The girder in the roof framing plan in Fig. 6.H is to be designed using the optimum cantilever length L_c. The girder is 20F Southern pine glulam. RDL = 16 psf. The top of the girder is laterally supported along the length of the member. Deflection need not be checked, but camber requirements are to be determined. CUF = 1.0.

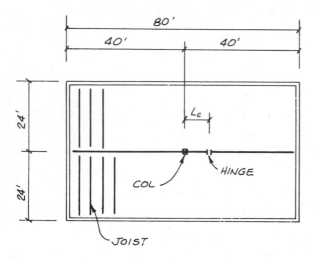

ROOF FRAMING PLAN

Fig. 6.H

    Find: *a.* The minimum required beam size if the girder is designed for a 20-psf roof LL with no reduction for trib. area.

          b. The minimum required beam size if the girder is designed for a basic 20-psf roof LL that is to be adjusted for trib. area. Roof LL is to be determined using UBC Table 23.C, Method 1 (consider the trib. area of the suspended portion of the cantilever system).

CHAPTER SEVEN

Axial Forces and Combined Bending and Axial Forces

7.1 Introduction

An axial force member has a load applied through the centroid of the cross section and parallel to the length of the member. The axial force may be either tension or compression. Because of the need to carry vertical gravity loads down through the structure into the foundation, columns are more often encountered than tension members. Both types of members, however, have widespread use in structural design in such items as trusses and diaphragms.

In addition to the design of axial force members, Chap. 7 covers the design of members with a more complicated loading condition. These include members with bending (beam action) occurring simultaneously with axial forces (tension or compression). This type of member is often referred to as a "combined stress" member. A combination of loadings is definitely more critical than if the same forces are applied individually. The case of compression combined with bending is probably more often encountered than tension plus bending, but both types of members may be used in typical wood-frame buildings.

To summarize, the design of the following types of members is covered in Chap. 7:

1. Axial tension
2. Axial compression
3. Combined bending and tension
4. Combined bending and compression

181

The design of axial-force tension members is relatively straightforward, and the required size of a member can be solved for directly. For the other three types of members, however, a trial-and-error solution is the typical design approach. Trial-and-error solutions may seem awkward in the beginning. However, with a little practice the designer will be able to pick an initial

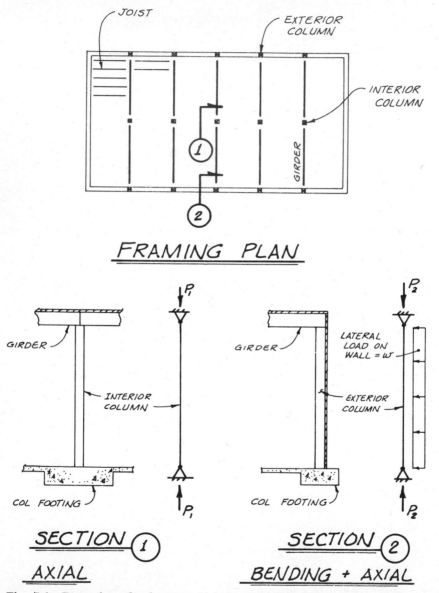

FRAMING PLAN

SECTION ①

AXIAL

SECTION ②

BENDING + AXIAL

Fig. 7.1 Examples of columns and beam-columns. Both of the members in this example are vertical, but horizontal or inclined members with this type of loading are also common.

trial size which will be relatively close to the required size. The final selection can often be made with very few trials. Several examples will illustrate the procedure used in design.

As noted, the most common axial-load member is probably the *column*, and the most common combined stress member is the *beam-column* (combined bending and compression). See Fig. 7.1. In this example an axial load is assumed to be applied to the interior column by the girder. For the exterior column there is both a lateral load that causes bending and a vertical load that causes axial compression.

The magnitude of the lateral load (wind or seismic) to the column depends on the unit design load and how the wall is framed. The wall may be framed horizontally to span between columns, or it may be framed vertically to span between story levels (Example 3.4).

Numerous other examples of axial-force members and combined load members can be cited. However, the examples given here are representative, and they adequately define the type of members and loadings that are considered in this chapter.

7.2   Axial Tension Members

Wood members are stressed in tension in a number of structural applications. For example, trusses have numerous axial-force members, and roughly half of these are in tension. It should be noted that unless the loads frame directly into the joints in the truss, bending stresses will be developed in addition to the axial stresses obtained in the standard truss analysis.

Axial tension members also occur in the chords of horizontal and vertical diaphragms. In addition, tension members are used in diaphragm design when the length of the horizontal diaphragm is greater than the length of the shearwall to which it is attached. This type of member is known as a strut, and it is considered in Chap. 9.

The check for the axial tension stress in a member of known size uses the formula

$$f_t = \frac{P}{A} \leq \text{adj. } F_t$$

This formula says that the actual (calculated) axial stress must be less than the adjusted allowable tension stress. The allowable tension stress is obtained by multiplying the tabulated allowable tension stress by the appropriate adjusting factors

$$\text{Adj. } F_t = \text{tab. } F_t \times \text{LDF} \times \text{CUF} \times \cdots$$

where adj. F_t = adjusted allowable axial tension
　　Tab. F_t = tabulated allowable axial tension
　　LDF　 = load duration factor

183

CUF　　= condition-of-use factor

× · · · = any other adjustment factors that may apply

The LDF and CUF are covered in detail in Chaps. 4 and 5. It should be pointed out that tabulated allowable tension stresses for both sawn lumber and glulam have been reduced in recent years. These reductions did not occur simultaneously for sawn lumber and glulam. The structural designer should be aware of the general tendency toward greater conservatism in the design of these members. It is also recommended that the allowable stresses be read from the most recently available NDS (or Code) tables. This practice will avoid the use of some allowable stress that may have been revised in later specifications.

The cross-sectional area to be used in the tension stress calculation is the *net area* of the member. This area is determined by subtracting the projected area of any bolt holes (or timber connector holes) from the gross cross-sectional area of the member. The arrangement and layout of bolt hole patterns in wood members is considered in Chap. 12 on bolted connections.

One adjusting factor for allowable tension stress is not identified by a special symbol, and it is relatively easy to overlook this reduction. A footnote to the NDS Supplement, Table 4A, *Design Values for Visually Graded Structural Lumber*, indicates that tabulated values of F_t for J&P sizes apply to members that are 5 in. and 6 in. wide. Multiplying factors (ranging from 0.9 to 0.6) are required to convert the tabulated F_t to an allowable stress for larger widths in the J&P category. This *adjustment for width* is used in the example in Example 7.8.

7.3　Design Problem: Tension Member

In this example the required size for the lower chord of a truss is determined. The loads are assumed to be applied to the top chord of the truss only. See Example 7.1.

In order to determine the axial forces in the truss members, it is necessary to have the loads applied to the joints. Loads for the truss analysis are obtained by taking the tributary width to one joint times the uniform load. Because the actual loads are applied uniformly to the top chord, these members will have combined stresses. Other members in the truss will have axial forces only.

The tension force in the bottom chord is obtained through a standard truss analysis (method of joints). The member size is determined by calculating the required net area and adding to it the area removed by the bolt hole.

EXAMPLE 7.1

Determine the required size of the lower (tension) chord in the truss shown below. The loads are (DL + snow), and the effects of the roof slope have already been taken into account. Joints are assumed to be pinned. Connections will be made with a single row of ¾-in.-diameter bolts. Trusses are 4 ft-0 in. o.c. Use No. 2 Southern pine surfaced dry. EMC ⩽ 19 percent. Allowable stresses and cross-sectional properties are to be taken from the NDS.

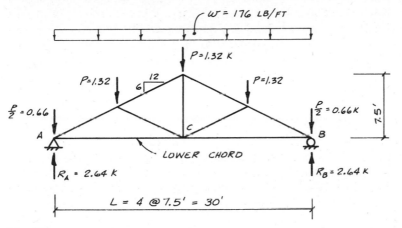

Fig. 7.2a

Loads

$$RDL = 14 \text{ psf} \qquad \text{horiz. plane}$$
$$\underline{\text{Reduced SL} = 30}$$
$$TL = \overline{44 \text{ psf}}$$
$$w_{TL} = 44 \times 4 = 176 \text{ lb/ft}$$

For truss analysis (load to joint)

$$P = 176 \times 7.5 = 1320 \text{ lb}$$

Force in Lower Chord
Use method of joints.

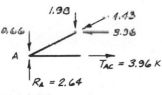

Fig. 7.2b

Determine Required Size of Tension Member
Assume that the member will be a J&P 1½ in. thick.

$$\text{Adj. } F_t = F_t \times LDF = 625 \times 1.15 = 719 \text{ psi*}$$

$$\text{Req'd net } A = \frac{P}{F_t} = \frac{3960}{719} = 5.51 \text{ in.}^2$$

---

* Because a 2 × 6 is adequate for the tension chord, no reduction of F_t for *width* is required. **185**

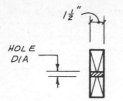

Fig. 7.2c

For net area calculations *arbitrarily* assume that bolt hole is ⅛ in. larger than the bolt diameter (for stress calculations only).

$$\text{Req'd gross } A = 5.51 + 1.5 \left(\frac{3}{4} + \frac{1}{8} \right)$$

$$= 6.82 \text{ in.}^2$$

> Use 2 × 6 No. 2 Southern pine
> $A = 8.25 > 6.82$ *OK*

NOTE: The simplified truss analysis used in this example applies only to trusses with pinned joints. If some form of toothed steel gussets are used for the connections, the design should conform to Ref. 35.

7.4   Column Design

The check on the capacity of an axially loaded wood column of known size uses the formula

$$f_c = \frac{P}{A} \leq \text{adj. } F_c'$$

The symbol F_c' is used to indicate that the effects of column instability must be considered. The symbol F_c is used for the allowable compressive stress parallel to the grain given in the NDS and UBC tables. It applies only to *short* columns.

In the calculation of actual stress f_c, the cross-sectional area to be used will be either the *gross area* of the column or the *net area* at some hole in the member. The area to be used depends on the location of the hole, and the tendency of that point in the member to buckle laterally. If the hole is located at a point which is braced, the gross area of the member may be used in the check for column stability. Another check of f_c at the reduced cross section (using the net area) should be compared with F_c with no reduction for stability. See Example 7.2. The other possibility is that some reduction of column area occurs in the laterally unbraced portion of the column. In the latter case, the net area is used directly in the stability check.

EXAMPLE 7.2 **Actual Stresses and Slenderness Ratios**

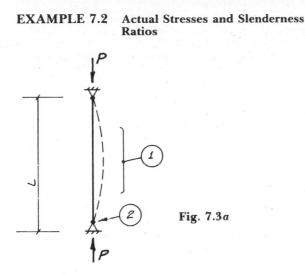

Fig. 7.3a

Actual Stresses

In Fig. 7.3a it is assumed that there are no holes in the column except at the supports (connections). Check the following column stresses:

1. Away from the supports stability must be considered.

$$f_c = \frac{P}{A_g} \leq F'_c$$

2. At the connections, buckling is not a factor.

$$f_c = \frac{P}{A_n} \leq F_c$$

Slenderness Ratio

Column stability is measured by the slenderness ratio.

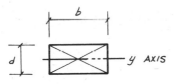

<u>RECTANGULAR</u>
<u>COLUMN SECTION</u>

Fig. 7.3b

$$\text{General slenderness ratio} = \frac{L}{r}$$

where L = unbraced length of column
r = least radius of gyration of column cross section

187

$$\text{Slenderness ratio for } \textit{rectangular columns} = \frac{L}{d}$$

where L = unbraced length of column
d = Least cross-sectional dimension of the column*

For a rectangular cross section the dimension d is directly proportional to the radius of gyration.

$$r_y = \sqrt{\frac{I_y}{A}} = \sqrt{\frac{bd^3/12}{bd}} = \sqrt{\frac{d^2}{12}} = d\sqrt{\frac{1}{12}}$$

$\therefore r \propto d$

The measure of the tendency of a column to buckle is the slenderness ratio. In its traditional form the slenderness ratio is expressed as the unbraced length of a column divided by the least radius of gyration (L/r).

For the design of rectangular wood columns, however, the slenderness ratio is modified to a form that is somewhat easier to apply. Here the slenderness ratio is the unbraced length of the column divided by the least dimension of the cross section (L/d). Use of this modified slenderness ratio is possible because the radius of gyration r can be expressed as a direct function of the width of a rectangular column. See Fig. 7.3b. The constant in the conversion of the modified slenderness ratio is simply incorporated into the allowable stress column design formulas.

The allowable stress formulas included in this chapter are for columns with rectangular cross sections. If columns of nonrectangular cross sections are considered, the formulas for allowable compression given in this section can be used by substituting $r\sqrt{12}$ for d.

There are three ranges of slenderness ratios for wood column design:

$$0 < L/d \le 11 \qquad \text{short column}$$
$$11 < L/d \le K \qquad \text{intermediate column}$$
$$K < L/d \le 50 \qquad \text{long column}$$

As one might expect, there is a different expression used to define the allowable column stress for each of these ranges. This is similar to the design of laterally unbraced beams (Sec. 6.3).

In the *short* column range, crushing of the wood fibers governs column strength. Here the full tabulated allowable compressive stress parallel to the grain is used with no reduction for stability. In the *long* column range buckling controls, and Euler's critical buckling stress divided by a factor of safety is used as the basis for design. In the *intermediate* column range, a transition expression is used between the Euler-based curve and the tabulated F_c value. See Example 7.3.

It should be noted that the intermediate-length column range formula is a fairly recent addition to the NDS column design criteria. Prior to its adoption,

* In beam design d is normally associated with the strong axis.

the short- and long-column range formulas were applied over the entire range of slenderness ratios. See dashed line in Fig. 7.4a. The intermediate-column range formula is obviously more conservative than the older criteria.

The method of applying the load duration factor in these column design formulas requires special attention. The expressions for F'_c are shown in the NDS without the LDF. The designer must then determine where the LDF is to be used. As a guide the LDF is shown at the appropriate points in the expressions in Example 7.3.

It should be noted that the adjustment for LDF in the column stability formulas is different than the adjustment used for unbraced beams. In the column formulas, the LDF is used to calculate the limit K. This has the effect of providing a transition from use of the full LDF in the short range to no LDF in the long range. This transition takes place throughout the intermediate column range (Fig. 7.4b).

EXAMPLE 7.3   Allowable Axial Compression

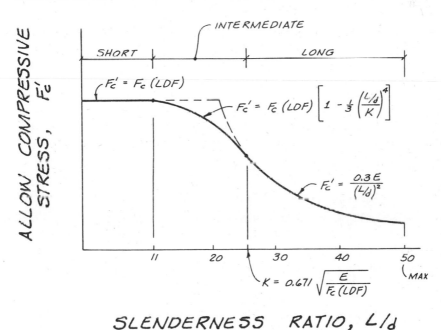

Fig. 7.4a

F'_c = allowable compressive stress considering the effects of column stability
F_c = tabulated allowable compressive stress parallel to the grain
The correct method of applying the load duration factor can easily be remembered. The tabulated allowable stress F_c is multiplied by the LDF in *all* column stress calculations.

189

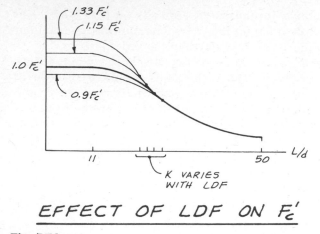

EFFECT OF LDF ON F_c'

Fig. 7.4b

7.5   Detailed Analysis of Slenderness Ratio

The concept of the slenderness ratio was briefly reviewed in Sec. 7.4. There it was stated that the *least* radius of gyration is used in the L/r ratio, and that the *least* dimension of the column cross section is used in the L/d ratio.

These statements assume that the unbraced length of the column is the same for both the x and the y axes. In this case the column, if loaded to failure, would buckle about the weak (y) axis. See Fig. 7.5a. Note that if buckling occurs about the y axis, the column moves in the x direction. Figure 7.5a illustrates this straightforward case of column buckling. Only the slenderness ratio about the weak axis need be calculated.

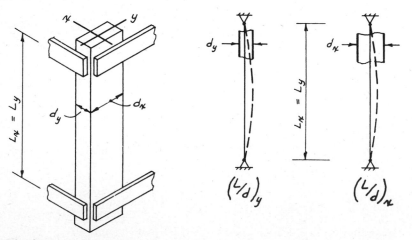

Fig. 7.5a   By inspection the slenderness ratio about the y axis is larger and is therefore critical.

Although this concept of column buckling applies in many situations, the designer should have a deeper insight into the concept of the slenderness ratio. Conditions can exist under which the column may actually buckle about the *strong* axis of the cross section rather than the weak axis.

In the more general case, the column can be viewed as having two slenderness ratios. One slenderness ratio would evaluate the tendency of the column to buckle about the *strong* axis of the cross section. The other would measure the tendency of buckling about the *weak* axis. For a rectangular column these slenderness ratios would be written

$(L/d)_x$ for buckling about the strong (x) axis (column movement in the y direction)

$(L/d)_y$ for buckling about the weak (y) axis (column movement in the x direction)

If the column is loaded to failure, buckling will occur about the axis that has the larger slenderness ratio. In design, the larger slenderness ratio is used to calculate the allowable compressive stress.

The reason that the strong axis can be critical can be understood from a consideration of the *bracing* and *end conditions* of the column. The *unbraced*

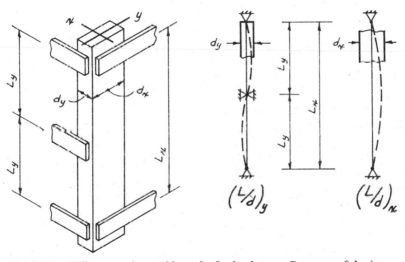

Fig. 7.5*b* Different unbraced lengths for both axes. Because of the intermediate bracing for the y axis, the critical slenderness ratio cannot be determined by inspection. Both slenderness ratios must be calculated, and the larger value is used to determine F'_c.

length is the length to be used in the calculation of the slenderness ratio. It is possible to have a column with different unbraced lengths for the x and y axes. See Fig. 7.5*b*. In this example the unbraced length for the x axis is twice as long as the unbraced length for the y axis. In practice bracing can occur at any interval. The effect of end conditions is explained below.

Another case where the unbraced lengths for the *x* and *y* axes are different occurs when sheathing is attached to a column. If the sheathing is attached to the column with an effective connection, buckling about an axis that is perpendicular to the sheathing is prevented. See Fig. 7.6. The most common example of this type of column is a stud in a bearing wall. The wall sheathing

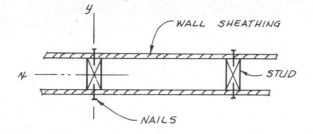

PLAN - STUD WALL

Fig. 7.6 Column braced by sheathing. Sheathing attached to the studs prevents column buckling about the weak *(y)* axis of the stud. Therefore, consider buckling about the *x* axis only.

can prevent column buckling about the weak axis of the stud, and only the slenderness ratio about the strong axis of the member needs to be evaluated.

The final item regarding the slenderness ratio is the effect of column *end conditions*. The length *L* used in the slenderness ratio is theoretically the unbraced length of a *pinned-end* column. For columns with other end conditions the length is taken as the distance between points of inflection (IPs) on a sketch of the buckled column. An inflection point corresponds to a point of reverse curvature on the deflected shape of the column and represents a point of zero moment. For this reason the inflection point is considered as a pinned end for purposes of column analysis. The *effective unbraced length* is taken as the distance between inflection points.

Typically, six "ideal" column end conditions are considered. Studies have resulted in recommended effective lengths for design purposes for these ideal columns. See Fig. 7.7. The effective unbraced length can be determined by multiplying the *effective length factor* times the actual unbraced length, giving the equation:

$$\text{Effective length} = \text{distance between inflection points*}$$
$$= \textit{effective length factor} \times \textit{unbraced length}$$

---

* When only one inflection point is on the sketch of the buckled column, the mirror image of the column is drawn to give a second inflection point. The effective length factors shown in Fig. 7.7 are the "recommended design values" from Ref. 12. These values equal or exceed the theoretical effective length factors. The larger values are recommended because practical end conditions can only approximate ideal pinned or fixed ends.

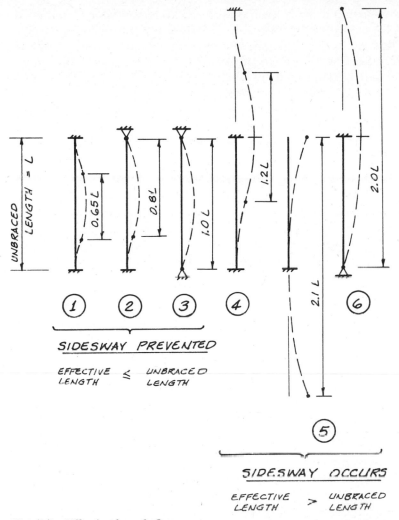

Fig. 7.7   Effective length factors.

In practice the designer must determine which "ideal" column most closely approximates the actual end conditions for a given column. Some degree of judgment is required for this evaluation, but several key items should be considered in making this comparison.

First of all, it should be noted that three of the ideal columns undergo sidesway, and the other three do not. *Sidesway* means that the top of the column is relatively free to move laterally with respect to the bottom of the column. The designer must be able to identify a column that will undergo sidesway and, on the other hand, what constitutes restraint against sidesway. In general the answer to this depends on the type of lateral-force-resisting system that is used (Sec. 3.3).

193

Generally speaking, if the column is part of a system in which lateral loads are resisted by *bracing* or by *shearwalls,* sidesway will be prevented. These types of LFRSs are relatively rigid, and the movement of one end of the column with respect to the other end is restricted. If an overload occurs in this case, column buckling will be symmetrical. See Fig. 7.8. However, if the column is part of a *frame*-type LFRS, the system is relatively flexible and sidesway can occur. Typical columns for the types of buildings considered in this text will have sidesway prevented.

It should also be noted that columns with *sidesway prevented* have an effective

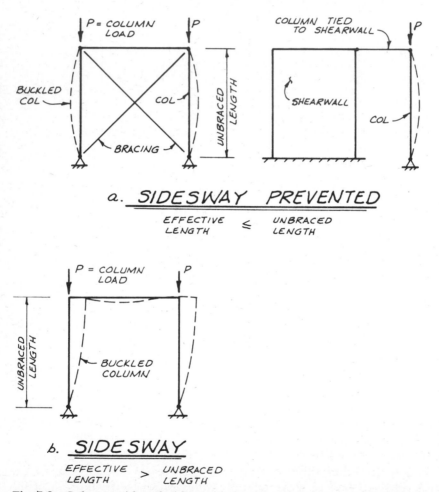

a. **SIDESWAY PREVENTED**

EFFECTIVE   ≤   UNBRACED
LENGTH   LENGTH

b. **SIDESWAY**

EFFECTIVE   >   UNBRACED
LENGTH   LENGTH

Fig. 7.8 Columns with and without sidesway. *(a)* Braced frames or buildings with shearwalls limit the displacement of the top end of the column so that sidesway does not occur. *(b)* Columns in rigid frames (without bracing) will undergo sidesway if the columns buckle.

length which is *less than or equal* to the actual unbraced length. A common and conservative practice is to consider the effective length equal to the unbraced length for these columns.

For columns where *sidesway* can occur, the effective length is *greater than* the actual unbraced length. For these types of columns, the larger slenderness ratio causes the allowable axial load to be considerably less than the allowable load on a braced column.

In addition to answering the question of sidesway, the comparison of an actual column to an ideal column should evaluate the effectiveness of the column connections. Practically all columns have square-cut ends. For structural design purposes, this type of column end condition is assumed to be pinned. This assumption is conservative because the square-cut end does offer some minor restraint against buckling. It does not function as an ideal frictionless pin.

It is possible to design moment-resisting connections in wood members, but they are the exception rather than the rule. The majority of connections in ordinary wood buildings are "simple" connections. Therefore, for many columns it is conservative to take the effective length equal to the unbraced length. However, the designer should examine the actual *bracing conditions* and *end conditions* for a given column and determine whether or not a larger effective length should be used.

7.6   Design Problem: Axially Loaded Column

The design of a column is a trial-and-error process because, in order to determine the allowable column stress F'_c, it is first necessary to know the slenderness ratio L/d. In the following example only two trials are required to determine the size of the column. See Example 7.4.

Several items in the solution should be emphasized. First of all, the importance of the size category should be noted. The initial trial size was in the Joist and Plank size category, and the second trial size was a Post and Timber. Tabulated allowable stresses are different for these two size categories.

The second item concerns the adjustment for duration of loading. This example involves (DL + roof LL). It will be remembered that RLL is an arbitrary minimum load required by the Code, and the LDF for this combination is 1.25. For many areas of the country the design load for a roof will be (DL + snow), and the corresponding LDF is 1.15. It is also important to note the locations where the LDF is applied *and* where it is not applied.

The design of a glulam column will follow essentially the same procedure used for a sawn column. The basic difference is in the stress grade of the material. For axial-force glulam members, the grade is given as a combination number. Calculations for a glulam column are illustrated in the example on combined bending and compression.

EXAMPLE 7.4   Sawn Lumber Column

Design the column in Fig. 7.19a using sawn lumber (No. 1 Douglas fir-larch). Bracing conditions are the same for buckling about both the x and y axes. The load is combined dead load and roof live load. Allowable stresses are to be in accordance with the NDS.

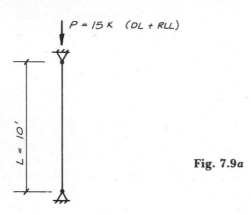

P = 15 K   (DL + RLL)

L = 10'

Fig. 7.9a

Trial 1

Try 4 × 6 (J&P size category). Tabulated values:

$$F_c = 1250 \text{ psi}$$
$$E = 1,800,000 \text{ psi}$$
$$A = 19.25 \text{ in.}^2$$

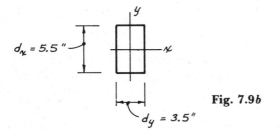

$d_x = 5.5''$

$d_y = 3.5''$

Fig. 7.9b

Determine column capacity:

$$\left(\frac{L}{d}\right)_{max} = \left(\frac{L}{d}\right)_y = \frac{10 \times 12}{3.5} = 34.3$$

$$K = 0.671 \sqrt{\frac{E}{F_c(\text{LDF})}} = 0.671 \sqrt{\frac{1,800,000}{1250\,(1.25)}} = 22.8 < 34.3$$

$$\left(\frac{L}{d}\right) > K$$

∴ Column is in *long* slenderness range, and the LDF does not apply

196

$$F'_c = \frac{0.3E}{\left(\frac{L}{d}\right)^2} = \frac{0.3(1,800,000)}{(34.3)^2} = 460 \text{ psi}$$

Allow. $P = F'_c A = (0.460)(19.25) = 8.84 \text{ k} < 15$ *NG*

Trial 2
Try 6 × 6 (P&T size category). Tabulated values:

$$d_x = d_y = 5.5 \text{ in.}$$
$$A = 30.25 \text{ in.}^2$$
$$F_c = 1000 \text{ psi}$$
$$E = 1,600,000 \text{ psi}$$

$$\left(\frac{L}{d}\right)_{max} = \frac{10 \times 12}{5.5} = 21.8 > 11$$

$$K = 0.671 \sqrt{\frac{E}{F_c(LDF)}} = 0.671 \sqrt{\frac{1,600,000}{1000\,(1.25)}} = 24.0 > 21.8$$

$$11 < \left(\frac{L}{d}\right) < K$$

∴ Column is in *intermediate* column range.

$$F'_c = F_c(LDF) \left[1 - \frac{1}{3}\left(\frac{L/d}{K}\right)^4\right] = (1000)(1.25)\left[1 - \frac{1}{3}\left(\frac{21.8}{24.0}\right)^4\right] = 965 \text{ psi}$$

Allow. $P = F'_c A = 0.965(30.25) = 29.2 \text{ k} > 15$ *OK*

> Use 6 × 6 column
> No. 1 DF-L

7.7 Design Problem: Capacity of a Bearing Wall

An axial load may be applied to a wood-frame wall. For example, an interior bearing wall may support the reactions of floor or roof joists (Figure 3.4c). Exterior bearing walls also carry reactions from joists and rafters, but, in addition, exterior walls usually must be designed to carry lateral wind loads. (The UBC requires a minimum 5-psf lateral load on interior walls. Theoretically both interior and exterior walls must be designed to carry lateral seismic forces perpendicular to the wall surface. However, for typical wood-frame walls, the dead load of the wall is so small that the seismic force is not critical— especially when an LDF of 1.33 is considered.)

In Example 7.5 the vertical load capacity of a wood frame wall is determined. Two main factors should be noted about this problem. The first item relates to the column capacity of a stud in a wood-frame wall. Because sheathing is attached to the stud throughout its height, continuous lateral support is provided in the x direction. Therefore, the possibility of buckling about the weak (y) axis is prevented. The column capacity is evaluated by the slenderness ratio about the strong axis of the stud $(L/d)_x$.

The second main factor to consider is the bearing capacity of the top and bottom wall plates. It is possible that the vertical load capacity of a bearing wall may be governed by compression perpendicular to the grain on the wall plates rather than by the column capacity of the studs. This is typically not a problem with major columns in a building because steel bearing plates can be used to distribute the load perpendicular to the grain on supporting members. However, in a standard wood-frame wall the stud bears *directly* on the horizontal wall plates.

In the analysis for bearing perpendicular to the grain on the wall plate, the adjustment factor based on the length of the bearing (BLF = 1.25) is used. If this adjustment factor had been ignored or did not apply, the capacity of the stud would be governed by bearing perpendicular to the grain.

EXAMPLE 7.5   Capacity of a Stud Wall

Determine the vertical load capacity of the stud shown in Fig 7.10a. No bending. Express the load in pounds per lineal foot of wall. Lumber is Standard grade Hem-Fir. Load is (DL + snow).

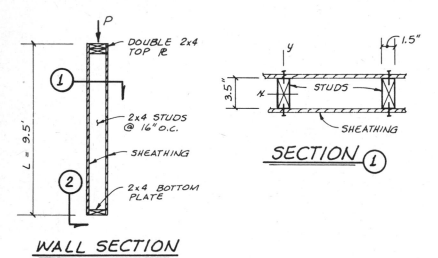

WALL SECTION

Fig. 7.10a

Tabulated properties, 2 × 4 (LF size category):

$$A = 5.25 \text{ in.}^2$$
$$F_c = 775 \text{ psi}$$
$$F_{c\perp} = 245 \text{ psi}$$
$$E = 1{,}200{,}000 \text{ psi}$$

Column Capacity of Stud

Buckling about the weak axis of the stud is prevented by the sheathing.

$$\left(\frac{L}{d}\right)_x = \frac{9.5 \times 12}{3.5} = 32.6$$

$$K = 0.671 \sqrt{\frac{E}{F_c(\text{LDF})}} = 0.671 \sqrt{\frac{1,200,000}{775(1.15)}} = 24.6 < 32.6$$

Stud is in *long* column range, and LDF does not apply.

$$F'_c = \frac{0.3E}{\left(\frac{L}{d}\right)^2} = \frac{0.3(1,200,000)}{(32.6)^2} = 339 \text{ psi}$$

Allow. $P = F'_c A = 339(5.25) = 1780$ lb/stud

Allow. $w = \dfrac{1780}{1.33} = \boxed{1340 \text{ lb/ft}}$

Bearing Capacity of Wall Plates

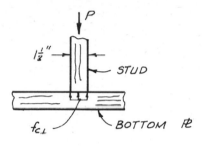

SECTION ②

Fig. 7.10b

A typical stud is located more than 3 in. from the end of the wall plate, and the bearing length of 1½ in. is less than 6 in. ∴ The bearing length factor (BLF) can be used to adjust $F_{c\perp}$ (Sec. 6.7).

$$\text{Adj. } F_{c\perp} = \text{tab. } F_{c\perp} \times \text{LDF} \times \text{BLF}$$

$$= 245 \times 1.15 \times \left(\frac{1\frac{1}{2} + \frac{3}{8}}{1\frac{1}{2}}\right)$$

$$= 352 \text{ psi} > 339$$

$$F_{c\perp} > F'_c$$

∴ Column capacity governs over bearing perpendicular to grain.

7.8 Built-up Columns

A *built-up column* is constructed from several parallel wood members which are nailed or bolted together to function as a composite column. These are to be distinguished from *spaced columns*, which have specially designed timber connectors to transfer shear between the separate parallel members.

199

DESIGN OF WOOD STRUCTURES

The NDS includes criteria for designing spaced columns. Spaced columns can be used to increase the allowable load in compression members in heavy wood trusses. They are, however, relatively expensive to fabricate and are not often used in ordinary wood buildings. For this reason spaced columns are not covered in this text.

Built-up columns may also have limited use, but they are fairly easy to fabricate, and their design is briefly covered here. The combination of several members in a built-up column results in a member with a larger cross-sectional dimension d and, correspondingly, a smaller slenderness ratio L/d.

With a smaller slenderness ratio, a larger allowable column stress can be used. Therefore, the allowable load on a built-up column is larger than the allowable load for the same members used individually. However, the fasteners connecting the members do not fully transfer the shear between the various pieces, and the capacity of a built-up column is less than the capacity of a *solid* sawn or glulam column of the same size and grade.

The capacity of a built-up column is determined by first calculating the column capacity of an equivalent solid column. This value is then reduced by an adjustment factor that is a function of the column slenderness ratio. This procedure is demonstrated in Example 7.6.

EXAMPLE 7.6   Strength of Built-up Columns

Determine the allowable axial load on the built-up column in Fig. 7.11. Lumber is No. 1 DF-L. Column length is 13 ft-0 in. LDF = 1.0, and CUF = 1.0.

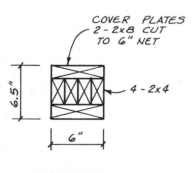

COVER PLATES
2 - 2x8 CUT
TO 6" NET

6.5"

4 - 2x4

6"

COLUMN
SECTION

Fig. 7.11

Tabulated values:

$$F_c = 1250 \text{ psi}$$
$$E = 1,800,000$$

Column Capacity

$$\left(\frac{L}{d}\right)_{\text{max}} = \frac{13 \times 12}{6.0} = 26$$

$$K = 0.671 \sqrt{\frac{E}{F_c(\text{LDF})}} = 0.671 \sqrt{\frac{1,800,000}{1250(1.0)}} = 25.5$$

$$K < \left(\frac{L}{d}\right)$$

∴ *Long* column, and LDF does not apply.

Capacity of a solid column:

$$F'_c = \frac{0.3E}{\left(\frac{L}{d}\right)^2} = \frac{0.3(1,800,000)}{(26)^2} = 800 \text{ psi}$$

Allow. $P = F_c A = (0.80)(6 \times 6.5) = 31.1$ k

Capacity of a built-up column:

Allow. $P = 0.82 \times 31.1 = \boxed{25.6 \text{ k}}$

The strength of a built-up column can be calculated as a percentage of the strength of a single-piece column. The reduction factors given in the following table apply only if the built-up section has *cover plates* and only if the members are *well nailed or bolted*. The percentages in the table below are obtained from Ref. 6.

Built-up column capacities	
Slenderness ratio L/d	Strength of one-piece column, percent
6	82
10	77
14	71
18	65
22	74
26	82

7.9 Combined Bending and Tension

When a bending moment occurs simultaneously with an axial tension force, the effects of combined stresses must be taken into account. Combined stresses are usually evaluated in an *interaction formula*.

The interaction formula for combined bending and tension is a straight-line formula which is made up of two separate terms. See Example 7.7. The first term measures the effects of axial tension, and the second term evaluates the effects of bending. In each case the actual stress is divided by the corresponding allowable stress.

201

EXAMPLE 7.7 Interaction Formula for Combined Bending and Tension

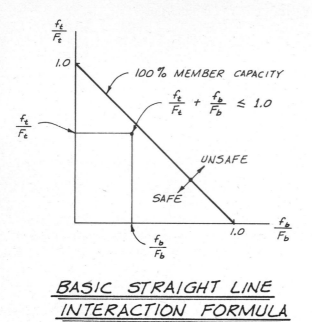

$$\frac{f_t}{F_t} + \frac{f_b}{F_b} \leq 1.0$$

BASIC STRAIGHT LINE INTERACTION FORMULA

Fig. 7.12a

The basic straight-line interaction formula is used for combined bending and tension. The terms in the interaction formula define a point on the graph (Fig. 7.12a). If the point lies *on* or *below* the line representing 100 percent of the member strength, the interaction formula is satisfied.

$$\frac{f_t}{F_t} = 1.0 \qquad \text{defines pure tension—no bending}$$

$$\frac{f_b}{F_b} = 1.0 \qquad \text{defines pure bending—no axial tension}$$

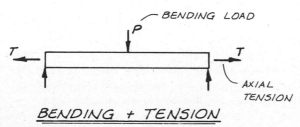

BENDING + TENSION

Fig. 7.12b

Definition of Terms

$$f_t = \frac{P}{A}$$

$F_t = $ adj. $F_t = $ tab. $F_t \times$ LDF $\times$ CUF $\times \cdots$

$$f_b = \frac{M}{S}$$

$F_b = $ adj. $F_b = $ tab. $F_b \times C_F \times$ LDF $\times$ CUF $\times \cdots$

or
$\qquad = F_b' \times$ CUF $\times \cdots$

The first allowable bending stress term includes the size factor, and the second term accounts for the effects of lateral instability. The allowable bending stress based on lateral stability, F_b', may or may not have an adjustment for load duration. This depends on the range of the beam slenderness factor (Example 6.6). The lower of the two allowable bending stress terms is to be used in the interaction formula.

It is convenient to think of the terms in the interaction formula as percentages or fractions of total member capacity. For example, the ratio of actual tension stress to allowable tension stress, f_t/F_t, can be viewed as the fraction of total member capacity that is used to resist axial tension. The ratio of actual bending stress to allowable bending stress, f_b/F_b, represents the fraction of total member capacity used to resist bending. The sum of these fractions must be less than the total member capacity (1.0).

In the calculation of allowable stresses for use in the interaction formula, it should be noted that all appropriate adjustment factors are to be applied. These include the load duration factor, the condition-of-use factor, and any other factor that may be required. For example, the allowable bending stress should reflect the *size effect* or a reduction for *lateral stability*, as appropriate.

7.10 Design Problem: Combined Bending and Tension

The truss in Example 7.8 is similar to the truss used in Example 7.1. The difference is that in the current example, an additional load is applied to the bottom chord of the truss. This load is uniformly distributed and represents the weight of a ceiling supported by the bottom chord of the truss.

The first part of the example deals with the calculation of the axial force in member *AC*. In order to analyze a truss it is necessary for the loads to be resolved into joint loads. The tributary width to the three joints along the top chord is 7.5 ft and the tributary width to the joint at the midspan of the truss on the bottom chord is 15 ft. The remaining loads (both top and bottom chord loads) are tributary to the joints at each support.

The design of a combined stress member is a trial-and-error procedure. In this example, a 2 × 8 bottom chord is the initial trial, and it proves satisfac-

203

tory. *Independent* checks on the tension and bending strengths are first completed. Finally, the combined effects are evaluated in the interaction formula.

It should be noted that the LDF used for the independent check for axial tension is the LDF for combined (DL + snow). Dead load plus snow causes the axial force of 4.44 k. The independent check of bending uses an LDF for DL only, because the DL of the ceiling alone cause the bending moment of 10.8 in.-k.

In the combined stress check, however, an LDF of 1.15 applies to *both* the axial and bending portions of the interaction formula. Recall that the LDF to be used in checking stresses caused by a *combination* of loads is the one associated with the shortest-duration load in the combination. For combined stresses, then, the same LDF applies to both terms.

EXAMPLE 7.8   Combined Bending and Tension

Design the lower chord of the truss in Fig. 7.13a. Use No. 1 Southern pine—surfaced dry. EMC ≤ 19 percent. Connections will be made with a single row of ¾-in.-diameter bolts. Trusses are 4 ft-0 in. o.c. Loads are applied to both the top and bottom chords. Lateral stability is not a consideration. Take allowable stresses from the NDS.

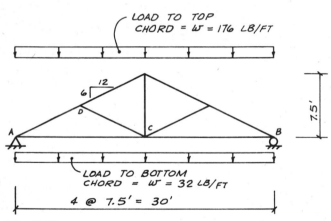

Fig. 7.13a

Loads

TOP CHORD:

$$RDL = 14 \text{ psf (horizontal plane)}$$
$$Snow = \underline{30} \quad \text{(reduced snow load based on roof slope)}$$
$$TL = 44 \text{ psf}$$
$$w_{TL} = 44 \times 4 = 176 \text{ lb/ft to truss}$$

Load to joint for truss analysis:

$$P_T = 176 \times 7.5 = 1320 \text{ lb/joint}$$

BOTTOM CHORD:

$$\text{Ceiling DL} = 8 \text{ psf}$$
$$w_{DL} = 8 \times 4 = 32 \text{ lb/ft}$$

Load to joint for truss analysis:

$$P_B = 32 \times 15 = 480 \text{ lb/joint}$$

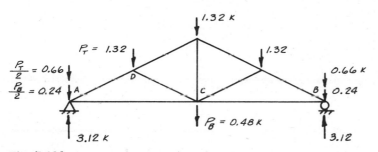

Fig. 7.13b

Force in lower chord (method of joints):

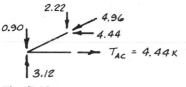

Fig. 7.13c

Load diagram for tension chord AC:

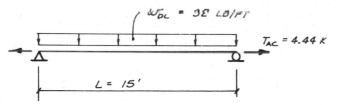

Fig. 7.13d

Member design
Try 2 × 8 (J&P). Tabulated values:

$$F_b = 1450 \text{ psi}$$
$$F_t = 975 \text{ psi}$$
$$S = 13.14 \text{ in.}^3$$

DESIGN OF WOOD STRUCTURES

TENSION:

1. Check tension at the net section. Because this truss is assumed to have pinned connections, the bending moment is theoretically zero at this point. Arbitrarily assume hole diameter is $\frac{1}{8}$ in. larger than bolt diameter (for stress calculations only). Actual clearance for bolt is less.

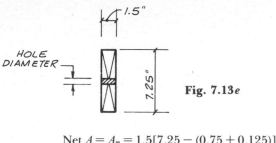

Fig. 7.13e

$$\text{Net } A = A_n = 1.5[7.25 - (0.75 + 0.125)]$$
$$= 9.56 \text{ in.}^2$$

$$f_t = \frac{T}{A} = \frac{4440}{9.56} = 464 \text{ psi}$$

$$\text{Adj. } F_t = \text{tab. } F_t \times \text{LDF} \times \text{(reduction for width—Sec. 7.2)}$$
$$= 975 \times 1.15 \times 0.80 = 897 \text{ psi}$$
$$897 > 464 \qquad OK$$

2. Determine tension stress at the point of maximum bending stress (midspan) for use in the interaction formula.

$$f_t = \frac{T}{A_g} = \frac{4440}{1.5(7.25)} = 408 \text{ psi} < 895 \qquad OK$$

BENDING:

For a simple span with a uniform load

$$M = \frac{wL^2}{8} = \frac{(32)(15)^2}{8} = 900 \text{ ft-lb} = 10,800 \text{ in.-lb}$$

$$f_b = \frac{M}{S} = \frac{10,800}{13.14} = 822 \text{ psi}$$

$$\text{Adj. } F_b = \text{tab. } F_b \times \text{LDF}$$
$$= 1450 \times 0.9 = 1305 \text{ psi} > 822 \qquad OK$$

COMBINED:

$$\frac{f_t}{F_t} + \frac{f_b}{F_b} = \frac{408}{897} + \frac{822}{1450 \times 1.15} = 0.46 + 0.49$$
$$= 0.95 < 1.0. \qquad OK$$

> Use 2 × 8 No. 1
> Southern pine*

NOTE: The simplified analysis used in this example applies only to trusses with pinned joints. For a truss connected with toothed steel gussets, a design approach is used which takes the continuity of the joints into account (Ref. 35).

---

* Shear and deflection should also be checked.

7.11   Combined Bending and Compression

When a bending moment occurs simultaneously with an axial compressive force, a more critical combined stress problem exists in comparison with combined bending and tension. In a beam-column an *additional* bending stress is created which is known as the *P-Δ* effect.

The *P-Δ* effect can be described in this way. The bending moment developed by the transverse loading causes a deflection Δ to occur. When the axial force *P* is applied, an additional bending moment of $P \times \Delta$ is generated. See Example 7.9. The *P-Δ* moment is known as a *second-order effect*, and the added bending stress is not calculated directly. Instead, it is taken into account by *modifying the straight-line* interaction formula.

The first term in the interaction formula represents the effects of compression and the second term represents the effects of bending. As with combined bending and tension, these terms can be thought of as being fractions of member capacity used to resist the respective types of stresses. The sum of

---

EXAMPLE 7.9   **Basic Interaction Formula for Beam—Columns**

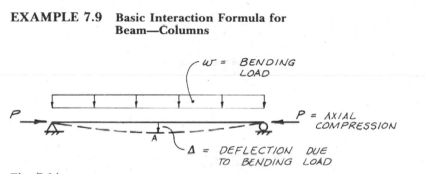

Fig. 7.14a

By cutting the member and summing moments at point *A*, it can be seen that a moment of $P \times \Delta$ is created which *adds* to the moment caused by the transverse load. In a beam with an axial tension force, this moment *subtracts* from the normal bending moment and is ignored in design.

Basic straight-line interaction formula:

$$\frac{f_c}{F_c'} + \frac{f_b}{F_b} \le 1.0$$

Modified interaction formula to account for *P-Δ* effect:

$$\frac{f_c}{F_c'} + \frac{f_b}{F_b - Jf_c} \le 1.0$$

where $f_c = P/A$
$F_c' =$ adjusted allowable compressive stress to account for stability effects and any other appropriate adjustments (Sec. 7.4)
$f_b = M/S$

207

F_b = adjusted allowable bending stress to account for size effect *or* stability (whichever governs) and any other appropriate adjustments (Sec. 6.2 and 6.3)

Jf_c = basic adjustment for P-Δ effect. The P-Δ effect is proportional to f_c (= P/A), and J measures the sensitivity of a member to the P-Δ effect. J essentially says that little deflection occurs in short members, and deflection is significant in long slender members.

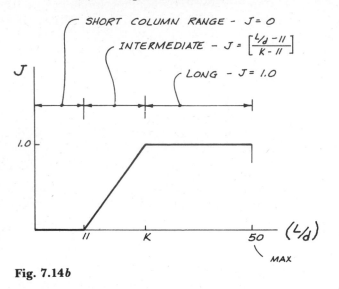

Fig. 7.14b

these fractions must be less than the total member strength (1.0 or 100 percent).

The modification of the straight-line formula comes in the denominator of the bending term. The P-Δ moment is taken into account by using a reduced allowable bending stress in members that are "sensitive" to this kind of action. The allowable bending stress is reduced by Jf_c. Here J reflects the sensitivity of the member to P-Δ effects. Hence, J is small for members with small slenderness ratios and large for members with large slenderness ratios.

The interaction formula given in Example 7.9 is used for beam-columns in which the bending moment is created by loads which are applied transverse (i.e., perpendicular) to the length of the member. It is possible for the moment, or part of the moment, to be generated by an eccentrically applied compressive force. In this case, the bending stress caused by the eccentrically applied load can be expressed as a function of the axial column stress f_c. See Example 7.10.

Obviously if any of the terms in the generalized interaction formula is zero, a simplified interaction formula can be written. Several of these are given in NDS Appendix H, *Formulas for Wood Columns with Side Loads and Eccentricity.*

EXAMPLE 7.10 **General Interaction Formula**

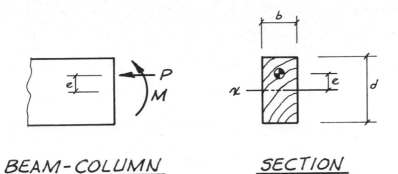

BEAM-COLUMN   SECTION

Fig. 7.15

The beam-column in Fig. 7.15 has a moment that is created by transverse bending loads M *and* a moment that is generated by an eccentrically applied compressive load Pe. The eccentric moment results from the centroid of the load not coinciding with the centroid of the member cross section. This latter moment is not to be confused with the P-Δ effect.

The development of the generalized interaction formula for beam-columns with *transverse* and *eccentric* bending stresses is shown below. Bending stresses caused by the eccentric moment are *amplified* for certain slenderness ratios. Greater P-Δ effects occur as a result of the eccentric moment, especially in the longer column ranges.

Section Properties

$$A = bd$$

$$S = \frac{bd^2}{6}$$

Bending Stresses

$$\text{Total moment} = \text{transverse load } M + \text{eccentric load } M$$
$$= M + Pe$$

Bending stress due to M:

$$f_b = \frac{M}{S}$$

Bending stress due to Pe:

$$\frac{Pe}{S} = \frac{Pe}{\left(\dfrac{bd^2}{6}\right)} = f_c\left(\frac{6e}{d}\right)$$

where $f_c = \dfrac{P}{A} = \dfrac{P}{bd}$

$$\text{Total bending stress} = f_b + f_c\left(\frac{6e}{d}\right)$$

209

General Interaction Formula

The portion of the total bending stress that is caused by Pe is increased to $f_c (6 + 1.5J) (e/d)$. With this amplified bending stress, the interaction formula becomes

$$\frac{f_c}{F'_c} + \frac{f_b + f_c(6 + 1.5J)\left(\frac{e}{d}\right)}{F_b - Jf_c} \leqslant 1.0$$

J and other terms are defined in Example 7.9.

Several examples of beam-column calculations are included in the remaining portion of this chapter.

7.12   Design Problem: Beam-Column

In this first example the top chord in the truss in Fig. 7.13a is considered. The top chord is subjected to bending loads caused by the DL and snow load being applied along the member, rather than framing directly into the joints. The top chord is also subjected to axial compression, which is obtained from a truss analysis using tributary loads to the truss joints. See Example 7.11.

A 2 × 8 was selected as the initial trial size. In a beam-column problem it is often convenient to divide the stress calculations into three subproblems. In this approach, somewhat independent checks on axial, bending, and combined stresses are performed.

The first check is on axial stresses. Because the top chord of the truss is attached directly to the roof sheathing, lateral buckling about the weak axis of the cross section is prevented. Bracing for the strong axis is provided at the truss joints by the members that frame into the top chord.

The compressive stress is calculated at two different locations along the length of the member. Column buckling (F'_c) is checked away from the joints using the gross area in the calculation of f_c. The second calculation involves the stress at the net section at a joint compared with the allowable stress F_c without any reduction for stability.

In the bending stress calculation, the moment is determined by use of the horizontal span of the top chord and the load on a horizontal plane. It was shown in Example 2.5 that the moment obtained in this manner is the same as the moment using the inclined span length and the normal component of the load.

The combined stress check is relatively straightforward. In this example the trial size is found to be inadequate. The 2 × 8 top chord is overstressed by roughly 6 percent, and a revised member size is required.

EXAMPLE 7.11 Beam—Column Design

Design the top chord of the truss shown in Fig. 7.13. The axial force and bending loads are reproduced in Fig. 7.16a. Use No. 1 Southern pine, surfaced dry, with EMC ≤ 19 percent. Connections will be made with a single row of ¾-in.-diameter bolts. The top chord is stayed laterally throughout its length by the roof sheathing. Trusses are 4 ft-0 in. o.c. Allowable stresses and section properties are to be obtained from the NDS.

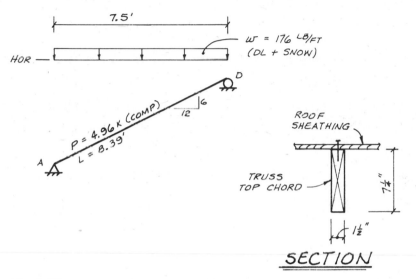

Fig. 7.16a

Try 2 × 8 (J&P):

$$A = 10.875 \text{ in.}^2$$
$$S = 13.141 \text{ in.}^2$$
$$F_c = 1250 \text{ psi}$$
$$F_b = 1450 \text{ (single member use)}$$
$$E = 1,700,000 \text{ psi}$$

Axial

1. *Stability check.* Column buckling occurs away from truss joints. Use gross area.

$$f_c = \frac{P}{A} = \frac{4960}{10.875} = 456 \text{ psi}$$

$$\left(\frac{L}{d}\right)_y = 0 \qquad \begin{array}{l}\text{because of lateral support} \\ \text{provided by roof diaphragm}\end{array}$$

$$\left(\frac{L}{d}\right)_x = \frac{8.39 \times 12}{7.25} = 13.9 > 11$$

$$K = 0.671 \sqrt{\frac{E}{F_c(\text{LDF})}} = 0.671 \sqrt{\frac{1,700,000}{1250(1.15)}}$$

$$23.1 > 13.9$$

211

$$11 < L/d < K \qquad \therefore \text{ intermediate column}$$

$$F_c' = F_c(\text{LDF})\left[1 - \frac{1}{3}\left(\frac{L/d}{K}\right)^4\right]$$

$$= 1250(1.15)\left[1 - \frac{1}{3}\left(\frac{13.9}{23.1}\right)^4\right] = 1375 \text{ psi} > 456 \qquad OK$$

2. *Check net section.* Arbitrarily assume hole diameter is ⅛ in. larger than bolt (for stress calculations only).

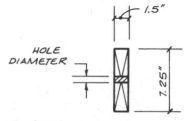

$$A_n = (7.25 - 0.875)1.5 = 9.56 \text{ in.}^2$$

$$f_c = \frac{P}{A_n} = \frac{4960}{9.56} = 518 \text{ psi}$$

Adj. $F_c = $ tab. $F_c \times \text{LDF} = 1250 \times 1.15$
$$= 1437 > 518 \qquad OK$$

Fig. 7.16b

Bending

Assume simple span (no end restraint).

$$M = \frac{wL^2}{8} = \frac{0.176(7.5)^2}{8} \qquad \text{(span and load on horizontal plane)}$$

$$= 1.24 \text{ ft-k} = 14.85 \text{ in.-k}$$

$$f_b = \frac{M}{S} = \frac{14,850}{13.14} = 1130 \text{ psi}$$

The beam has full lateral support.

$$\therefore \text{ Adj. } F_b = \text{tab. } F_b \times \text{LDF} \qquad (d < 12 \text{ in., } \therefore C_F = 1.0)$$
$$= 1450 \times 1.15 = 1667 \text{ psi} > 1130 \qquad OK$$

Combined Stresses

Because the maximum bending is assumed to occur at the midspan, it is reasonable to combine the bending stress with the axial compression away from the support (i.e., by using the gross area). For intermediate range columns,

$$J = \frac{(L/d) - 11}{K - 11} = \frac{13.9 - 11}{23.1 - 11} = 0.24$$

$$\frac{f_c}{F_c'} + \frac{f_b}{F_b - Jf_c} = \frac{456}{1375} + \frac{1130}{1667 - (0.24)456}$$

$$= 0.332 + 0.725$$

$$= 1.057 > 1.0$$

$$(1.057 - 1.0)100\% \approx 6 \text{ percent overstress}$$

Because of the inaccuracy involved in estimating design loads, and because of variations in material properties, a calculated overstress of 1 or 2 percent is considered by many designers to be still within the "spirit" of the specifications. However, the calculated overstress in this problem is unacceptable.

A 2 × 10 will prove satisfactory. Calculations are not repeated here for this revised member.

NOTE: The simplified analysis used in this example applies only to trusses with pinned joints. For a truss connected with toothed steel gussets, a design approach is used which takes the continuity of the joints into account (Ref. 35).

It should be noted that all stress calculations in this example are for loads caused by the design load of (DL + snow). For this reason the LDF for snow (1.15) is applied to each *individual* stress calculation as well as to the *combined* stress check.

In some cases of combined loading, the individual stresses (axial and bending) may not both be caused by the same load. In this situation the respective LDFs are used in evaluating the *individual* stresses. However, in the *combined* stress calculation, the same LDF is used for both components. Again, the reason for this is that the LDF applies to the *shortest* duration load in the *combination*. The appropriate rules for applying the LDF must be followed in checking both individual and combined stresses. The application of different LDFs in the individual stress calculations is illustrated in the following examples.

7.13 Design Problem: Beam-Column Action in a Stud Wall

A common occurrence of beam-column action is found in an exterior bearing wall. Axial column stresses are developed in the wall studs by vertical dead and live loads. Bending stresses are caused by lateral wind or seismic loads.

The seismic load to be used in the design of a wall for bending is given by the Code expression for F_p (Sec. 2.13). For loads normal to a wall, C_p is taken as 0.3.

For typical wood-frame walls, the wall DL W_p is so small that the design wind load usually exceeds F_p. This applies to loads normal to the wall only, and seismic loads can be critical for other design items. In buildings with large wall dead loads, F_p often exceeds the wind load. Large wall dead loads usually occur in concrete and masonry (concrete block and brick) buildings.

In the two-story building of Example 7.12, the stud carries a number of axial compressive loads. These include dead, roof live, and floor live loads. In the first load case, the various possible combinations of vertical loads are considered.

A statement should be made at this point regarding the analysis of these axial column loads. In Example 4.10, part *b*, a "system" was introduced to determine the *critical load combination* for certain members. This system breaks down in the analysis of laterally unbraced beams and columns because of the treatment of the LDF. It will be remembered that the LDF is applied somewhat differently for beams and columns, *and* the LDF has a varied effect depending on the slenderness. For these reasons, the idea of dividing out the duration load is inappropriate for these members.

As a result, each vertical load combination theoretically should be checked using the appropriate LDF. In this example DL alone can be eliminated by inspection. The results of the other two combinations are close. This can be determined by evaluating f_c/F_c' for each combination. The combination with the larger value for this ratio is critical. However, only the calculations for the critical combination of DL + RLL + FLL are shown in the example.

The second load combination involves both vertical and lateral loads. Roof LL need not be considered simultaneously with lateral loads. (If snow load occurs instead of RLL, it must be considered with lateral loads although some reduction may be possible.) Thus, for the combined stress check, the vertical load effects are recalculated. The reduction of f_c from 123 to 97.3 psi has a relatively minor effect on the final outcome of this problem. For larger axial forces the difference could be of importance.

Notice that the LDF for wind (the shortest-duration load in the combination) applies to both components of stress in the interaction formula.

EXAMPLE 7.12   Combined Bending and Compression in a Stud Wall

Check the 2 × 6 stud in the first floor bearing wall in the building shown in Fig. 7.17*a*. Consider both vertical and lateral loads. Lumber is Stud grade DF-L. EMC ≤ 19 percent. Design loads are given below. Take allowable stresses from the NDS.

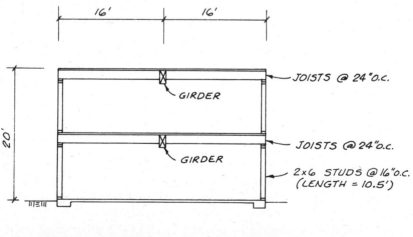

TRANSVERSE   SECTION

Fig. 7.17*a*

Roof DL = 10 psf
Roof LL = 20 psf
Wall DL = 7 psf
Floor DL = 8 psf
Floor LL = 40 psf
Wind = 25 psf horizontal
Seismic $F_p = ZIC_pW_p = 1.0(1.0)(0.3)W_p = 0.3W_p$
$= 0.3(7) = 2.1$ psf < 25 ∴ wind governs

Tabulated stresses (J&P size category):

$F_b = 850$ psi (repetitive member)
$F_c = 675$ psi
$E = 1,500,000$ psi

Load Case 1 (Vertical loads only)

Tributary width of roof and floor to exterior bearing wall is 8 ft. Dead loads:

Roof DL = 10 × 8 = 80 lb/ft
Wall DL − 7 × 20 − 140
Floor DL = 8 × 8 = 64
$w_{DL} = \overline{284}$ lb/ft

Live loads:

Roof LL = 20 × 8 = 160 lb/ft
Floor LL = 40 × 8 = 320 lb/ft

LOAD COMBINATIONS:

Calculate the axial load on a typical stud:

DL alone = 284(1.33 ft) = 378 lb
DL + FLL = (284 + 320)1.33 = 803 lb
DL + FLL + RLL = (284 + 320 + 160)1.33 = 1016 lb

The combination of DL alone can be eliminated by inspection. When the LDFs are considered for the second two combinations, the net effects are roughly the same. However, DL + FLL + RLL is the critical vertical loading, and stress calculations for this combination only are shown. The axial stress in the stud and the bearing stress on the wall plate are equal

$$f_c = f_{c\perp} = \frac{P}{A} = \frac{1016}{8.25} = 123 \text{ psi}$$

COLUMN CAPACITY:

Sheathing provides lateral support about the weak axis of the stud. Therefore, check column buckling about the x axis only:

$$\left(\frac{L}{d}\right)_x = \frac{10.5 \times 12}{5.5} = 22.9 > 11$$

$$K = 0.671\sqrt{\frac{E}{F_c(\text{LDF})}} = 0.671\sqrt{\frac{1,500,000}{675(1.25)}}$$

$$= 28.3 > 22.9 \qquad \therefore \text{ intermediate column}$$

$$F_c' = F_c(\text{LDF})\left[1 - \frac{1}{3}\left(\frac{L/d}{K}\right)^4\right]$$

$$= 675(1.25)\left[1 - \frac{1}{3}\left(\frac{22.9}{28.3}\right)^4\right] = 723 \text{ psi}$$

215

BEARING OF STUD ON WALL PLATES:

$$\text{Adj. } F_{c\perp} = \text{tab. } F_{c\perp} \times \text{LDF} \times \text{BLF}$$
$$= 385 \times 1.25 \times 1.25 = 601 \text{ psi} < 723$$

∴ Bearing governs over stability.

$$601 > 123 \text{ psi}$$
$$\text{Vertical loads} \qquad OK$$

Load Case 2 (Vertical + lateral loads)

BENDING:

Wind governs.

$$Wind = 25 \text{ psf}$$
$$w = 25 \times 1.33 = 33.3 \text{ lb/ft}$$
$$M = \frac{wL^2}{8} = \frac{33.3(10.5)^2}{8} = 458 \text{ ft-lb}$$
$$= 5.5 \text{ in.-k}$$

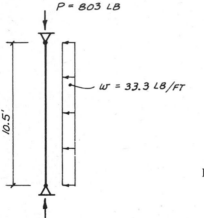

P = 803 LB

w = 33.3 LB/FT

10.5'

Fig. 7.17b

$$f_b = \frac{M}{S} = \frac{5500}{7.56} = 728 \text{ psi}$$

$$F_b = \text{tab. } F_b \times \text{LDF} = 850 \times 1.33 = 1130 > 728 \qquad OK$$

AXIAL:

Roof LL need not be considered simultaneously with lateral loads—use DL + FLL.

$$f_c = \frac{P}{A} = \frac{803}{8.25} = 97.3 \text{ psi}$$

In Load Case 1 it was noted that DL + FLL + RLL is the critical combination for vertical loads. Therefore, a separate allowable axial stress for DL + FLL alone (i.e., using an LDF of 1.0) is not required. In the combined stress calculation the same LDF is used throughout. Hence, F_c' is determined here using an LDF of 1.33.

$$\left(\frac{L}{d}\right)_x = 22.9$$

$$K = 0.671\sqrt{\frac{E}{F_c(\text{LDF})}} = 0.671\sqrt{\frac{1,500,000}{675(1.33)}}$$

$$= 27.4 > 22.9 \qquad \therefore \text{intermediate column}$$

$$F_c' = F_c(\text{LDF})\left[1 - \frac{1}{3}\left(\frac{L/d}{K}\right)^4\right]$$

$$= 675(1.33)\left[1 - \frac{1}{3}\left(\frac{22.9}{27.4}\right)^4\right]$$

$$= 752 \text{ psi} > 97.3$$

For combined stresses in the stud, the use of F_c' (instead of F_{c1}) is appropriate.

COMBINED STRESS:

Intermediate-length column:

$$J = \frac{(L/d) - 11}{K - 11} = \frac{22.9 - 11}{27.4 - 11} = 0.726$$

$$\frac{f_c}{F_c'} + \frac{f_b}{F_b - Jf_c} = \frac{97.3}{752} + \frac{728}{1130 - (0.726)(97.3)}$$

$$= 0.82 < 1.0 \qquad OK$$

2 × 6 Stud grade DF-L Exterior bearing wall   OK

7.14 Design Problem: Glulam Beam-Column

In this example a somewhat more complicated bracing condition is considered. The column is a glulam that supports both roof dead and live loads as well as lateral wind loads. See Example 7.13.

In the first load case the vertical loads are considered. (DL + RLL) is the critical loading. The interesting aspect of this problem is that there are different unbraced lengths for the x and y axes. Lateral support for the strong axis is provided at the ends only. However, for the weak axis the unbraced length is the height of the window.

In the second load case the vertical DL and lateral wind load are considered. The bending analysis includes a check of lateral stability using the window height as the unbraced length. Bending takes place about the strong axis and if lateral buckling occurs the beam will buckle about the weak axis of the member.

In checking combined stresses the LDF of 1.33 for wind applies to all components of the interaction formula. The limit K is reevaluated for the combined stress check because K depends on the LDF. If the column had been in the intermediate range, the revised value of K would be used to determine J as well as F_c'.

EXAMPLE 7.13   Glulam Beam-Column

Check the column in the building shown in Fig. 7.18*a* for the given loads. The column is Combination 2 DF glulam. The member supports the tributary (DL + roof LL) and the lateral wind load. The wind load is transferred to the column by the window framing in the wall. Seismic is not critical. Stresses are from Refs. 1, 2, 3, or 28.

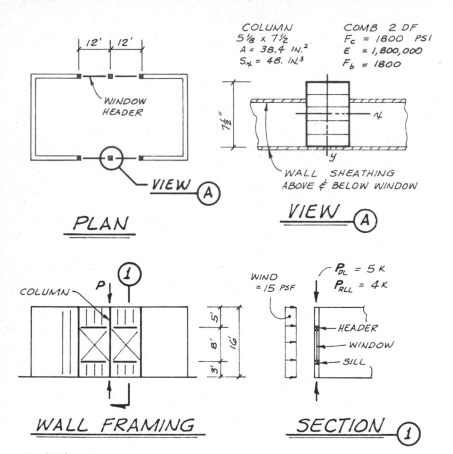

Fig. 7.18*a*

Load Case 1 (Vertical Loads)

$$DL = 5 \text{ k}$$
$$DL + RLL = 5 + 4 = 9 \text{ k}$$

By inspection DL + RLL is the critical vertical loading condition.

$$f_c = \frac{P}{A} = \frac{9000}{38.4} = 234 \text{ psi}$$

Neglect the column end restraint offered by wall sheathing for column buckling about the *y* axis. Assume an effective length factor (Fig. 7.7) of 1.0 for both the *x* and *y* axes.

$$\left(\frac{L}{d}\right)_x = \frac{1(16 \times 12)}{7.5} = 25.6$$

$$\left(\frac{L}{d}\right)_y = \frac{1(8 \times 12)}{5.125} = 18.7 < 25.6$$

Strong axis of column governs.

$$K = 0.671\sqrt{\frac{E}{F_c(LDF)}} = 0.671\sqrt{\frac{1,800,000}{1800(1.25)}}$$

$$= 19.0 < 25.6 \qquad \therefore \text{long column}$$

$$F'_c = \frac{0.3E}{(L/d)^2} = \frac{0.3(1,800,000)}{(25.6)^2} = 824 \text{ psi} > 234 \qquad OK$$

The member is adequate for vertical loads.

Load Case 2 (DL + Wind)

Roof LL need not be considered simultaneously with wind. Snow load (if it occurs) must be considered with wind although some reduction may be allowed. The LDF is taken as 1.33 throughout.

AXIAL (DL ALONE):

$$\left(\frac{L}{d}\right)_{max} = 25.6 \text{ from Load Case 1}$$

$$K = 0.671\sqrt{\frac{E}{F_c(LDF)}} = 0.671\sqrt{\frac{1,800,000}{1800(1.33)}}$$

$$= 18.4 < 25.6 \qquad \therefore \text{long column}$$

F'_c remains the same as in Load Case 1.

$$f_c = \frac{P}{A} = \frac{5000}{38.4} = 130 \text{ psi} < 824 \qquad OK$$

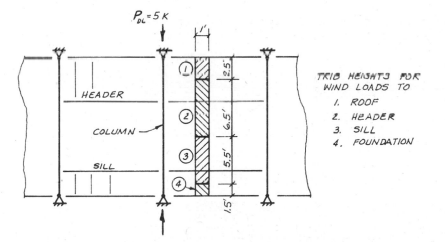

WALL FRAMING & TRIB HEIGHTS

Fig. 7.18b

219

DESIGN OF WOOD STRUCTURES

BENDING (WIND):

Window header and sill span horizontally as beams between columns. Uniform wind load to *header* and *sill* is calculated using a 1-ft section of wall and the trib. heights to the members (Fig. 7.18b).

$$\text{Wind on header} = w_1 = (15)(6.5) = 97.5 \text{ lb/ft}$$

Horizontal reaction of headers on center column (Fig. 7.18c):

$$P_1 = 97.5 \times 12 = 1170 \text{ lb}$$

$$\text{Wind on sill} = w_2 = (15)(5.5) = 82.5 \text{ lb/ft}$$

Horizontal reaction of sills on center column:

$$P_2 = 82.5 \times 12 = 990 \text{ lb}$$

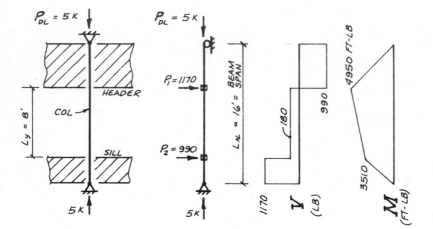

LOAD & SHEAR & MOMENT

Fig. 7.18c

From the moment diagram,

$$M = 4950 \text{ ft-lb} = 59.4 \text{ in.-k}$$

$$f_b = \frac{M}{S_x} = \frac{59,400}{48} = 1240 \text{ psi}$$

Unbraced length $= L_u = 8 \text{ ft} = 96 \text{ in.}$

Effective length $= L_e = 1.92 L_u = 1.92(96) = 184$

$$C_s = \sqrt{\frac{L_e d}{b^2}} = \sqrt{\frac{184(7.5)}{(5.125)^2}} = 7.25 < 10$$

∴ Short unbraced length for bending and lateral stability does not govern (*dng*).
Check size effect:

$$\text{Adj. } F_b = \text{tab. } F_b \times \text{LDF} \times C_F = 1800 \times 1.33 \times 1.0$$
$$= 2390 \text{ psi} > 1240 \qquad OK$$

COMBINED STRESSES (DL + WIND):

$$J = 1.0 \quad \text{(long column)}$$

$$\frac{f_c}{F_c'} + \frac{f_b}{F_b - Jf_c} = \frac{130}{824} + \frac{1240}{2390 - (1)(130)}$$

$$= 0.158 + 0.549 = 0.71 < 1.0 \quad OK$$

$5\frac{1}{8} \times 7\frac{1}{2}$ Combination 2 DF glulam is
OK for bending and compression.

7.15 Design for Minimum Eccentricity

The design procedures for an axially loaded column were covered in detail in Secs. 7.4 and 7.5. A large number of interior columns and some exterior columns qualify as axial-load-carrying members. That is, the applied load is *assumed* to pass directly through the centroid of the column cross section, and, in addition, no transverse bending loads are involved.

Although many columns can theoretically be classed as axial-load members, there may be some question as to whether the load in practical columns is truly an axial load. In actual construction there may be some misalignment or nonuniform bearing in connections that cause the load to be applied eccentrically.

Some eccentric moment probably develops in columns which are thought to support axial loads only. The magnitude of the eccentric moment, however, is unknown. Many designers simply ignore the possible eccentric moment and design for axial stresses only. This practice may be justified because practical columns typically have square-cut ends. In addition, the ends are attached with connection hardware such that the column end conditions do not exactly resemble the end conditions of an "ideal" pinned-end column. With the restraint provided by practical end conditions, the effective column length is somewhat less than the actual unbraced length.

However, Ref. 7 states that the possible eccentric moment should not be ignored, and it suggests that columns should be designed for some minimum eccentricity. The *minimum eccentricity* recommended is similar to that required in the design of axially loaded reinforced concrete columns. In this approach the moment used is that caused by considering the compressive load at an eccentricity of 1 in. or one-tenth the width of the column ($0.1d$), whichever is larger. The moment is considered independently about both principal axes.

In the design of wood columns there is no Code requirement to design for a minimum eccentric moment. The suggestion that some designers may provide for an eccentric moment in their column calculations is presented here for information only. Including an eccentric moment in the design column is definitely a more conservative design approach. Whether or not eccentricity should be included in a design is left to the judgment of the designer.

221

7.16 Design Problem: Column with an Eccentric Load

Example 7.14 is provided to demonstrate the use of the interaction formula for eccentric loads. The load is theoretically an axial load, but the calculations are expanded to include a check for the minimum eccentricity discussed in Sec. 7.15. The same interaction formula would be used in the case of a known actual eccentricity.

EXAMPLE 7.14 Column Design for Minimum Eccentricity

The column in Fig. 7.19a is an interior column in a large auditorium. The design roof DL and snow load is theoretically an axial load on the column. Because of the importance of the column, it is desired to provide a conservative design, and a minimum eccentricity of 0.1d or 1 in. will be used in design. Bracing conditions are shown. Lumber is Select Structural DF-L. CUF = 1.0. Loads:

$$DL = 20 \text{ k}$$
$$\underline{SL = 60}$$
$$P_{TL} = 80 \text{ k} \qquad \text{total load governs over DL alone}$$

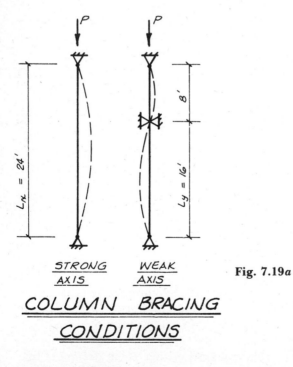

Fig. 7.19a

STRONG   WEAK
AXIS   AXIS

COLUMN BRACING

CONDITIONS

Try 10 × 14 (B&S):

$b = 9\frac{1}{2}$ in.	$A = 128.25$ in.^{2}	$F_b = 1600$ psi
$d = 13\frac{1}{2}$ in.	$S_x = 288.6$ in.^{3}	$F_c = 1100$ psi
	$S_y = 203.1$	$E = 1{,}600{,}000$ psi

Axial

$$\left(\frac{L}{d}\right)_x = \frac{24 \times 12}{13.5} = 21.3$$

$$\left(\frac{L}{d}\right)_y = \frac{16 \times 12}{9.5} = 20.2 < 21.3 \qquad \therefore \text{ strong axis governs}$$

$$K = 0.671 \sqrt{\frac{E}{F_c(\text{LDF})}} = 0.671 \sqrt{\frac{1,600,000}{1100(1.15)}} = 23.9$$

$$11 < \frac{L}{d} < K \qquad \therefore \text{ intermediate-length column}$$

$$F'_c = F_c(\text{LDF}) \left[1 - \frac{1}{3}\left(\frac{L/d}{K}\right)^4\right] = 1100\,(1.15)\left[1 - \frac{1}{3}\left(\frac{21.3}{23.9}\right)^4\right] = 995 \text{ psi}$$

$$f_c = \frac{P}{A} = \frac{80,000}{128.25} = 624 \text{ psi} < 995 \qquad OK$$

Check for minimum eccentricity.

Eccentric Load about Strong Axis

BENDING:

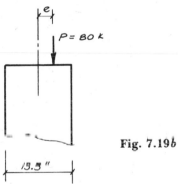

Fig. 7.19b

$$e = 0.1d = 0.1(13.5) = 1.35 \text{ in.} > 1.0$$

$$\text{Ecc. } M = Pe = 80(1.35) = 108 \text{ in.-k}$$

$$\text{Ecc. } f_b = \frac{Pe}{S_x} = f_c\left(\frac{6e}{d}\right) = 624\left(\frac{6 \times 1.35}{13.5}\right)$$

$$= 374 \text{ psi}$$

Check lateral stability

$$L_u = 16 \text{ ft} = 192 \text{ in.}$$

$$C_s = \sqrt{\frac{L_e d}{b^2}} = \sqrt{\frac{(1.92 \times 192)13.5}{(9.5)^2}} = 7.43 < 10$$

$\therefore$ Use full F_b.

$\quad$ Adj. F_b = tab. $F_b \times$ LDF $\times C_F = 1600 \times 1.15 \times 0.99 = 1822$ psi $> 374 \qquad OK \qquad$ **223**

COMBINED STRESSES:

General interaction formula:

$$\frac{f_c}{F_c'} + \frac{f_b + f_c(6 + 1.5J)\left(\frac{e}{d}\right)}{F_b - Jf_c} \le 1.0$$

Bending stress due to transverse loads is zero ($f_b = 0$). For intermediate column,

$$J = \frac{(L/d) - 11}{K - 11} = \frac{21.3 - 11}{23.9 - 11} = 0.798$$

$$\frac{f_c}{F_c'} + \frac{f_c(6 + 1.5J)\left(\frac{e}{d}\right)}{F_b - Jf_c} = \frac{624}{995} + \frac{624[6 + 1.5(0.798)]\left(\frac{1.35}{13.5}\right)}{1822 - (0.798)(624)}$$

$$= 0.627 + 0.339$$

$$= 0.97 < 1.0 \quad OK$$

Eccentricity about Weak Axis

BENDING:

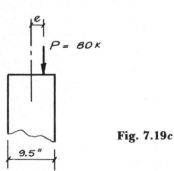

Fig. 7.19c

$$e = 0.1d = 0.1(9.5) = 0.95 < 1.0$$
$$\text{Ecc. } M = Pe = 80(1.0) = 80 \text{ in.-k.}$$

Calculate directly the amplified eccentric bending stress:

$$f_c(6 + 1.5J)\left(\frac{e}{d}\right) = 624[6 + (1.5)(0.798)]\left(\frac{1.0}{9.5}\right)$$

$$= 473 \text{ psi}$$

Check lateral stability.

$$L_u = 24 \text{ ft} = 288 \text{ in.}$$

$$C_s = \sqrt{\frac{L_e d}{b^2}} = \sqrt{\frac{1.92(288)9.5}{(13.5)^2}} = 5.4 < 10$$

224 ∴ Use full F_b, and bending alone is *OK* ($F_b = 1600 \times 1.15 = 1840$ psi).

COMBINED STRESSES:

$$\frac{f_c}{F'_c} + \frac{f_c(6 + 1.5J)(e/d)}{F_b - Jf_c} = \frac{624}{995} + \frac{473}{1840 - (0.798)(624)}$$

$$= 0.627 + 0.352$$

$$= 0.98 < 1.0 \qquad OK$$

$$\boxed{\begin{array}{c} \text{Use } 10 \times 14 \text{ Sel. Str.} \\ \text{DF-L column} \end{array}}$$

The column has different unbraced lengths for the x and y axes. These different lengths are considered in both the axial and bending calculations. Column buckling is found to be governed by the strong axis of the cross section. The slenderness factor is less than 10 for both axes, and lateral stability for bending is not critical for either eccentric loading.

(DL + snow) is the critical load combination, and an LDF of 1.15 is used throughout the problem.

7.17 Problems

Allowable stresses for the following problems are to be in accordance with the NDS.

7.1 A 3 × 8 tension member in a horizontal diaphragm resists a tension load of 20 k. This load is caused by the lateral wind load. Lumber is Select Structural DF-L. A single line of ⅞-in.-diameter bolts is used to make the connection of the member to the diaphragm. CUF = 1.0.

Find: Check the tension stress in the member.

7.2 A 5⅛ × 15 glulam is used as the tension member in a large roof truss. A single row of 1-in.-diameter bolts occurs at the net section of the member. Material is Combination 3 Southern pine glulam. Loads are a combination of (DL + snow). Joints are assumed to be pin connected.

Find: The allowable axial tension load.

7.3 The truss in Fig. 7.A has a 2 × 4 lower chord of Sel. Str. Idaho white pine. The loads shown are the result of (DL + snow). No reduction of area for fasteners. CUF = 1.0. Joints are assumed to be pin connected.

Find: Check the tension stress in the member.

7.4 The truss in Fig. 7.A has a 2 × 6 lower chord of No. 1 DF–larch. In addition to the loads shown on the sketch, the lower chord supports a ceiling load of 5 psf (20 lb/ft). There is no reduction of member area for fasteners. Lateral stability is not a factor. Joints are assumed to be pin connected. CUF = 1.0.

Find: Check combined stresses in the lower chord.

225

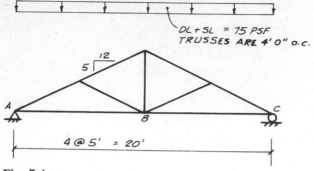

Fig. 7.A

7.5 The truss in Fig. 7.B supports the roof DL of 16 psf shown in the sketch. Trusses are spaced 24 in o.c., and the roof LL is to be in accordance with UBC Table 23C, Method 1. Lumber is No. 2 DF-L. Fasteners do not reduce the area of the members. Truss joints are assumed to be pin connected. CUF = 1.0.

Find: The required member size for the tension chord.

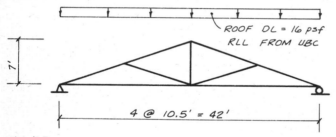

Fig. 7.B

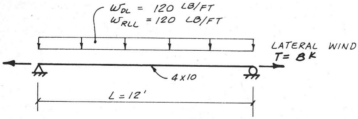

Fig. 7.C

7.6 Problem 7.6 is the same as Prob. 7.5 except that in addition there is a ceiling load applied to the bottom chord of 8 psf (16 lb/ft). Neglect deflection. CUF = 1.0.

7.7 The door header in Fig. 7.C supports the loads shown. Lumber is No. 2 Hem-Fir. CUF = 1.0. No bolt holes at maximum moment. Lateral stability is not a problem.

Find: Check the given member size under the following loading conditions.
   a. Vertical loads only
   b. UBC-required vertical and lateral loads
   Which loading condition is the more severe?

7.8 A 4 × 4 carries an axial compressive force caused by DL + FLL. Lumber is No. 1 DF–larch.

   Find: The allowable column load if the unbraced length of the member is
      a. 3 ft
      b. 6 ft
      c. 9 ft
      d. 12 ft

7.9 Problem 7.9 is the same as Prob. 7.8 except that the force is the result of DL + RLL.

7.10 A 6 × 6 carries an axial compressive force caused by DL + FLL. Lumber is No. 1 DF–larch.

   Find: The allowable column load if the unbraced length of the member is
      a. 5 ft
      b. 9 ft
      c. 11 ft
      d. 15 ft
      e. 19 ft

7.11 Problem 7.11 is the same as Prob. 7.10 except that the force is the result of DL + snow.

7.12 An 8¾ × 15 glulam column is made from Combination 1 California redwood. The bracing conditions for the column are shown in Fig. 7.D. The design load

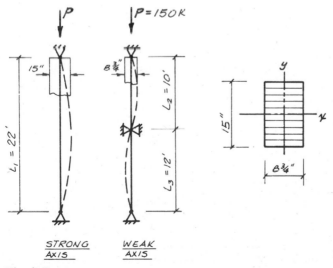

Fig. 7.D

is shown on the sketch and is the result of DL + floor LL + roof LL. Effective length factor = 1.0.

Find: Is the column adequate to support the design load?

7.13 Problem 7.13 is the same as Prob. 7.12 except that the following lengths are to be used in Figure 7.23:

$$L_1 = 24 \text{ ft}$$
$$L_2 = 10 \text{ ft}$$
$$L_3 = 14 \text{ ft}$$

7.14 A sawn lumber column is used to support an axial load (DL + snow) of 75 k. Use No. 1 DF–Larch. The unbraced length is the same for both the x and y axis of the member. The effective length factor is 1.0 for both axes.

Find: The minimum column size if the unbraced length is
   a. 8 ft
   b. 10 ft
   c. 14 ft
   d. 18 ft
   e. 22 ft

7.15 An 8 × 12 column of No. 1 Ponderosa pine has an unbraced length for buckling about the strong *(x)* axis of 16 ft, and an unbraced length for buckling about the weak *(y)* axis of 8 ft.

Find: The allowable axial DL + FLL.

7.16 Problem 7.16 is the same as Prob. 7.12 except that $P = 50$ k (instead of 150 k) and the minimum eccentricity described in Sec. 7.15 is to be considered.

7.17 A stud wall is to be used as a bearing wall in a wood-frame building. The wall carries an axial load of DL + RLL. Studs are 2 × 4 Construction grade Hem-Fir and are located 16 in. o.c. Studs have sheathing attached.

Find: The allowable load per lineal foot of wall if the wall height is
   a. 8 ft
   b. 10 ft
   c. 12 ft

7.18 A stud wall is to be used as a bearing wall in a wood-frame building. The wall carries an axial load of DL + snow. Studs are 2 × 6 No. 2 Southern pine (surfaced dry) and are 24 in. o.c. Studs have sheathing attached.

Find: The allowable load per lineal foot of wall if the wall height is
   a. 10 ft
   b. 14 ft

7.19 The exterior column in Fig. 7.E carries a vertical load due to a girder reaction and a lateral load due to horizontal wall framing. The vertical load is the result of DL + snow. The lateral load is the tributary wind load. CUF = 1.0.

Find: Check the trial column for combined stresses.

7.20 Problem 7.20 is the same as Prob. 7.19 except that the length is 21 ft and the uniform wind load is 100 lb/ft.

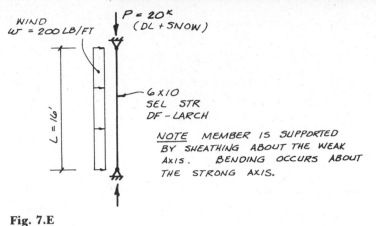

Fig. 7.E

7.21 The truss in Fig. 7.A has a 2 × 10 top chord of No. 2 Hem-Fir. The top of the truss is fully supported along its length by roof sheathing. No reduction of area for fasteners. CUF = 1.0. Joints are assumed to be pin connected.

Find: Check combined stresses in the top chord.

7.22 A truss is similar to the one shown in Fig. 7.A except that the span is 36 ft and the load is 30 psf. The top chord is a 2 × 10 of No. 2 Hem-Fir, and it is laterally supported along its length by roof sheathing. No reduction of member area for fasteners. CUF = 1.0. Joints are assumed to be pin connected.

Find: Check combined stresses in the top chord.

7.23 A 2 × 6 exterior stud wall is 14 ft tall. Studs are 16 in. o.c. The stud supports the following vertical loads per foot of wall:

$$w_{DL} = 800 \text{ lb/ft}$$
$$w_{FLL} = 800 \text{ lb/ft}$$
$$w_{RLL} = 400 \text{ lb/ft}$$

In addition, the wall carries a uniform wind load of 15 psf (horizontal). Lumber is No. 1 DF–larch. CUF = 1.0. Sheathing provides lateral support in the weak direction.

Find: Check the stud using the UBC required load combinations.

CHAPTER EIGHT

Plywood

8.1 Introduction

Plywood is a widely used building material with a variety of structural and nonstructural applications. Some of the major structural uses are

1. Roof, floor, and wall sheathing
2. Horizontal and vertical (shearwall) diaphragms
3. Structural components
   a. Lumber and plywood beams
   b. Stressed skin panels
   c. Curved panels
   d. Folded plates
   e. Sandwich panels
4. Gusset plates
   a. Trusses
   b. Rigid frame connections
5. Wood foundation systems
6. Concrete formwork

Numerous other uses of plywood can be cited, including a large number of industrial, commercial, and architectural applications.

As far as the types of buildings covered in this text are concerned, the first two items in the above list are of primary interest. The relatively high allowable loads and the ease with which sheets of plywood can be installed have made plywood widely accepted for use in these applications. The other topics listed above are beyond the scope of this text. Information on these and other subjects is available from the American Plywood Association (APA).

Chapter 8 will essentially serve as a turning point from the design of the vertical-load-carrying system (beams and columns) to the design of the lateral-force-resisting system (horizontal diaphragms and shearwalls). Plywood provides this transition because it is often used as a structural element in *both* the vertical and lateral systems.

In the vertical system, plywood functions as the *sheathing* material. As such, it directly supports the roof and floor loads and distributes these loads to the beams in the framing system. See Example 8.1. Wall sheathing, in a similar manner, distributes the normal wind load to the studs in the wall. In the lateral-force-resisting system (LFRS), plywood serves as the *shear*-resisting element.

The required *thickness* of the plywood is often determined by *sheathing-type loads* (loads normal to the surface of the plywood). On the other hand, the *nailing* requirements for the plywood are determined by the *unit shears* in the horizontal or vertical diaphragm. When the shears are high, the required thickness of plywood may be determined by the diaphragm unit shears instead of by the sheathing loads.

It should be noted that the required thickness of plywood for roof, floor, and wall sheathing may usually be determined from *design aids* provided in Code tables or APA literature. It is important to realize that the bases for the design aids are beam calculations or concentrated load considerations, whichever is more critical. The need may arise for beam calculations of this nature if the design aids are found not to cover a particular situation. However, structural calculations for plywood are usually necessary only for the design of a structural-type component (e.g., a lumber and plywood beam or stress-skin panel).

Chapter 8 introduces plywood properties and grades and reviews the procedures used to determine the required thickness and grade of plywood for *sheathing applications*. The aids for designing sheathing are included, but the calculation of stresses in plywood is beyond the scope of this text. However, the basic structural behavior of plywood is explained, and some of the unique design aspects of plywood are introduced. Understanding these basic principles is necessary for the proper use of plywood.

The concept of plywood is basically simple, but some effort is required in order to understand the grades of plywood and its structural behavior. Chapter 8 is basically descriptive in nature, and little is required in the form of design calculations. However, several readings may be necessary for this material to be assimilated.

Chapter 9 deals with the problem of *diaphragm design* and Chap. 10 covers *shearwalls*. There the calculations necessary for the design of the LFRS are treated in considerable detail.

EXAMPLE 8.1 Plywood Sheathing

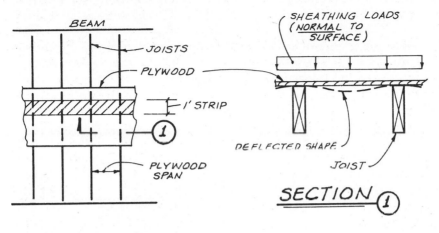

FRAMING PLAN

Fig. 8.1

The term *sheathing load,* as used in this book, refers to loads that are normal to the surface of the sheathing. See Fig. 8.1. Sheathing loads for floors and roofs include DL and LL. For walls, the wind load is the sheathing load. Typical plywood sheathing applications use the plywood continuously over two or more spans. For common joist spacings and typical loads, *design aids* have been developed so that the required plywood sheathing can be chosen without having to perform beam design calculations. A number of these design aids are included in Chap. 8. When required, cross-sectional properties for a 1-ft-wide section of plywood can be obtained from Ref. 14.1.

8.2 Plywood Construction and Panel Dimensions

There are a number of variables in the construction of plywood which affect its use as a structural material. These variables include:

1. Panel dimensions and thickness
2. Number and thickness of layers and plies
3. Species of wood
4. Orientation of grain in the plies
5. Veneer grade
6. Type of glue
7. Surface condition of the panel

When the various combinations of these items are considered, it can be seen why there are such a large number of plywood grades. Each type of plywood has an appropriate use, and only by a thorough understanding of the above factors can the designer make the proper choice for a given application.

The fabrication of plywood used in structural applications is covered in Ref. 14.11. Voluntary product standards are published by the National Bureau of Standards under the Department of Commerce. The standards are made effective by incorporating them by reference into other specifications and buildings codes.

Plywood is manufactured in sheets or *panels*. The standard size of these panels is 4 ft × 8 ft. Certain manufacturers are capable of producing larger sizes, but the standard 4 ft × 8 ft dimensions should be assumed in design unless the availability of other sizes is known.

The tolerance for the length and width dimensions for a plywood panel

---

EXAMPLE 8.2   Plywood Tolerances and Clearances

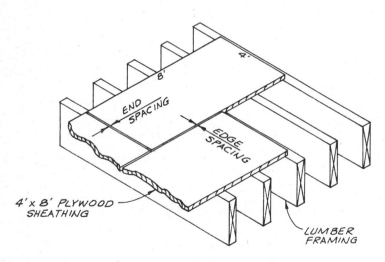

PLYWOOD   SHEATHING

Fig. 8.2

Many plywood sheathing applications for roofs, floors, and walls recommend an *edge spacing* as much as ⅛ in. and an *end spacing* of ¹⁄₁₆ in. to permit panel movement with changes in moisture content. Other spacing provisions may apply, depending on the type of plywood, application, and moisture-content conditions. Refer to APA publications given in Ref. 14 for specific recommendations.

is plus 0 in. and minus $\frac{1}{16}$ in. Because of its cross-laminated construction, plywood is dimensionally stable. However, some small change in dimensions can be expected under varying moisture conditions, especially during the early stages of construction when the material is adjusting to the local atmospheric conditions. For this reason installation instructions for many roof and floor applications recommend a clearance between panel edges and panel ends. See Example 8.2.

These clearance recommendations explain the negative tolerance (+0, $-\frac{1}{16}$) for the panel dimensions. By having the panel dimension slightly less than the stated size, the clearances between panels can be provided while maintaining the basic 4-ft module that the use of plywood naturally implies.

Plywood is available in a number of standard thicknesses from $\frac{1}{4}$ to $1\frac{1}{8}$ in. The tolerances for thickness vary, depending on the thickness of the panel and the surface condition of the panel.

For sanded panels, a tolerance of $\pm\frac{1}{64}$ inch applies to thicknesses of $\frac{3}{4}$ in. or less. For larger thicknesses, a tolerance of ± 3 percent of the panel thickness is used. For *unsanded, touch-sanded,* and *overlaid* panels a tolerance of $\pm\frac{1}{32}$ in. applies to thicknesses of $1\frac{3}{16}$ inch or less. For larger thicknesses, a tolerance of ± 5 percent of the panel thickness is specified. The different surface conditions are described in Sec. 8.8 and 8.9.

8.3 Plywood Makeup

A plywood panel is made up of a number of veneers (thin sheets or pieces of wood). Veneer is obtained by rotating *peeler logs* (approximately $8\frac{1}{2}$ ft long) in a lathe. A continuous veneer is obtained as the log is forced into a long knife. The log is simply unwound or "peeled." See Example 8.3. The veneer is then clipped to the proper size, dried to a low moisture content (2 to 5 percent), and graded according to quality.

The veneer is spread with glue and cross-laminated (adjacent layers have the wood grain at right angles) into a plywood panel with an *odd* number of *layers.* See Example 8.4. The panel is then cured under pressure in a large hydraulic press. The glue bond obtained in this process is stronger than the wood in the plies. After curing, the panels are trimmed and finished (e.g., sanded) if necessary. Finally the appropriate grade-trademark is stamped on the panel.

In summary, *veneer* is the thin sheet of wood obtained from the peeler log. When veneer is used in the construction of plywood it becomes a *ply.* The cross-laminated pieces of wood in a plywood panel are known as *layers.* Layers are often simply an individual ply, but they can consist of more than one ply.

It is the cross-laminating that provides plywood with its unique strength characteristics. It provides increased dimensional stability over wood that is not cross-laminated. Cracking and splitting are reduced, and fasteners, such **235** as nails and staples, can be placed close to the edge without a reduction in load capacity.

If structural calculations are required, the cross laminations make stress analysis somewhat more involved. Wood is stronger parallel to the grain than perpendicular to the grain. This is especially true in tension, where

---

EXAMPLE 8.3   Fabrication of Veneer

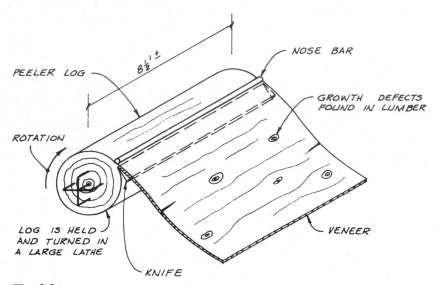

Fig. 8.3

The log is unwound or peeled into a continuous sheet of *veneer*. Thicknesses range between $\frac{1}{16}$ in. and $\frac{5}{16}$ in. As with sawn lumber, the veneer is graded visually by observing the size and number of defects. Most veneers may be repaired or patched to improve the grade to which it is assigned. Veneer grades are discussed in Sec. 8.5.

---

wood has little strength across the grain; it is also true in compression but to a lesser extent. In addition, wood is much stiffer parallel to the grain compared with perpendicular to the grain. The modulus of elasticity across the grain is approximately $\frac{1}{35}$ of the modulus of elasticity parallel to the grain.

Because of the differences in strength and stiffness, the *plies that have the grain parallel to the stress are much more effective* than those that are perpendicular to the stress. As a result of the odd number of layers used in plywood construction, different strength properties result when the applied stress is in one

236

EXAMPLE 8.4   **Plywood Cross-Laminated Construction**

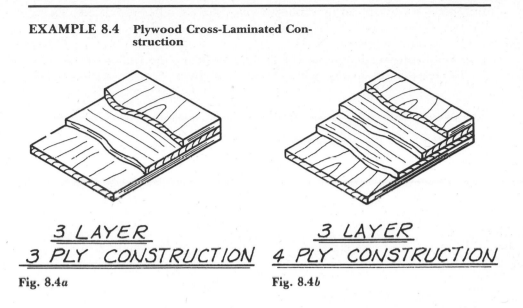

3 LAYER
3 PLY CONSTRUCTION

Fig. 8.4a

3 LAYER
4 PLY CONSTRUCTION

Fig. 8.4b

In its simplest form plywood consists of 3 plies. Each ply has wood grain at right angles to the adjacent veneer (Fig. 8.4a).

An extension of the simple three-ply construction is the three-layer four-ply construction (Fig. 8.4b). The center two plies have the grain running in the same direction. However, the basic concept of cross-laminating is still present because the two center plies are viewed as a single layer. It is the *layers* which are *cross-laminated*.

Three-layer construction is used in the thinner plywood panels. Depending on the thickness and grade of the plywood, five- and seven-layer constructions are also fabricated.

Detailed information on plywood panel makeup is contained in Ref. 14.11.

---

direction (say parallel to the 8-ft dimension), compared with the section properties that result when the load is applied in the other direction (parallel to the 4-ft dimension).

Thus, two sets of cross-sectional properties apply to plywood. One set is used for stresses parallel to the 8-ft dimension, and the other is used for stresses parallel to the 4-ft dimension.

Even if the sheathing thickness and allowable load is read from a table (structural calculations not required), the orientation of the panel and its directional properties are important to the proper use of the plywood. Obviously the tables giving the design values were prepared using the section properties for the two different directions.

The direction of the grain in a panel must be clearly understood. See Example 8.5. The names assigned to the various layers in the makeup of a plywood panel are

1. Face—outside ply. If the outside plies are of different veneer quality, the face is the better veneer grade.
2. Back—the other outside ply.
3. Crossband—inner layer(s) laid at right angles to the face and back plies.
4. Centers—inner layer(s) parallel with outer plies.

EXAMPLE 8.5   Direction of Grain

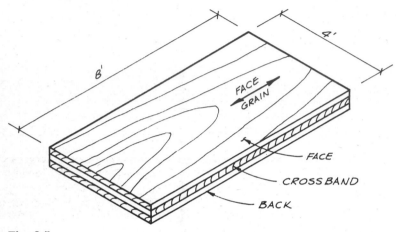

FACE GRAIN

FACE

CROSS BAND

BACK

Fig. 8.5

In standard plywood construction, the face and back plies have the grain running parallel to the 8-ft dimension of the panel. Crossbands are inner plies that have the grain at right angles to the face and back (i.e., parallel to the 4-ft dimension). If a panel has more than three layers, some inner plies (centers) will have grain that is parallel to the face and back.

When the stress in plywood is parallel to the 8-ft dimension, there are more "effective" plies (i.e., there are more plies with grain parallel to the stress). The designer should be aware that different section properties are involved, depending on how the panel is turned. This is important even if stress calculations are not performed.

To illustrate the effects of panel orientation, two different panel layouts are considered. See Fig. 8.6. With each panel layout, the corresponding 1-ft-wide beam cross section is shown. The bending stresses in these beams are parallel to the span. For simplicity, the plywood in this example is three-layer construction.

In the first example, the 8-ft dimension of the plywood panel is parallel to the span (sheathing spans between joists). When the plywood is turned this way (face grain perpendicular to the supports), it is said to be used in the strong direction. In the second example, the 4-ft dimension is parallel to the span of the plywood (face grain parallel to the supports). Here the panel is used in the weak direction.

From these sketches it can be seen that the cross section for the strong direction has more plies with the grain running parallel to the span. In addi-

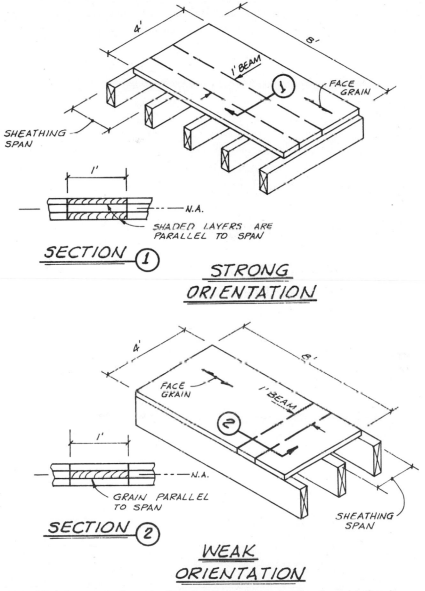

Fig. 8.6 Strong and weak directions of plywood. Wood grain that is parallel to the span and stress is more effective than the wood grain that is perpendicular to the span.

239

tion, these plies are located at a larger distance from the neutral axis. These two facts explain why the effective cross-sectional properties are larger for plywood oriented with the face grain parallel to the span.

8.4   Species Groups

There are a large number of species of wood that can be used to manufacture plywood. See Fig. 8.7. The various species are assigned, according to strength

Group 1	Group 2	Group 3	Group 4	Group 5
Apitong	Cedar, Port	Alder, Red	Aspen	Basswood
Beech,	Orford	Birch, Paper	Bigtooth	Poplar,
American	Cypress	Cedar, Alaska	Quaking	Balsam
Birch	Douglas	Fir,	Cativo	
Sweet	Fir 2 (a)	Subalpine	Cedar	
Yellow	Fir	Hemlock,	Incense	
Douglas	Balsam	Eastern	Western	
Fir 1 (a)	California	Maple,	Red	
Kapur	Red	Bigleaf	Cottonwood	
Keruing	Grand	Pine	Eastern	
Larch,	Noble	Jack	Black	
Western	Pacific	Lodgepole	(Western	
Maple, Sugar	Silver	Ponderosa	Poplar)	
Pine	White	Spruce	Pine	
Caribbean	Hemlock,	Redwood	Eastern	
Ocote	Western	Spruce	White	
Pine, South.	Lauan	Engelmann	Sugar	
Loblolly	Almon	White		
Longleaf	Bagtikan			
Shortleaf	Mayapis			
Slash	Red			
Tanoak	Tangile			
	White			
	Maple, Black			
	Mengkulang			
	Meranti,			
	Red (b)			
	Mersawa			
	Pine			
	Pond			
	Red			
	Virginia			
	Western			
	White			
	Spruce			
	Black			
	Red			
	Sitka			
	Sweetgum			
	Tamarack			
	Yellow-			
	Poplar			

(a) Douglas Fir from trees grown in the states of Washington, Oregon, California, Idaho, Montana, Wyoming, and the Canadian Provinces of Alberta and British Columbia shall be classed as Douglas Fir No. 1. Douglas Fir from trees grown in the states of Nevada, Utah, Colorado, Arizona and New Mexico shall be classed as Douglas Fir No. 2.
(b) Red Meranti shall be limited to species having a specific gravity of 0.41 or more based on green volume and oven dry weight.

240   **Fig. 8.7**   Species of wood used in plywood. *(Courtesy of APA.)*

and stiffness, to one of five different species groups. Group 1 species have the highest strength characteristics, and Group 5 species have the lowest strength properties. Allowable stresses have been determined for species Groups 1 through 4, and plywood using these species can be used in structural applications. Group 5 has not been assigned allowable stresses.

The specifications for the fabrication of plywood allow the mixing of various species of wood in a given plywood panel. This practice allows the more complete usage of raw materials. If it should become necessary to perform stress calculations, the allowable stresses for plywood calculations have been simplified for use in design. This is accomplished by providing allowable values based on the species group of the face and back plies. The species group of the *outer* plies is included in the grade stamp of certain grades of plywood (Secs. 8.8 and 8.9). Tabulated section properties are calculated for Group 4 *inner* plys (the weakest species group allowed in structural plywood). The assumption of Group 4 inner plies is made regardless of the actual makeup. Allowable stresses and cross-sectional properties are given in the APA publication *Plywood Design Specification* (PDS—Ref. 14.1).

Although plywood grades have not been specifically discussed up to this point, it should be noted that there are some designations that can be added to the plywood grade which limit the species used in the plywood. For example, the term STRUCTURAL I can be added to certain plywood grades to provide increased strength properties. The addition of the STRUCTURAL I designation restricts all veneers in the plywood to Group 1 species. Thus the greatest section properties apply to plywood with this designation, because species Group 1 (rather than Group 4) inner plies are used in calculations. (It should also be noted that there is another term known as STRUCTURAL II which limits the species to those in Groups 1, 2, and 3. STRUCTURAL II, however, is not generally available, and its use in design and building specifications is not recommended.)

Besides limiting the species of wood used in the manufacture of plywood, the STRUCTURAL I designation requires the use of exterior glue and provides further restrictions on knot sizes and repairs over the same grades without the designation. The STRUCTURAL I designation should be added to the plywood grade specification when strength requirements are not otherwise met.

A separate grade of plywood known as "marine" uses the same section properties as the STRUCTURAL I. It is limited to Douglas fir and Western larch species and has stringent limitations on defects. It is a more expensive plywood grade, and, as its name implies, it may be used for boat hull construction. It is not normally used in building construction.

Before the methods for determining the required plywood grade and thickness can be reviewed, there are some additional topics that should be covered. These include veneer grades, glue types, plywood grades, and the identification index system.

8.5 Veneer Grades

The method of producing the veneers which are used to construct a plywood panel was described in Sec. 8.3. Before a panel is manufactured the individual veneers are graded according to quality. The *grade of the veneers* is one of the factors that determines the *grade of the panel*.

The six basic veneer grades are designated by a letter name:

N   Special order "natural finish" veneer. Not used in ordinary structural applications.

A   Smooth paintable surface. Solid surface veneer without knots but may contain a limited number (18 in a 4-ft × 8-ft veneer) of neatly made repairs.

B   Solid-surface veneer. May contain knots up to 1 in. in width across the grain if they are both sound and tight-fitting. May contain an unlimited number of repairs.

C-plugged   A grade of C veneer which meets more stringent limitations on defects than the normal C veneer. For example, open defects such as knotholes may not exceed ¼ in. by ½ in. Further restrictions apply.

C   May contain open knotholes up to 1 in. in width across the grain and occasional knotholes up to 1½ in. across the grain. This is the minimum grade veneer allowed in exterior-type plywood.

D   Allows open knotholes to 2½ in. in width across the grain and occasional knotholes up to 3 in. across the grain. This veneer is not allowed in exterior-type plywood.

The veneer grades in this list are given in order of decreasing quality. Detailed descriptions of the defects and patching provisions for each veneer grade can be found in Ref. 14.11.

Although A and B veneer grades have better surface qualities than C and D veneers, on a structural basis A and C grades are more similar. Likewise, B and D grades are similar, structurally speaking. The reason for these structural similarities is that C veneers can be upgraded through patching and other repairs to qualify as A veneers. See Example 8.6. On the other hand, a B veneer grade can be obtained by upgrading a D veneer. The result is more unbroken wood fiber with A and C veneers.

Except for the special "marine" exterior grade of plywood, A and B grade veneers are used only for the face and back veneers. They may be used for the inner plies, but, in general, C-plugged, C, and D veneers will be the grades used for the inner plies. It should be noted that D veneers represent a large percentage of the total veneer production, and their use, where appropriate, constitutes an efficient use of materials.

EXAMPLE 8.6   Veneer Grades and Repairs

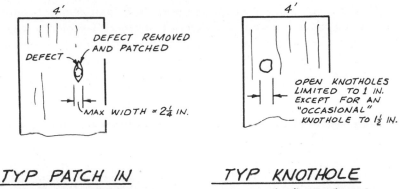

TYP PATCH IN
AN "A" VENEER

TYP KNOTHOLE
IN A "C" VENEER

Fig. 8.8a

A veneers are smooth and firm and free from knots, pitch pockets, open splits, and other open defects. A-grade veneers can be obtained by upgrading (repairing) C grade veneers.

Another upgraded C veneer is C plugged veneer. Although it has fewer open defects than C, it does not qualify as an A veneer.

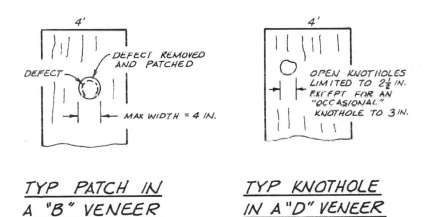

TYP PATCH IN
A "B" VENEER

TYP KNOTHOLE
IN A "D" VENEER

Fig. 8.8b

B veneers are solid and free from *open* defects with some minor exceptions. B veneers can be obtained by upgrading (repairing) D-grade veneers. A and B veneers have similar surface qualities, but A and C are *structurally* similar. Likewise, B and D grades have similar strength properties.

8.6   Types of Plywood

The two basic *types* of plywood are classifications of their "exposure capability." *Exterior-type* plywood is glued with an insoluble "waterproof" glue *and* is constructed using a minimum of C-grade veneers. It will retain its glue bond when repeatedly wetted and dried. Exterior plywood is required when it will be permanently exposed to the weather or when the moisture content in use will exceed 18 percent, either continuously or in repeated cycles.

Interior-type plywood may be used if it is not exposed to the weather and if the EMC in service does not continuously or repeatedly exceed 18 percent. Interior plywood can be manufactured with interior glue, but it is generally available with exterior glue. Thus, *plywood manufactured with exterior glue is not necessarily classified as exterior-type plywood.* If a plywood panel contains a D-grade veneer, it cannot qualify as an exterior panel even if it contains exterior glue. The reason for this veneer grade restriction is that the knotholes allowed in the D veneer grade are so large that the glue bond, even with exterior glue, does not stand up under exposure to the weather. Exposure may result in localized delamination in the area of the knothole.

Interior plywood with any glue type is intended for use in interior (protected) applications. Interior plywood bonded with exterior glue is intended for use where protection against moisture exposure due to construction delays may occur. In addition, the UBC requires that the plywood used for roof sheathing be bonded with exterior (or intermediate) glue. Although the roofing materials provide protection to the plywood, the Code specifies the added glue bond requirements to protect against leakage and possible higher moisture contents in roofing systems.

8.7   Plywood Grades

There are a fairly large number of *plywood grades*. Plywood grades are based on a number of the important factors that affect both strength and appearance. Many of these factors have been discussed in previous sections including wood species, veneer grades, and glue type. Broadly speaking, plywood is divided into

1.   Appearance grades
2.   Engineered grades

These two categories are used to describe the *surface qualities* of the plywood and indicate their normal use. Both appearance grades and engineered grades can be used in structural work. However, appearance grades are more costly and are typically not used unless their surface qualities are necessary for a particular job. Appearance and engineered grades are available in both interior and exterior types.

8.8   Appearance Grades

In general, appearance grades will have smooth, solid surfaces. For these grades the face ply will normally be either an A- or B-grade veneer. The back ply may be an A, B, C, or D veneer. To obtain a smooth surface on appearance grades, faces of A and B panels are sanded.

The sanding operation improves the surface condition of the panel, but in doing so it reduces the thickness of the outer veneers by a measurable amount. In fact, the relative thickness of the layers is reduced by such an amount that different cross-sectional properties are used in strength calculations for sanded and unsanded panels.

Although most appearance grades have sanded exterior plies of A- or B-grade veneers, some appearance grades can have C or better face veneers. These are used where special appearance qualities are desired such as in decorative panels and plywood sidings. These appearance grades may include surface qualities such as rough-sawn, brushed, and grooved surfaces.

Another type of surface available is known as "overlay." This is a finish which produces a hard, smooth surface that has a high resistance to abrasion and to chemicals. The finish is produced by overlaying exterior plywood with a resin-impregnated fiber. Two different grades of overlaid plywood are produced

Medium-density overlay (MDO), a paintable surface for general use

High-density overlay (HDO), for use in more severe conditions such as concrete forms

Each plywood panel is stamped with a grade-trademark which allows it to be fully identified. Grade-trademarks for appearance grades and engineered grades may differ in a number of respects. The grade-trademark on a typical appearance grade will identify the following items:

1.   Veneer grade of face and back

2.   Minimum species group of the outer plies

3.   Type of plywood (exterior or interior) and/or type of glue.

These items essentially identify the plywood. See Example 8.7. Other information in the grade stamp indicates the product standard, manufacturer mill number, and the abbreviation of the "qualified inspection and testing agency." The American Plywood Association (APA) is the agency that provides this quality assurance for most of the plywood manufacturers in the United States.

The recommended uses for the appearance grades of plywood are given in the *Guide to Appearance Grades of Plywood.* See Fig. 8.9.

8.9   Engineered Grades

Engineered grades are used when appearance is not a major concern. Normally in these applications the plywood is covered with some finish material so

Interior Type

Grade Designation (2)	Description and Most Common Uses	Typical (3) Grade-trademarks	Veneer Grade Face	Veneer Grade Inner Plies	Veneer Grade Back	Most Common Thicknesses (inch)					
						1/4	5/16	3/8	1/2	5/8	3/4
N-N, N-A N-B INT-APA	Cabinet quality. For natural finish furniture, cabinet doors, built-ins, etc. Special order items.	N-N, N-A INT-APA PS 1-74 000 / N-B, N-A INT-APA PS 1-74 000	N	C	N,A, or B						3/4
N-D-INT-APA	For natural finish paneling. Special order item.	N-D·1/2 INT-APA PS 1-74 000	N	D	D	1/4					
A-A INT-APA	For applications with both sides on view, built-ins, cabinets, furniture, partitions. Smooth face; suitable for painting.	A-A·1/2 INT-APA PS 1-74 000	A	D	A	1/4		3/8	1/2	5/8	3/4
A-B INT-APA	Use where appearance of one side is less important but where two solid surfaces are necessary.	A-B·1/2 INT-APA PS 1-74 000	A	D	B	1/4		3/8	1/2	5/8	3/4
A-D INT-APA	Use where appearance of only one side is important. Paneling, built-ins, shelving, partitions, flow racks.	A-D GROUP 1 INTERIOR PS 1 000 APA	A	D	D	1/4		3/8	1/2	5/8	3/4
B-B INT-APA	Utility panel with two solid sides. Permits circular plugs.	B-B·1/2 INT-APA PS 1-74 000	B	D	B	1/4		3/8	1/2	5/8	3/4
B-D INT-APA	Utility panel with one solid side. Good for backing, sides of built-ins, industry shelving, slip sheets, separator boards, bins.	B-D GROUP 2 INTERIOR PS 1 000 APA	B	D	D	1/4		3/8	1/2	5/8	3/4
DECORATIVE PANELS—APA	Rough-sawn, brushed, grooved, or striated faces. For paneling, interior accent walls, built-ins, counter facing, displays, exhibits.	DECORATIVE B-D G1 INT APA PS 1-74 000	C or btr.	D	D		5/16	3/8	1/2	5/8	
PLYRON INT-APA	Hardboard face on both sides. For counter tops, shelving, cabinet doors, flooring. Faces tempered, untempered, smooth, or screened.	PLYRON INT-APA PS 1-74 000		C & D					1/2	5/8	3/4
A-A EXT-APA	Use where appearance of both sides is important. Fences, built-ins, signs, boats, cabinets, commercial refrigerators, shipping containers, tote boxes, tanks, ducts. (4)	A-A·G1 EXT-APA PS 1-74 000	A	C	A	1/4		3/8	1/2	5/8	3/4
A-B EXT-APA	Use where the appearance of one side is less important. (4)	A-B·G1 EXT-APA PS 1-74 000	A	C	B	1/4		3/8	1/2	5/8	3/4

Fig. 8.9 Guide to appearance grades of plywood. (*Courtesy of APA.*)

Exterior Type

Grade	Use	Typical Grade-trademark	Face	Inner plies	Back	1/4	3/8	1/2	5/8	3/4
A-C EXT-APA	Use where the appearance of only one side is important. Soffits, fences, structural uses, boxcar and truck lining, farm buildings. Tanks, trailers, commercial refrigerators. (4)	A-C GROUP EXTERIOR APA PS 1 74 000	A	C	C	1/4	3/8	1/2	5/8	3/4
B-B EXT-APA	Utility panel with solid faces. (4)		B	C	B	1/4	3/8	1/2	5/8	3/4
B-C EXT-APA	Utility panel for farm service and work buildings, boxcar and truck lining, containers, tanks, agricultural equipment. Also as base for exterior coatings for walls, roofs. (4)	B-C GROUP 2 EXTERIOR APA PS 1 74 000	B	C	C	1/4	3/8	1/2	5/8	3/4
HDO EXT-APA	High Density Overlay plywood. Has a hard, semi-opaque resin-fiber overlay both faces. Abrasion resistant. For concrete forms cabinets, counter tops, signs, tanks. (4)	HDO · A·A · G·1 · EXT·APA · PS 1 74 · 000	A or B	C or C plgd	A or B		3/8	1/2	5/8	3/4
MDO EXT-APA	Medium Density Overlay with smooth, opaque resin-fiber overlay one or both panel faces. Highly recommended for siding and other outdoor applications, built-ins, signs, displays. Ideal base for paint. (4)(6)	MDO · B·B · G·2 · EXT·APA · PS 1 74 · 000	B	C	B or C		3/8	1/2	5/8	3/4
303 SIDING EXT-APA	Proprietary plywood products for exter or siding, fencing, etc. Special surface treatment such as V-groove, channel groove, striated, brushed, rough-sawn and texture-embossed MDC. Stud spacing (Span Index) and face grade classification indicated on grade stamp.	303 SIDING 6-S GROUP 24 o·c SPAN EXTERIOR PS 1·74 000	(5)	C	C		3/8	1/2	5/8	
T 1-11 EXT-APA	Special 303 panel having grooves 1/4" deep, 3/8" wide, spaced 4" or 8" o.c. Other spacing optional. Edges shiplapped. Available unsanded, textured and MDO.	303 SID·NG 6·S·W T 1·11 GROUP 2 16·o·c SPAN EXTERIOR PS 1·74 000	C or btr.	C	C			19/32	5/8	
PLYRON EXT-APA	Hardboard faces both sides, tempered, smooth or screened.	PLYRON EXT·APA 000		C				1/2	5/8	3/4
MARINE EXT-APA	Ideal for boat hulls. Made only with Douglas fir or western larch. Special solid jointed core construction. Subject to special limitations on core gaps and number of face repairs. Also available with HDO or MDO faces.	MARINE · A·A · EXT·APA · PS 1·74 · 000	A or B	B	A or B	1/4	3/8	1/2	5/8	3/4

(1) Sanded both sides except where decorative or other surfaces specified.

(2) Can be manufactured in Group 1, 2, 3, 4 or 5.

(3) The species groups, certification Indexes and Span Indexes shown in the typical grade-trademarks are examples only. See "Group," "Identification Index" and "Span Index" for explanations and availability.

(4) Can also be manufactured in Structural I (all plies limited to Group 1 species) and Structural II (all plies limited to Group 1, 2, or 3 species).

(5) C or better for 5 plies. C Plugged or better for 3 plies.

(6) Also available as a 303 siding.

247

EXAMPLE 8.7 Appearance Grade-Trademark

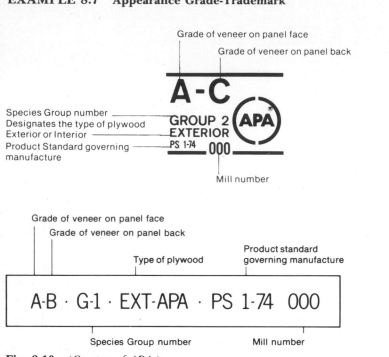

Fig. 8.10   *(Courtesy of APA.)*

Two types of grade stamps for appearance grades of plywood are shown in Fig. 8.10. The first type is found on the back of a panel, and the second type is branded into the edge. In general, appearance grades have A or B veneers for the face ply and may have A, B, C, or D veneer grades for the back and inner plies.

Exterior type requires exterior glue *and* a minimum of C-grade veneers throughout. *Interior type* is generally available with exterior glue and inner plies are usually D-grade veneers.

that the surface qualities of the plywood are not objectionable. Most engineered grades, therefore, do not have smooth solid surfaces.

If a smooth solid surface is required, an appearance grade will ordinarily be used. One exception to this is a special engineered grade that is used for concrete formwork. B-B Plyform is intended to provide a smooth surface for pouring concrete.

The engineered grades are used in the fabrication of plywood components. However, the major use of this type of plywood is for sheathing. The engineered grades of C-C and C-D plywood are also known as *sheathing grades.* Other grades of plywood that are commonly used for sheathing are UNDER-

248

LAYMENT and APA's Sturd-I-Floor. Specific considerations for floor, roof, and wall sheathing are covered in Secs. 8.11 through 8.13.

Except for B-B Plyform, the face plies of engineered panels are either C– or C-plugged–grade veneers. The back and inner plies are generally C- or D-grade veneers, but in some cases, improved C- or D-veneer grades are required for these plies. Engineered grades are available in several different surface conditions:

1. Unsanded—rough surface

2. Touch sanded—a sizing operation that provides an intermediate surface. Sander skips may occur.

3. Sanded—smooth surface

Most engineered grades will be unsanded. This generally applies to grades with C or D outer plies. Touch sanding is a light surface sanding and is

---

EXAMPLE 8.8 **Engineered (Sheathing) Grade Stamp**

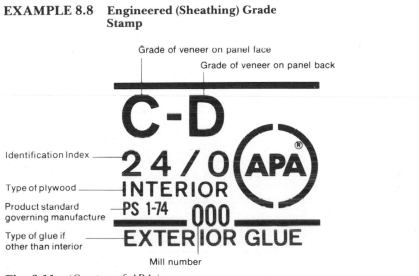

Fig. 8.11 *(Courtesy of APA.)*

Engineered grades of plywood usually have C or C-plugged face veneers and C or D backs and inner plies. *Exterior type* requires exterior glue and a minimum of C-grade veneers throughout. *Interior type* is generally available with exterior glue and inner plies are usually D-grade veneers.

Panel identification index (PII) indicates allowable roof span (inches)/allowable floor span (inches). In this stamp the panel can span a maximum of 24 in. when used for roof sheathing and is not recommended for floor sheathing (allowable span = 0 in.).

The sheathing grades of plywood can be "modified" by adding the STRUCTURAL I designation to the grade. Higher structural properties result.

Interior Type

Grade Designation	Description and Most Common Uses	Typical [1] Grade-trademarks	Veneer Grade — Face	Inner Plies	Back	Most Common Thicknesses (inch) — 1/4	5/16	3/8	1/2	5/8 (19/32)	3/4 (23/32)
C-D INT-APA	For wall and roof sheathing, subflooring, industrial uses such as pallets. Most commonly available with exterior glue (CDX). Specify exterior glue where construction delays are anticipated and for treated-wood foundations. (7)	C-D 32/16 INTERIOR PS 1-74 000 (APA) / C-D 24/0 INTERIOR PS 1-74 000 EXTERIOR GLUE (APA)	C	D	D		5/16	3/8	1/2	5/8	3/4
STRUCTURAL I C-D INT-APA and STRUCTURAL II C-D INT-APA	Unsanded structural grades where plywood strength properties are of maximum importance: structural diaphragms, box beams, gusset plates, stressed-skin panels, containers, pallet bins. Made only with exterior glue. See (6) for species group requirements. Structural I more commonly available. (7)	STRUCTURAL I C-D 24/0 INTERIOR PS 1-74 000 EXTERIOR GLUE (APA)	C(3)	D(3)	D(3)		5/16	3/8	1/2	5/8	3/4
STURD-I-FLOOR INT-APA	For combination subfloor-underlayment. Provides smooth surface for application of resilient floor covering. Possesses high concentrated- and impact-load resistance during construction and occupancy. Manufactured with exterior glue only. Touch-sanded. Available square edge or tongue-and-groove. (7)	STURD-I-FLOOR 24oc T&G INTERIOR PS 1-74 000 (APA)	C Plugged	(4)	D					19/32	23/32
STURD-I-FLOOR 48 O.C. (2-4-1) INT-APA	For combination subfloor-underlayment on 32- and 48-inch spans. Provides smooth surface for application of resilient floor coverings. Possesses high concentrated- and impact-load resistance during construction and occupancy. Manufactured with exterior glue only. Unsanded or touch-sanded. Available square edge or tongue-and-groove. (7)	STURD-I-FLOOR 48oc T&G 24 t T&G 1 1/8 INCH INTERIOR 000 EXTERIOR GLUE (APA)	C Plugged	C(5) & D	D	*1-1/8* (spanning thickness columns)					
UNDERLAY-MENT INT-APA	For application over structural subfloor. Provides smooth surface for application of resilient floor coverings. Touch-sanded. Also available with exterior glue. (2)(6)	UNDERLAYMENT GROUP 1 INTERIOR PS 1-74 000 (APA)	C Plugged	C(5) & D	D	1/4		3/8	1/2	19/32 5/8	23/32 3/4
C-D PLUGGED INT-APA	For built-ins, wall and ceiling tile backing, cable reels, walkways, separator boards. Not a substitute for Underlayment or Sturd-I-Floor as it lacks their indentation resistance. Touch-sanded. Also made with exterior glue. (2)(6)	C-D PLUGGED GROUP 2 INTERIOR PS 1-74 000 (APA)	C Plugged	D	D	1/4		3/8	1/2	19/32 5/8	23/32 3/4

250

Fig. 8.12 Guide to engineered grades of plywood. (Courtesy of APA.)

Type	Grade	Description	Typical grade-trademark	Face	Inner	Back	1/4	5/16	3/8	1/2	19/32	5/8	23/32	3/4
Exterior Type	C-C EXT-APA	Unsanded grade with waterproof bond for sub-flooring and roof decking, siding on service and farm buildings, crating, pallets, pallet bins, cable reels, treated-wood foundations. (7)	C-C 42/20 (APA) EXTERIOR PS 1-__ 000	C	C	C		5/16	3/8	1/2		5/8		3/4
	STRUCTURAL I C-C EXT-APA and STRUCTURAL II C-C EXT-APA	For engineered applications in construction and industry where full Exterior type panels are required. Unsanded. See (6) for species group requirements. (7)	STRUCTURAL I C-C 32/16 (APA) EXTERIOR PS 1-__ 000	C	C	C		5/16	3/8	1/2		5/8		3/4
	STURD-I-FLOOR EXT-APA	For combination subfloor-underlayment under resilient floor coverings where severe moisture conditions may be present, as in balcony decks. Possesses high concentrated-and impact-load resistance during construction and occupancy. Touch-sanded. Available square edge or tongue-and-groove. (7)	STURD-I-FLOOR 20OC 16 INCH (APA) EXTERIOR 000	Plugged	C	C(5)					19/32	5/8	23/32	3/4
	UNDERLAYMENT C-C PLUGGED EXT-APA	For application over structural subfloor. Provides smooth surface for application of resilient floor coverings where severe moisture conditions may be present. Touch-sanded. (2)(6)	UNDERLAYMENT GROUP 2 (APA) EXTERIOR PS 1-__ 000	Plugged	C	C(5)	1/4		3/8	1/2	19/32	5/8	23/32	3/4
	C-C PLUGGED EXT-APA	For use as tile backing where severe moisture conditions exist. For refrigerated or controlled atmosphere rooms, pallet fruit bins, tanks, box car and truck floors and linings, open soffits. Touch-sanded. (2)(6)	C-C PLUGGED GROUP 2 (APA) EXTERIOR PS 1-__ 000	Plugged	C	C	1/4		3/8	1/2	19/32	5/8	23/32	3/4
	B-B PLYFORM CLASS I & CLASS II EXT-APA	Concrete form grades with high reuse factor. Sanded both sides. Mill-oiled unless otherwise specified. Special restrictions on species. Available in HDO and Structural 1. Class 1 most commonly available. (8)	B-B PLYFORM CLASS 1 (APA) EXTERIOR PS 1-__ 000	B	C	B						5/8		3/4

(1) The species groups, Identification Indexes and Span Indexes shown in the typical grade-trademarks are examples only. See "Group," "Identification Index" and "Span Index" for explanations and availability.

(2) Can be manufactured in Group 1, 2, 3, 4, or 5.

(3) Special improved grade for structural panels.

(4) Special veneer grade to resist indentation from concentrated loads, or other solid wood base materials.

(5) Special construction to resist indentation from concentrated loads.

(6) Can also be manufactured in Structural I (all plies limited to Group 1 species) and Structural II (all plies limited to Group 1, 2, or 3 species).

(7) Specify by Identification Index for sheathing and Span Index for Sturd-I-Floor panels.

(8) Made only from certain wood species to conform to APA specifications.

251

normally applied to C-plugged faces. Plywood with B or better faces and backs is typically sanded.

Production of these different surface conditions changes the relative thicknesses of the layers in the make-up of a plywood panel. Thus, different cross-sectional properties apply for each of the three surface conditions. In addition, a number of the engineered grades can be modified with the STRUCTURAL I designation as described in Sec. 8.4. This grade modification affects allowable stresses and effective section properties for the plywood.

The grade-trademark used on engineered plywood grades can take several different forms. One type is similar to the grade mark used on appearance grades (Figure 8.10). With this type of grade-trademark an engineered grade will be recognized by the fact that a C or C-plugged veneer will be used for the face and C or D veneers for the back. Generally appearance grades will have a minimum of an A or B face veneer.

Another type of grade-trademark is used on the *sheathing* grades. See Example 8.8. The sheathing grade-trademarks include the veneer grade of the face and back and the plywood type (exterior or interior), as do the appearance grades. However, instead of giving the group number of the outer plies, the sheathing grade-trademark includes an *identification index* (Sec. 8.10). Other items in the grade stamp are similar to those in the appearance grade-trademark.

The typical uses and recommended applications of the engineered grades are summarized in the *Guide to Engineered Grades of Plywood*. See Fig. 8.12.

8.10   Panel Identification Index

The panel identification index (abbreviated PII in this text) is a set of two numbers which is included in the grade-trademark (Fig. 8.11) of the sheathing grades (i.e., unsanded C-C EXT and C-D INT).

The first number in the PII is the allowable span for the plywood in inches when used as roof sheathing with face grain across supports. The second number is the allowable span in inches when the plywood is used as subflooring. For example, an identification index of 48/24 indicates that the panel can be used to span 48 in. in a roof system and 24 in. as floor sheathing. The purpose of the identification index is to allow the selection of a proper plywood panel for sheathing applications without the need for structural calculations.

The use of the PII to *directly* determine the allowable span for a given panel, requires that the plywood be oriented in the strong direction (i.e., face grain parallel to the span). In addition, there are certain plywood *edge support* requirements which must be satisfied in order to apply the PII without a reduction in allowable load or allowable span.

The allowable uniform loads that apply when the identification index is used are, in most cases, conservative. As a general rule, the PII will provide an allowable load of 35 psf for roofs and 160 psf for floors. Specific allowable loads and spans for roofs, floors, and walls are given in Secs. 8.11 through 8.13. The edge support requirements for these applications are also covered in these sections. The basic Code reference for the PII is UBC Table 25R-1, and this table applies to both roof and floor applications. The fact to note at this time is that a given identification index may be found on panels of different thicknesses.

The PII theoretically accounts for the equivalent strength of panels fabricated from different species of wood. Thus, the same PII may be found on a thin panel that is fabricated from a strong species of wood *and* a thicker panel that is manufactured from a weaker species. However, practically speaking, plywood is usually constructed so that the thinner (or thinnest) of the panel thicknesses given in the table will be generally available in the market.

This fact is significant because of the dual function of plywood in many buildings. The *identification index* should be specified for sheathing loads *and* the *plywood thickness* should be specified for diaphragm load design. Thus both the PII and panel thickness are required in specifying a sheathing grade of plywood.

To summarize, panels can be manufactured with different thicknesses for a given PII. Generally speaking the thickness that *is available* is the smaller of those listed for a given Index in UBC Table 25R-1. A panel thickness should be specified that is compatible with the required PII.

8.11   Roof Sheathing

In standard sheathing applications plywood is assumed to be continuous over two or more spans. Plywood is normally used in the strong direction, but in panelized roof systems (Sec. 3.2), the plywood is often used in the weak direction. In this latter case, a thicker panel is required, but a savings in labor results in this type of construction. In addition, higher diaphragm shears can be carried with the increased plywood thickness.

The identification index system can be used as described in Sec. 8.10 to determine *directly* the panel requirements for plywood used in the strong direction. The minimum roof live load with this system is 35 psf, and in most cases tabulated allowable loads are considerably higher. See Fig. 8.13a. With this table the PII can be used *indirectly* to determine sheathing requirements for larger design loads. For example, if plywood is used to span 24 in. on a roof with a 90-psf snow load, the required PII of 32/16 can be read from the table. In this case the numbers in the PII do not agree with the span length.

The allowable loads and spans given in this table are accepted by the Code and appear in abbreviated form in UBC Table 25R-1. The UBC Table, however, includes the allowable total load as well as the live load. Where controlled by bending stress, allowable loads have been adjusted for the duration of snow loads.

Allowable Uniform Live Loads for Roof Decking

(C-D INT-APA, C-C EXT-APA, Structural I and II C-D INT-APA and Structural I and II C-C EXT-APA)

Identification Index	Plywood Thickness (in.)	Maximum Span (in.)	Unsupported Edge-Max. Length (in.)	Allowable Live Loads (psf) Spacing of Supports Center-to-Center (in.)									
				12	16	20	24	30	32	36	42	48	60
12/0	5/16	12	12	150									
16/0	5/16, 3/8	16	16	160	75								
20/0	5/16, 3/8	20	20	190	105	65							
24/0	3/8, 1/2	24	20, 24 (b)	250	140	95	50						
32/16	1/2, 5/8	32	28	385	215	150	95	50	40				
42/20	5/8, 3/4, 7/8	42	32		330	230	145	90	75	50	35		
48/24	3/4, 7/8	48	36			300	190	120	105	65	45	35	
48/24 (a)	3/4, 7/8	48	36				225	125	105	75	55	40	
2-4-1 (c)	1-1/8	72	48				390	245	215	135	100	75	45
1-1/8" Grp. 1 & 2	1-1/8	72	48				305	195	170	105	75	55	35
1-1/4" Grp. 3 & 4	1-1/4	72	48				355	225	195	125	90	65	40

(a) Loads apply only to C-C EXT-APA, Structural I C-D INT-APA, and Structural I C-C EXT-APA. Check availability of these grades before specifying.

(b) Maximum unsupported length 20 inches for 3/8 inch plywood, 24 inches for 1/2 inch plywood.

(c) 2-4-1 is synonymous with Sturd-I-Floor 48 o.c.

Fig. 8.13a Allowable roof LL on plywood sheathing. *(Courtesy of APA.)*

EXAMPLE 8.9   Plywood Roof Sheathing

Allowable loads given in Fig. 8.13a assume that the face grain is parallel to the span (perpendicular to the supports). Plywood is also assumed to be continuous over two or more spans. UBC Table 25R-1 is similar.

A roof dead load of 5 psf applies in addition to roof live load. Deflection criteria are L/180 under total load (TL) and L/240 under live load (LL) only. UBC Table 25R-1 gives higher roof dead loads where appropriate.

For blocking and edge support requirements see Example 8.10.

For 20-year *bonded built-up roofs* the plywood spans (spacing of supports) should be reduced to those shown in Figure 8.13b.

Spans for 20-Year Bonded Roofs

Allowable grades for all Indexes	Panel Identifi- cation Index	Plywood thickness (in.)	Spacing of supports (in.)	Edge support Plyclips (number as shown) (a)
C-D INT-APA, C-D INT-APA w/ext. glue, C-C EXT-APA, STR. I & II C-D INT APA, STR. I & II C-C EXT-APA	24/0	3/8, 1/2	16	1
	32/16	1/2, 5/8	24	1
	42/20	5/8, 3/4, 7/8	32	1
	48/24	3/4, 7/8	48	2

(a) May also be blocking or tongue-and-groove edges.

Fig. 8.13b   Plywood sheathing for bonded built-up roofs. *(Courtesy of APA.)*

The spans given in the allowable load table of Fig. 8.13a have proven satisfactory for the support of various roofing materials, including built-up roofs. However, many roofing companies require that the plywood spans be reduced in order to qualify for a 20-year bond on built-up roofs. See Fig. 8.13b. Built-up roofing is made up of alternate layers of roofing felt and mopped-on hot tar or asphalt. Often a layer of gravel is applied as a finish surface coating.

The allowable loads and maximum allowable spans in Fig. 8.13a and UBC Table 25R-1 are basically in agreement. However, the Code indicates that the maximum allowable spans can be used only if the plywood edges parallel to the span are supported by *blocking, tongue-and-groove edges,* or *panel clips.* See Example 8.10. Generally if this edge support is not provided, the Code requires that a reduced roof sheathing span be used. In some cases these reduced spans are less than those given in the column titled "Unsupported edge—maximum length" in Fig. 8.13a. The reduced spacing of supports is to limit the differential movement between adjacent plywood panels that do not have all edges supported.

EXAMPLE 8.10   **Roof Sheathing Edge Support Requirements**

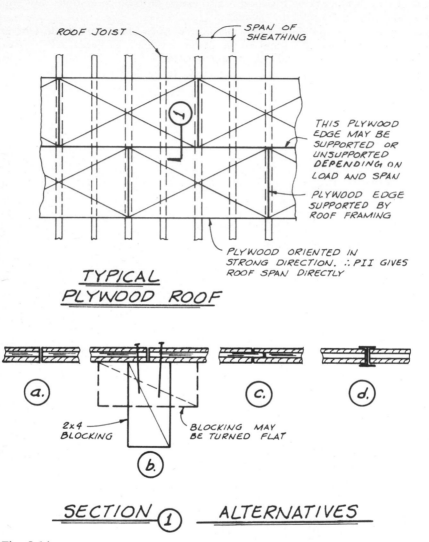

ROOF JOIST

SPAN OF SHEATHING

THIS PLYWOOD EDGE MAY BE SUPPORTED OR UNSUPPORTED DEPENDING ON LOAD AND SPAN

PLYWOOD EDGE SUPPORTED BY ROOF FRAMING

PLYWOOD ORIENTED IN STRONG DIRECTION. ∴ PII GIVES ROOF SPAN DIRECTLY

TYPICAL PLYWOOD ROOF

a.

2×4 BLOCKING

b.

BLOCKING MAY BE TURNED FLAT

c.

d.

SECTION ① ALTERNATIVES

Fig. 8.14

a. Unsupported edge—UBC Table 25R-1 requires closer roof joist spacing than given by the PII when all panel edges are not supported. UBC unsupported edge lengths are less than those given in the APA recommendations in Fig. 8.13a. Note that plywood thickness may be increased as an alternative to providing support to all edges (or reducing the roof beam spacing).

256

b. *Lumber blocking*—cut and fitted between roof joists.

c. *Tongue-and-groove (T&G) edges*

d. *Panel clips*—Metal H-shaped clips placed between plywood edges. For the number of panel clips refer to Fig. 8.13*b*.

Use of lumber blocking, T&G edges, and panel clips constitutes *edge support* for vertical loads. T&G edges and panel clips cannot be used in place of blocking *if* blocking is required for diaphragm action. There is one exception to this: 1⅛-in.-thick plywood with stapled T&G edges qualifies as a blocked diaphragm. Diaphragms are covered in Chap. 9.

A point about the support of plywood edges should be emphasized. The use of tongue-and-groove edges or panel clips is accepted as an alternative to lumber blocking for sheathing loads only. If blocking is required for diaphragm action (Chap. 9), panel clips or T&G edges (except 1⅛-in.-thick 2-4-1 plywood with stapled T&G edges) *cannot* be substituted for lumber blocking.

It was noted at the beginning of this section that plywood is often oriented in the weak direction in panelized roof systems. The panel identification index does not apply directly when plywood is used in this manner, and a different method of determining the required type of plywood must be used. For the common roof support spacing of 24 in. o.c., the plywood thickness can be determined from the UBC Table 25R-2. There is a more detailed table giving allowable loads for several different support spacings and types of plywood. See Fig. 8.15.

8.12 Design Problem: Roof Sheathing

Example 8.11 considers a common roof sheathing problem in which the roof supports are spaced 24 in. o.c. Part 1 of the example considers the plywood oriented in the strong direction and uses a sheathing grade of plywood.

Several alternate plywood sheathing systems are suggested in part 1(*f*). The edge support provisions that are given relate to vertical load requirements only. When the diaphragm shears (Chap. 9) are considered, *blocking* of plywood edges may be required. If blocking is required for lateral loads, the alternate edge supports (i.e., T&G edges, or panel clips) are not permitted.

In part 2 the plywood is oriented in the weak direction. In this case the number of plies used in the construction of the plywood panels is significant. If five-ply construction is used, the effective section properties in the weak direction are larger. In four-ply (three-layer) construction, the section properties are smaller, and either stronger wood (STRUCTURAL I) or a thicker panel is required.

257

Allowable Loads (psf) for Plywood Roof Sheathing with Face Grain Parallel to Supports

Grade	Thickness (in.)	No. of Plies (a)	Identification Index	Max. Span (in.)	Number and Length of Spans							
					Four @ 12"		Three @ 16"		Two @ 24"		One @ 48"	
					Live Ld.	Total Ld.	Live Ld.	Total Ld.	Live Ld.	Total Ld.	Live Ld.	Total Ld.
STRUCTURAL I	3/8	3	24/0	12	35	45	75	130	25	35		
	1/2	4	32/16	24	185	230	115	150	40	55		
	1/2	5	32/20	24	260	265	220	235	80	85		
	5/8	5&6	42/20	24			325	330	110	115		
	3/4	7	48/24	24							45	55
	1-1/8		—	48								
Other than STRUCTURAL I	1/2	3&4	24/0, 32/16	12	45	80	65	85	25	30		
	1/2	5	24/0, 32/16	24 (b)	160	180	80	110	30	40		
	5/8	5	32/16, 42/20	24	195	250	125	160	45	55		
	3/4	4	42/20, 48/24	24	280	285	130	175	50	65		
	3/4	5&6	42/20, 48/24	24			185	220	70	80		
	7/8	5&7	42/20, 48/24	24			345	350	120	125		
	1-1/8 (2-4-1)	7	—	48							20 (c)	30 (c)

(a) Number of layers equal to number of plies, except 4 ply is 3 layer and 6 ply is 5 layer.
(b) Solid blocking recommended at 24-inch span.
(c) 25 psf live and 35 psf total load with solid blocking at panel ends.

Fig. 8.15 Allowable roof loads for plywood turned in the weak direction. (*Courtesy of APA.*)

EXAMPLE 8.11 Roof Sheathing with 24-in. Span

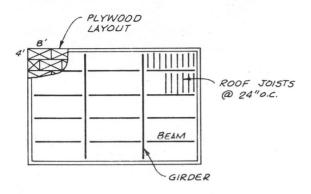

ROOF FRAMING PLAN

Fig. 8.16

Loads:

$$\text{Roof DL} = 8 \text{ psf}$$
$$\text{Roof LL} = 20 \text{ psf} \qquad \text{(no snow load)}$$
$$\text{TL} = 28 \text{ psf}$$

Part 1

For the roof layout shown, determine the plywood sheathing requirements using a sheathing grade.

a. Sheathing grades have the PII in the grade trademark. These grades are

    C-C EXT
    C-C EXT—STR I
    C-D INT*
    C-D INT—STR I

The UBC permits the use of interior-type plywood for roof sheathing protected from the weather by roofing materials. Exterior glue is required. STRUCTURAL I should be used when the added shear capacity is necessary for lateral loads.

b. From roof framing plan, plywood is spanning in the strong direction and the design loads are well within the allowable loads for the PII. Therefore PII can be used directly.

$$\text{Plywood span} = 24 \text{ in.}$$
$$\text{Req'd PII} = \text{roof span/floor span}$$
$$= 24/0$$

From UBC Table 25R-1:

$$\text{Allow. roof LL} = 50 \text{ psf} > 20$$
$$\text{Allow. roof TL} = 65 \text{ psf} > 28 \qquad OK$$

---

* Interior plywood is *generally available* with exterior glue. This type of plywood is abbreviated C-DX (X for exterior glue).

c. UBC Table 25R-1 indicates that 24/0 plywood may be either ⅜ in. or ½ in. thick. Normally, however, plywood is constructed so that the thinner thickness qualifies for a given PII.
Therefore, the *minimum* plywood requirement for this building is

$$⅜\text{-in. C-DX} \qquad \text{and} \qquad \text{PII} = 24/0$$

d. UBC Table 25R-1 also indicates that with unblocked edges the maximum allowable span for 24/0 plywood is 16 in. Therefore, *plywood edges* that are perpendicular to the roof joists *must be supported*. Edge supports may be provided by:
1. Blocking
2. T&G edges (available in plywood thicknesses ½ in. and larger)
3. H-shaped metal clips (panel clips)
See Fig. 8.14.

e. Although ⅜-in. plywood normally qualifies for a 24/0 identification index, ½-in. plywood is often used to span 24 in. for this type of building. From UBC Table 25R-1, ½-in. plywood normally qualifies for a PII = 32/16. If 32/16 plywood is used, the maximum unsupported edge length is 28 in. > 24 in. Therefore, edge support is not required for this alternative. If this building is to have a bonded roof, a PII of 32/16 is required *and* some form of edge support is necessary. See Fig. 8.13*b*.

f. Summary of minimum plywood sheathing requirements:
1. ⅜-in. C-DX (24/0)—edge support required.
2. ½-in. C-DX (32/16)—edge support not required.
3. ½-in. C-DX (32/16)—edge support required for a bonded roof.
Blocking may be required for lateral loads.

Part 2

In the above building, assume that the plywood panels are turned 90 degrees so that the face grain is perpendicular to the 24-in. span of the sheathing. What plywood can be used?

a. PII cannot be used directly because panels are oriented in the weak direction. UBC Table 25R-2 indicates that *any plywood panel* ½ in. thick with five-ply construction when used in this manner has

$$\text{Allow. roof LL} = 25 \text{ psf} > 20$$
$$\text{Allow. roof TL} = 30 \text{ psf} > 28 \qquad OK$$

All plywood edges must be supported.

b. Panels of four-ply (three-layer) construction require ½-in. STRUCTURAL I or a ⅝-in. thickness in other plywood grades.
(Figure 8.15 covers a variety of additional spans and loading conditions for plywood oriented in this manner.)

8.13 Floor Sheathing

When plywood is used for floors, it can serve several different functions. The terms used to refer to these different functions are

1. Subfloor

2. Underlayment

3. Combined subfloor-underlayment (or APA Sturd-I-Floor)

In some types of floor construction, the floor is made up of *two layers* of material plus a finish floor such as tile or carpeting. See Fig. 8.17a. The first or bottom layer is the basic structural sheathing material and is known as the *subfloor*. The sheathing grades of plywood (C-D INT and C-C EXT) are normally used for subflooring. These grades are also available in STRUC-TURAL I. If the floor is protected from the weather, interior-type plywood can be used.

EXAMPLE 8.12   Two-Layer Floor Construction

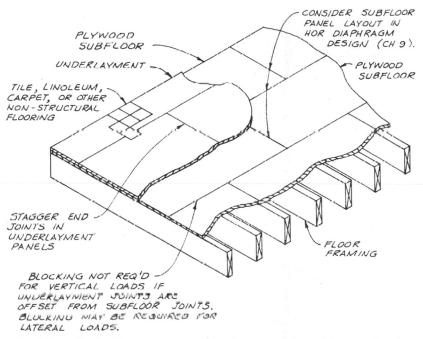

CONSIDER SUBFLOOR PANEL LAYOUT IN HOR DIAPHRAGM DESIGN (CH 9).

PLYWOOD SUBFLOOR

UNDERLAYMENT

PLYWOOD SUBFLOOR

TILE, LINOLEUM, CARPET, OR OTHER NON-STRUCTURAL FLOORING

STAGGER END JOINTS IN UNDERLAYMENT PANELS

FLOOR FRAMING

BLOCKING NOT REQ'D FOR VERTICAL LOADS IF UNDERLAYMENT JOINTS ARE OFFSET FROM SUBFLOOR JOINTS. BLOCKING MAY BE REQUIRED FOR LATERAL LOADS.

Fig. 8.17a   Typical floor construction using a separate subfloor and under-layment.

UNDERLAYMENT-grade plywood has C-plugged face veneer and special C-grade inner-ply construction to resist indentations. Typical underlayment thickness is ¼ in. for use over plywood subfloor and ⅜ in. for use over lumber subfloor.

When flooring has some structural capacity, the underlayment layer is not required. Wood strip flooring and lightweight concrete are examples of flooring which do not require the use of underlayment over the subfloor.

Plywood subfloor must be oriented in the strong direction and must be continuous over two or more spans. Plywood edges must be supported by blocking or T&G edges, or underlayment joints must be offset (see Fig. 8.17a). For heavy floor loads the plywood requirements can be obtained *indirectly* from the PII in Fig. 8.17b.

Plywood Recommended for Uniform Loads for Heavy Duty Floors (a)(b)

(Deflection limited to 1/240th of span)

Uniform Live Load (psf)	Center-to-Center Support Spacing (inches) (Nominal 2-Inch-Width Supports Unless Noted)					
	12 (c)	16 (c)	20 (c)	24 (c)	32	48 (d)
50	32/16	32/16	42/20	48/24	2-4-1	2-4-1
100	32/16	32/16	42/20	48/24	2-4-1	1½ (e)
150	32/16	32/16	42/20	48/24	2-4-1	1¾ (f) 2 (e)
200	32/16	42/20	42/20	2-4-1	1⅛ (f) 1⅜ (e)	2 (f) 2½ (e)
250	32/16	42/20	48/24	2-4-1	1⅜ (f) 1½ (e)	2¼ (f)
300	32/16	48/24	2-4-1	2-4-1	1½ (f) 1⅝ (e)	2¼ (f)
350	42/20	48/24	2-4-1	1⅛ (f) 1⅜ (e)	1½ (f) 2 (e)	
400	42/20	2-4-1	2-4-1	1¼ (f) 1⅜ (e)	1⅝ (f) 2 (e)	
450	42/20	2-4-1	2-4-1	1⅜ (f) 1½ (e)	2 (f) 2¼ (e)	
500	48/24	2-4-1	2-4-1	1½ (e)	2 (f) 2¼ (e)	

(a) Use plywood with T & G edges, or provide structural blocking at panel edges, or install a separate underlayment.
(b) 2-4-1 is synonymous with Sturd-I-Floor 48 o.c.
(c) A-C Group 1 sanded panels may be substituted for Identification Index panels (1/2-inch for 32/16; 5/8-inch for 42/20; 3/4-inch for 48/24).
(d) Nominal 4-inch-wide supports.
(e) Group 1 face and back, any species inner plies, sanded or unsanded, single layer.
(f) Structural I, sanded or unsanded, single layer.

Fig. 8.17b Plywood subfloor requirements for heavy loads. (*Courtesy of APA.*)

In most cases the plywood requirements for subflooring can be determined *directly* using the panel identification index. Recall that the second number in the PII represents the allowable floor span in inches. The allowable uniform floor load for plywood sheathing is 160 psf when the PII is used. A deflection limit of L/360 is used for floors. If heavier loads are encountered, the identification index can be used *indirectly* to determine the plywood requirements. See Fig. 8.17b. This table relates the load and span to the required PII. In this approach the allowable span for a given panel will probably be different from that given in the PII number.

In the type of floor construction that uses two layers of material, the top layer of plywood is known as the *underlayment.* Underlayment lies under the finish floor covering (on top of the subfloor) and is typically either ¼ in. or ⅜ in. in thickness. It is used to provide a solid surface for the direct application of nonstructural floor finishes.

UNDERLAYMENT-grade plywood (Fig. 8.12) is designed specifically for this application. Underlayment panels are touch-sanded to provide a reasonably smooth surface to support the nonstructural finish floor.

In another type of floor construction, thicker panels of the UNDERLAYMENT or APA Sturd-I-Floor grades are used in a *single-layer* floor construction. When a single layer of plywood serves as both the sheathing and the underlayment, it is known as *combination subfloor-underlayment.* In this case, plywood edges must be T&G or blocked with lumber. UBC Table 25S gives the required thickness of UNDERLAYMENT plywood required to span various floor joist spacings when used as combination subfloor-underlayment. This table also applies to C–C-plugged and all grades of sanded exterior-type plywood. (Note that C–D-plugged is not acceptable for this application.) The allowable uniform load associated with the use of this table is 125 psf.

The APA proprietary product known as Sturd-I-Floor has a grade-trademark that contains a span index (abbreviated SI in this text). The span index gives the allowable floor span in inches for a given panel. See Example 8.13. The allowable uniform load for this single-plywood-layer system is 140 psf.

The plywood that has the capacity to span 48 in. in a floor system deserves some additional explanation. This plywood is 1⅛ in. thick and has been available for some time under the name 2–4–1. Actually, if this panel is used to span 48 in., the supporting beams should be 4-in.-nominal or wider members, and the allowable uniform load is 70 psf. If 2-in.-nominal framing members are used, the recommended plywood span is reduced to 32 in. but the allowable load increases. With these large spans it is likely that 4-in.-nominal framing would be used for the floor supports.

It was noted earlier that T&G edges for most plywood construction provide an alternative to lumber blocking for vertical loads, but they do not qualify for blocking in diaphragms. The one exception to this general rule is the 1⅛-in.-thick 2–4–1 plywood. If the T&G edges on interlocking 2–4–1 panels are properly stapled, blocked diaphragm values can be used for lateral loads (Chap. 9).

263

EXAMPLE 8.13   APA Sturd-I-Floor System

Panel grade — STURD-I-FLOOR®
Span Index — 24oc
Tongue-and-groove — T&G
Thickness — 23/32 INCH
Type of plywood — INTERIOR
Mill number — 000
Type of glue if other than interior — EXTERIOR GLUE
National Research Board report number — NRB-108

STURD-I-FLOOR®
48oc T&G
2·4·1®
1-1/8 INCH
INTERIOR
000
EXTERIOR GLUE
NRB-108

Fig. 8.18a   Typical Grade-trademarks. *(Courtesy of APA.)*

These panels can be used in single-layer floor construction to span the distance given in the span index in the grade-trademark.

Panel thicknesses and installation requirements are given in Fig. 8.18b. Installation may be by nails only or by a combination of field gluing and nailing. The latter method is known as the *APA Glued Floor System.* It provides increased stiffness and reduces squeaking in floors caused by nail popping. Additional information is given in Ref. 14.7.

8.14   Design Problem: Floor Sheathing

In this example a typical floor sheathing problem for an office building is considered. See Example 8.14. Part 1 utilizes a two-layer floor system with a separate subfloor and underlayment. The subfloor was chosen from the sheathing grades, and a ¼-in. UNDERLAYMENT panel is used over the subfloor. If the joints of the underlayment are staggered with respect to the joints in the subfloor, no special edge support is required for the subfloor panels.

Part 2 of the example considers a single-layer-plywood floor construction. Several different plywood grades can be used for this application. The example first uses the Sturd-I-Floor system. The allowable single-layer floor span is included as the span index in the grade-trademark. With this system the plywood edges must be supported by blocking or T&G edges. As an alternative to these methods of edge support, a second layer of plywood UNDERLAY-MENT can be added similar to that in part 1. This may be desirable to ensure a smooth surface for certain finish floor coverings. Blocking may be required for diaphragm action.

APA Sturd-I-Floor (a)

Span Index (Maximum Joist Spacing) (in.)	Panel Thickness (b) (in.)	Fastening: Glue-Nailed (c)			Fastening: Nailed-Only		
		Nail Size and Type	Spacing (in.)		Nail Size and Type	Spacing (in.)	
			Panel Edges	Intermediate		Panel Edges	Intermediate
16	19/32, 5/8	6d deformed-shank (d)	12	12	6d deformed-shank	6	10
20	19/32, 5/8 23/32, 3/4	6d deformed-shank (d)	12	12	6d deformed-shank	6	10
24	23/32, 3/4	6d deformed-shank (d)	12	12	6d deformed-shank	6	10
	7/8	8d deformed-shank (d)	12	12	8d deformed-shank	6	10
48 (2-4-1)	1-1/8	8d deformed-shank (e)	6	(f)	8d deformed-shank (e)	6	(f)

(a) Special conditions may impose heavy traffic and concentrated loads that require construction in excess of the minimums shown. See page 18 or heavy duty floor recommendations.

(b) As indicated above, panels in a given thickness may be manufactured in more than one Span Index. Panels with a Span Index greater than the actual joist spacing may be substituted for panels of the same thickness with a Span Index matching the actual joist spacing. For example, 19/32-inch-thick Sturd-I-Floor 20 o.c. may be substituted for 19/32-inch-thick Sturd-I-Floor 16 o.c. over joists 16 inches on center.

(c) Use only adhesives conforming to APA Specification AFG-01 applied in accordance with the manufacturer's recommendations.

(d) 8d common nails may be substituted if deformed-shank nails are not available.

(e) 10d common nails may be substituted with 1-1/8-inch panels if supports are well seasoned.

(f) Space nails 6 inches for 48-inch spans and 10 inches for 32-inch spans.

Fig. 8.18b Panel thickness and installation requirements. *(Courtesy of APA.)*

EXAMPLE 8.14   Floor Sheathing with 24-in. Span

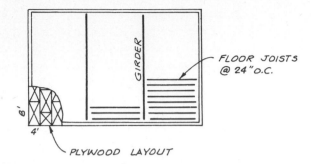

PLYWOOD LAYOUT

2ᴺᴰ FLOOR FRAMING PLAN

Fig. 8.19

Loads:

$$Floor\ DL = 12\ psf$$
$$Partition\ DL = 20\ psf$$
$$Floor\ LL = \underline{50\ psf}$$
$$TL = 82\ psf$$

Part 1

For the floor layout, determine the plywood requirements for vertical loads assuming a separate subfloor and underlayment construction. A thin resilient-tile finish floor will be used.

a. Sheathing grades of plywood normally provide the greatest economy for subfloor applications. These grades are

      C-C EXT
      C-C EXT—STR I
      C-D INT*
      C-D INT—STR I

Interior plywood is acceptable for the protected floor sheathing application above. STRUCTURAL I should be used when the added shear capacity is necessary for lateral loads.

b. From the floor framing plan, the plywood is oriented in the strong direction. Therefore the PII applies.

$$Plywood\ span = 24\ in.$$
$$Req'd\ PII = roof\ span/floor\ span$$
$$= 48/24$$

From UBC Table 25R-1,

$$Allow.\ floor\ TL = 165\ psf > 82 \quad OK$$

---

* Interior plywood is generally available with exterior glue. This type of plywood is abbreviated C-DX (X for exterior glue).

c. UBC Table 25R-1 indicates that 48/24 plywood may be either ¾ in. or ⅞ in. thick. Normally plywood is constructed so that the thinner thickness is generally available. Therefore, the minimum plywood requirement for subflooring in this problem is

$$\text{¾-in. C-D} \qquad \text{and} \qquad \text{PII} = 48/24$$

d. UBC Table 25R-1 indicates that plywood edges not supported directly by the floor framing must have some other type of vertical support. This support can be provided by one of the following:
1. Blocking
2. T&G edges
Alternatively a minimum of ¼-in. UNDERLAYMENT-grade plywood may be installed with panel edges offset from the subfloor panel joints (Fig. 8.17a). A layer of UNDERLAYMENT plywood is required to provide an acceptable surface for the installation of the finish floor tile.

e. Summary of plywood floor sheathing requirements:
1. Subfloor—¾-in. C-D with 48/24 PII
2. Underlayment—¼-in. UNDERLAYMENT INT with panel edges offset from subfloor panel joints below. Blocking is not required for vertical loads, but it may be required for diaphragm loading.
Although interior-type plywood is generally available with exterior glue, exterior glue should be specified for plywood floors where moisture may be present (such as bathrooms and utility rooms).

Part 2

For the same floor layout and loads, select a single-layer-plywood floor sheathing.
a. Plywood is used in the strong direction and the span index (SI) in the Sturd-I-Floor system can be used directly.

$$\text{Req'd SI} = 24$$

Sturd-I-Floor panels are available in either exterior or interior types. Interior panels all have exterior glue. Figure 8.18b indicates that this span index may be found on panels of several different thicknesses.

b. All panel edges not supported directly by floor framing must be supported by one of the following:
1. Blocking
2. T&G edges
Although Sturd-I-Floor panels are suitable for the direct application of nonstructural finish flooring, an additional layer of ¼-in. UNDERLAYMENT plywood may be desirable in some areas. Use of a second layer of plywood ensures a smooth, flush surface for tile or linoleum. If the panel edges are staggered as described in part 1, additional edge support is not required.

c. As an alternative to Sturd-I-Floor panels, a single-layer combination subfloor-underlayment can be obtained from UBC Table 25S. Acceptable grades are given in footnote 1 in the table.

8.15 Plywood for Wall Sheathing and Siding

Plywood can be used in wall construction in *two basic ways*. In *one method* the plywood serves a structural purpose only. It is attached directly to the framing **267** and serves as sheathing to distribute the normal wind load to the studs, and it may also function as the basic shear-resisting element if the wall is a shearwall. See Example 8.15. Finished siding of wood or other material is then attached to the outside of the wall.

Typical sheathing grades (C-D interior and C-C exterior) are used when finished siding will cover the plywood. In wall construction the face grain of the plywood can either be parallel or perpendicular to the studs (supports). For shearwall action all edges of the plywood must be supported. This is provided by studs in one direction and wall plates or blocking between the studs in the other direction.

Finish siding materials can either be attached directly to the wall by nailing through the plywood into the wall framing, or it may be attached by nailing directly to the plywood sheathing. The maximum allowed spacing of studs is related to the panel thickness and the PII by UBC Table 25N. It was noted earlier that for a given identification index the thinner plywood panel is generally available.

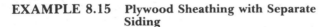

EXAMPLE 8.15   Plywood Sheathing with Separate Siding

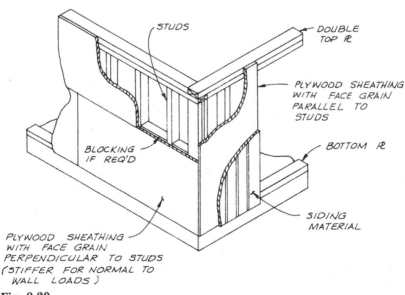

Fig. 8.20

In this system the plywood sheathing is the basic structural wall element. Wind loads normal to the wall are carried by the sheathing to the wall studs. UBC Table 25N gives the plywood sheathing requirements. Certain siding materials offer little structural resistance, and in some cases the minimum plywood requirements are increased if the face grain is not perpendicular to the studs.

If the plywood sheathing also functions as a shearwall (lateral loads parallel to wall), panel edges not supported by wall framing must be blocked and nailed (Chap. 10). *Minimum plywood nailing* is 6d common or galvanized box nails at 6 in. o.c. at supported edges and 12 in. o.c. along intermediate supports (studs). Heavier nailing may be required for shearwall action.

The *second main method* of using plywood in wall construction is to use a single layer as a combined sheathing-siding system. See Example 8.16. If the plywood is to serve as the siding as well as the sheathing, an appearance grade of plywood will probably be required.

A type of plywood often used for this application is known as 303 siding. This is an exterior-type plywood and is available with a variety of textured surface finishes. A special 303 panel is known as Texture 1–11 and is manufactured in $^{19}\!/_{32}$-in. and $\frac{5}{8}$-in. thicknesses only. It has shiplap edges to aid in waterproofing and to maintain surface pattern continuity. T 1–11 panels also have $\frac{3}{8}$-in.-wide grooves cut into the finished side for decorative purposes. A net panel thickness of $\frac{3}{8}$ in. is maintained at the groove. The required thickness of the plywood for use as combined sheathing and siding is given in UBC Table 25M. APA 303 siding includes a span index in the grade-trademark. Additional information on 303 siding is available from APA.

EXAMPLE 8.16   **Plywood Combined Sheathing/ Siding**

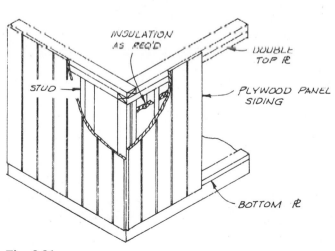

Fig. 8.21*a*

DESIGN OF WOOD STRUCTURES

In the combined sheathing-siding system, the plywood usually has a textured surface finish. These finishes include rough sawn, brushed, and a smooth finish for painting (medium density overlay—MDO). In addition to different surface textures, most siding panels are available with grooving for added decorative effect.

APA 303 siding is a plywood grade typically used for this application. UBC Table 25M gives the thickness requirements, and 303 siding grade-trademarks include the allowable stud spacing (e.g., 16 in. o.c.).

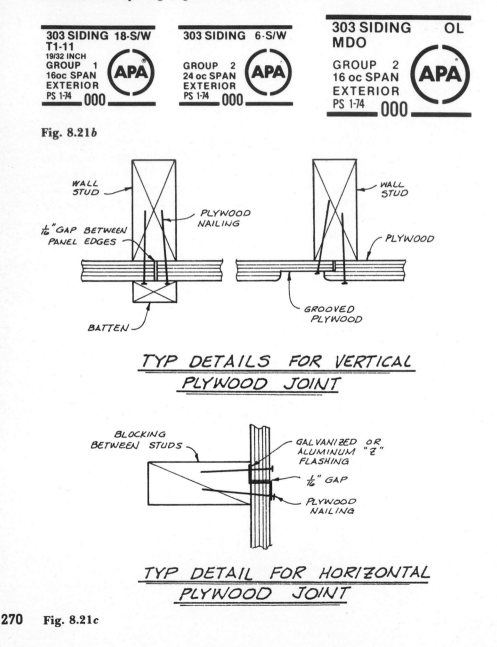

Fig. 8.21b

TYP DETAILS FOR VERTICAL PLYWOOD JOINT

TYP DETAIL FOR HORIZONTAL PLYWOOD JOINT

270   **Fig. 8.21c**

Plywood panel siding is usually installed with the 8-ft panel dimension running vertically. However, these panels can also be installed horizontally. Various panel joint details can be used for protection against the weather.

Nailing requirements are similar to those for plywood sheathing (Example 8.15). However, galvanized nails are normally used to reduce staining. Casing nails may be used where the presence of a common or box nail head is objectionable. Additional information on plywood nailing for walls is included in Chap. 10. For panels greater than ½ in. thick use a minimum of 8d nails.

Heavier nailing may be required for shearwall action. For shearwall design the thickness to be used is the *thickness where nailing occurs*. If grooves are nailed (see sketch) the net thickness at the groove is used.

8.16 Beam Design Calculations for Plywood

The design aids which allow the required thickness of plywood sheathing to be determined without detailed design calculations have been described in the previous sections. For most practical sheathing problems, these methods are adequate to determine the required grade and thickness of plywood.

If it becomes necessary to perform structural calculations for plywood, the designer must become familiar with a number of factors which interrelate somewhat to define plywood structurally. Calculations involving plywood are more often required in the design of plywood components rather than for sheathing.

The basic document which gives effective cross-sectional properties and allowable stresses for plywood is the *Plywood Design Specification* (PDS—Ref. 14.1). A designer who must perform structural calculations for plywood should obtain a copy of the PDS. A thorough study of it will provide a working knowledge of the basic design properties and allowable stresses for plywood.

It will be helpful, however, to review here several factors which are unique to plywood structural calculations. These will provide a useful background to the designer even if structural calculations are not required. Although some of these points have been introduced previously, they will be summarized here.

Cross-Sectional Properties. The cross-sectional properties for plywood are tabulated in the PDS. The variables which affect the cross-sectional properties are

1. Direction of stress
2. Surface condition
3. Species makeup

The *direction of stress* relates to the two-directional behavior of plywood because of its cross-laminated construction. Because of this type of construction, two sets of cross-sectional properties are tabulated. One applies when

271

plywood is stressed parallel to the face grain, and the other applies when it is stressed perpendicular to the face grain.

The *basic surface* conditions for plywood are

1. Sanded
2. Touch-sanded
3. Unsanded

It will be recalled that the relative thickness of the plies is different for these different surface conditions. Thus different section properties apply for these three surface conditions.

Finally, the *species makeup* of a panel also affects its cross-sectional properties. A special table of section properties applies to STRUCTURAL I (and II) and marine grades. Plywood panels made up of all other combinations of species are assigned a different set of cross-sectional properties.

The product standard (PS 1) allows some variation in the veneer and layer thicknesses used in the makeup of a plywood panel. Therefore, the tabulated cross-sectional properties are based on veneer thickness combinations which produce minimum properties. Thus for a given panel the properties for the weak direction and for the strong direction are not complementary.

Stress Calculations. Although plywood can be made up of a variety of wood species having different strengths and different values for modulus of elasticity, stress calculations are carried out on the assumption of uniform stress properties. This is made possible by use of the effective cross-sectional properties and the appropriate allowable stresses given in the PDS.

The use of effective cross-sectional properties differs somewhat from ordinary beam design calculations. For example, for sheathing-type loads, the

---

EXAMPLE 8.17   Plywood Beam Loadings and Section Properties

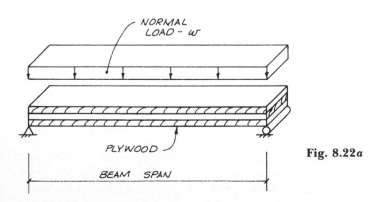

Fig. 8.22*a*

Plywood under Normal (Sheathing) Loads

For loads normal to the surface of the plywood the effective section modulus KS is used for *bending* stress calculations and the moment of inertia I is used for *deflection* calculations. For shear requirements see Fig. 8.23.

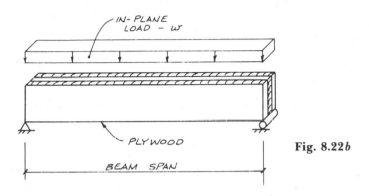

Fig. 8.22*b*

Plywood under In-Plane Loads

Procedures for calculating cross-sectional properties for plywood loaded in its plane are given in the *Plywood Design Specification (PDS) Supplement 2.* Plywood used in this manner is typically found in fabricated beams using lumber flanges and plywood webs.

FABRICATED BEAM

Fig. 8.22*c*

---

bending stress in the plywood is to be calculated using the effective section modulus KS and not I/c (the normal definition of section modulus). See Figure 8.22*a* (Example 8.17). The tabulated value for I is calculated by normalizing all ply areas relative to the face plies (transformed area technique), and I is to be used in stiffness (deflection) calculations only. Values for effective section modulus are then further adjusted based on the number and direction of plies using experimentally determined factors.

Another factor that is unique to plywood structural calculations is that there are two different allowable shear stresses. The different allowable shear stresses are a result of the cross laminations. The type and direction of the

273

EXAMPLE 8.18 Types of Shear in Plywood

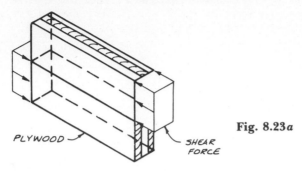

Fig. 8.23*a*

Shear through the Thickness

Shear through the thickness can occur from the type of loading shown in Fig. 8.23*a* (as in diaphragm and shearwall action) or from flexural shear (horizontal shear) caused by the type of loading shown in Fig. 8.22*b*. The stress calculated for these types of loading conditions is based on the *effective thickness for shear*. See PDS.

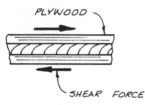

Fig. 8.23*b*

Rolling Shear

Rolling shear occurs in the ply (or plies) that are at right angles to the applied stress. This type of stress develops when plywood is loaded as shown in Fig. 8.22*a*. The stress is shear due to bending (horizontal shear) and is calculated from the rolling shear constant *Ib/Q*.

The allowable stress for rolling shear is considerably less than the allowable shear through the thickness. A test for rolling-shear strength is shown in Fig. 8.23*c*.

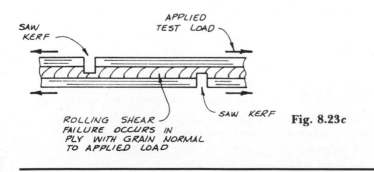

Fig. 8.23*c*

loading determines the type of shear involved, and the appropriate allowable shear must be used in checking the stress.

The PDS refers to these shear stresses as:

1. Shear in a plane perpendicular to the plies, or *shear through the thickness* of the plywood.

2. Shear in the plane of the plies, or *rolling shear.*

The first type of shear occurs when the load is in the plane of the plywood, as in a diaphragm. See Fig. 8.23a (Example 8.18). This same type of stress occurs in fabricated beams using plywood webs. In the latter case, shear through the thickness of the plywood is the result of flexural (horizontal) shear. The design procedures for fabricated plywood beams is covered in PDS Supplement 2 (Ref. 14.2).

Rolling shear can also be visualized as the horizontal shear in a beam, but in this case the loads are normal to the surface of plywood (as with sheathing loads). See Fig. 8.23b. The shear is seen to be "in the plane of the plies" rather than "through the thickness." With this type of loading the wood fibers that are at right angles to the direction of the stress tend to slide or *roll* past one another. Hence the name "rolling shear." If the stress is parallel to the face plies, the fibers in the inner crossband(s) are subjected to rolling shear.

The PDS provides different cross-sectional properties for the calculation of shear through the thickness and rolling shear. For shear through the thickness the "effective thickness for shear" is used, and for rolling shear the "rolling shear constant" Ib/Q is used. In addition to different cross-sectional properties, there are different allowable stresses for the two types of shear stress. The allowable rolling shear stress is considerably less than the allowable shear through the thickness.

For a further study of structural calculations for plywood the designer is referred to the PDS and its supplements.

8.17   Problems

8.1   What two major types of loading are considered in designing a plywood roof or floor system that also functions as part of the LFRS?

8.2   Regarding the fabrication of plywood panels, distinguish between (1) veneer, (2) plies, and (3) layers.

8.3   In the cross section of a plywood panel shown in Fig. 8.A, label the names used to describe the five plies.

8.4   Plywood panels (4 ft × 8 ft) of five-ply construction (similar to that shown in Fig. 8.A) are used to span between roof joists that are spaced 16 in. o.c.

   Find: *a.* Sketch a plan view of the framing and the plywood showing the plywood oriented in the strong direction.

275

        b. Sketch a 1-ft-wide cross section of the sheathing. Shade the plies that are effective in supporting the "sheathing" loads.

8.5 Problem 8.5 is the same as Prob. 8.4 except that the plywood is oriented in the weak direction.

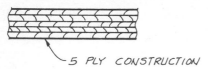

5 PLY CONSTRUCTION

Fig. 8.A

8.6 How are the species of wood used in the fabrication of plywood classified?

8.7 The plywood in Fig. 8.A has a grade stamp that indicates *group 2.*

    Find: *a.* What plies in the panel contain group 2 species?
            b. If all plies are not of group 2 species, what are the others?

8.8 What is the meaning of the term STRUCTURAL I? How is it used?

8.9 What are the veneer grades? Which veneers are the more similar in appearance and surface qualities? Which veneers are structurally more similar? State the reasons for the above similarities.

8.10 What is the most common type of glue used in the fabrication of plywood (interior, intermediate, or exterior)?

8.11 What is the difference between the construction of interior and exterior plywood?

8.12 *a.* What veneer grades are normally used for the face and back of appearance grades of plywood?
       b. What surface treatment (sanded, touch-sanded, or unsanded) is given to an appearance grade of plywood?

8.13 What veneer grades are normally used for the face and back of engineered grades of plywood?

8.14 What are the sheathing grades of plywood?

8.15 The spacing of rafters in a roof is 48 in. o.c. Roof DL = 5 psf. Snow load = 30 psf.

    Find: The required PII, thickness, grade, and edge support for plywood roof sheathing. Plywood is oriented in the strong direction. Use UBC criteria.

8.16 The spacing of joists in a roof is 24 in. o.c. Roof DL = 5 psf. Snow load = 100 psf.

    Find: The required PII, thickness, grade, and edge support for plywood sheathing. Plywood is oriented in the strong direction. Use APA recommendations.

8.17 The spacing of joists in a floor system is 16 in. o.c. and the design floor LL is 200 psf. DL is minimal.

Find: The required PII, thickness, grade, and edge support for plywood sub-flooring.

8.18 Problem 8-18 is the same as Prob. 8-17 except that the joist spacing is 24 in. o.c.

8.19 Describe the construction of UNDERLAYMENT-grade plywood. How is it used in floor construction?

8.20 Single-layer-plywood floor sheathing (glued and nailed) is to be used to support a floor DL of 10 psf and a floor LL of 75 psf. Floor joists are 24 in. o.c.

Find: a. The required APA plywood grade and span index (SI) and edge support requirements.
b. Describe the glued and nailed floor system. Why it is used?

8.21 Describe the following types of plywood used on walls (include typical grades):
a. Plywood sheathing
b. Plywood siding
c. Combined sheathing/siding

8.22 Grooved plywood (Texture 1–11) is used for combined sheathing/siding on a shearwall. Nailing to the studs is similar to that in Fig. 8.21c, the detail for a vertical plywood joint.

Find: What thickness plywood is to be used in the shearwall design calculations?

8.23 Regarding the calculation of stresses in plywood,
a. What is the effective section modulus KS and when is it used?
b. What is shear through the thickness?
c. What is rolling shear?

CHAPTER NINE

Horizontal Diaphragms

9.1 Introduction

The lateral loads that act on *box*-type buildings were described in Chap. 2, and the distribution of these forces was covered in Chap. 3. In the typical case the lateral loads were seen to be carried by the wall framing to the horizontal diaphragms at the top and bottom of the wall sections. A horizontal diaphragm acts as a beam in the plane of a roof or floor that spans between shearwalls. See Fig. 9.1.

The examples in Chap. 3 were basically load calculation and load distribution problems. In addition, the calculation of the unit shear in a diaphragm was illustrated. Although the unit shear is a major factor in a diaphragm design, there are a number of additional items that must be considered.

The basic design considerations for a horizontal diaphragm are

1. Sheathing thickness
2. Diaphragm nailing
3. Chord design
4. Strut design
5. Diaphragm deflection
6. Tie and anchorage requirements

The first item is often governed by loads normal to the surface of the sheathing (i.e., by sheathing loads). This subject is covered in Sec. 6.16 for lumber diaphragms and in Chap. 8 for plywood diaphragms. The nailing requirements, on the other hand, are a function of the unit shear. The sheathing thickness and nailing requirements may, however, both be governed by the unit shears *when the shears are large.*

In Chap. 9 the general behavior of a horizontal diaphragm is described and the functions of the various items mentioned above are explained. This

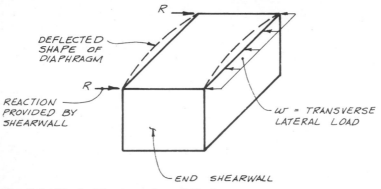

Fig. 9.1 Typical horizontal roof diaphragm.

is followed by detailed design considerations for the individual elements. Tie and anchorage requirements are touched upon in Chap. 9, but these are treated more systematically in Chap. 14. Shearwalls (vertical diaphragms) are covered in Chap. 10.

9.2 Basic Horizontal Diaphragm Action

A horizontal diaphragm can be defined as a large, thin structural element that is loaded in its plane. It is an assemblage of elements which typically includes:

1. Roof or floor sheathing
2. Framing members supporting the sheathing
3. Boundary or perimeter members

When properly designed and connected together, this assemblage will function as a horizontal beam that spans between the vertical resisting elements in the lateral-force-resisting system (LFRS).

The diaphragm must be considered for lateral loads in both the transverse and longitudinal directions. See Example 9.1. Like all beams, a horizontal diaphragm must be designed to resist both shear and bending.

In general, a horizontal diaphragm can be thought of as being made up of a shear-resisting element (the roof or floor sheathing) and some boundary members. There are two types of boundary members in a horizontal diaphragm (chords and struts), and the direction of the applied load determines

EXAMPLE 9.1   **Horizontal Diaphragm Loads and
Boundary Members**

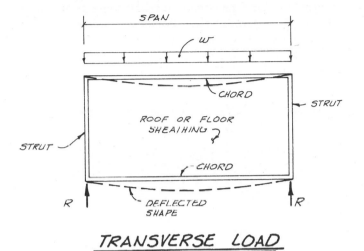

TRANSVERSE LOAD

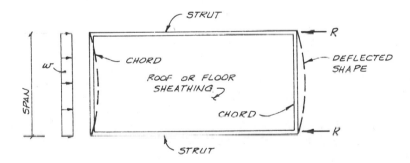

LONGITUDINAL LOAD

Fig. 9.2

Diaphragm *boundary members* change functions depending on the direction of the lateral load.

Chords are boundary members that are perpendicular to the applied load.
Struts are boundary members that are parallel to the applied load.

the function of the members (Fig. 9.2). The *chords* are designed to *carry the moment* in the diaphragm.

An analogy is often drawn between a horizontal diaphragm and a steel wide-flange (W-shape) beam. In a steel beam the flanges resist most of the moment and the web essentially carries the shear. In a horizontal diaphragm the sheathing corresponds to the web and the chords are assumed to be the flanges. The chords are designed to carry axial forces created by the moment. These forces are obtained by resolving the internal moment into a couple (tension and compression forces). See Fig. 9.3. The shear is assumed

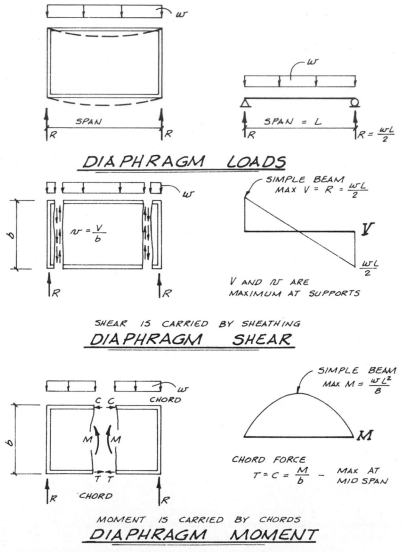

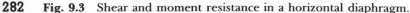

282   Fig. 9.3 Shear and moment resistance in a horizontal diaphragm.

EXAMPLE 9.2   **Function of Drag Strut**

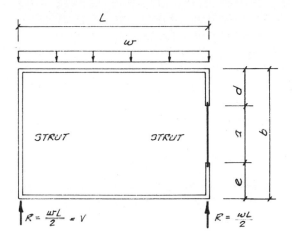

STRUT   STRUT

$$R = \frac{wL}{2} = V$$

$$R = \frac{wL}{2}$$

PLAN

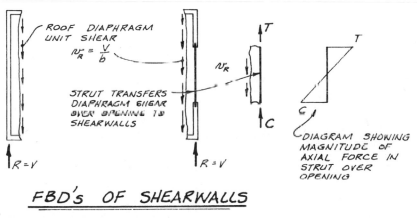

ROOF DIAPHRAGM
UNIT SHEAR
$$n_R = \frac{V}{b}$$

STRUT TRANSFERS
DIAPHRAGM SHEAR
OVER OPENING TO
SHEARWALLS

n_R

$R = V$

$R = V$

T

T

C

C

DIAGRAM SHOWING
MAGNITUDE OF
AXIAL FORCE IN
STRUT OVER
OPENING

FBD's OF SHEARWALLS

Fig. 9.4

The left wall has no openings and roof shear is transferred directly to the wall. The right wall has a large opening, and the strut transfers the roof diaphragm shear over the opening to the shearwalls. The strut here is assumed to function in both tension and compression. When shearwall lengths d and e are equal, the strut forces T and C are equal. Strut forces for other ratios of d and e are covered in Sec. 9.6.

to be carried entirely by the sheathing material. Proper nailing of the sheathing to the framing members is essential for this resistance to develop.

The *struts* are designed to transmit the horizontal diaphragm *loads to the shearwalls*. This becomes a design consideration only when the supporting shearwalls are shorter in length than the horizontal diaphragm. See Example 9.2. Essentially, the unsupported horizontal diaphragm unit shear over an opening must be transmitted to the shearwalls by the strut. The strut load is also known as a *drag force* or *collector force* because the strut collects or drags the diaphragm shear into the shearwall. When the supporting shearwalls are of wood-frame construction, the unit shear in the shearwalls is assumed to be uniform throughout the length of the wall.

9.3 Shear Resistance

The shear resisting element in the horizontal diaphragm assemblage is the roof or floor sheathing. This can be either lumber sheathing or plywood. The majority of wood horizontal diaphragms use plywood sheathing because of the economy of installation and the relatively high allowable unit shears it provides. For this reason, Chap. 9 deals primarily with *plywood diaphragms*. Lumber-sheathed diaphragms are covered in Ref. 4.

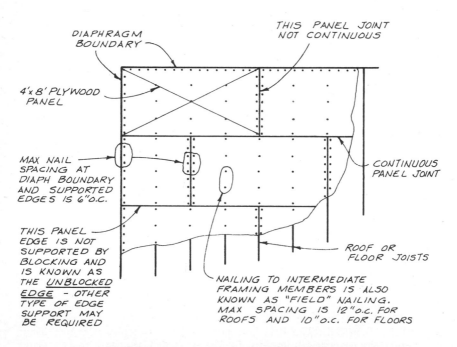

PARTIAL ROOF OR FLOOR PLAN

Fig. 9.5a Unblocked diaphragm.

For plywood diaphragms the starting point is the determination of the required thickness for sheathing loads (Chap. 8). It has been previously noted that these loads often determine the final thickness of the plywood. This is especially true for floors because the sheathing loads are larger and the deflection criteria are more stringent than for roofs.

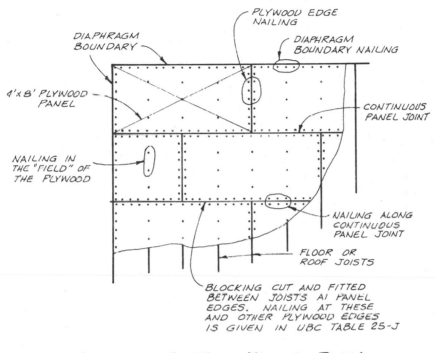

PARTIAL ROOF or FLOOR PLAN

Fig. 9.5b Blocked diaphragm

The shear capacity of the plywood for diaphragm action is the shear "through the thickness" (Fig. 8.23a). However, the shear capacity of the plywood is typically not the determining factor. Unit diaphragm shears are usually based on the *nail capacity* in the wood, rather than the strength of the plywood panel.

The nail spacing required for a horizontal diaphragm may be different at various points in the diaphragm. Because of the importance of the plywood nailing requirements, it is necessary for the designer to clearly understand the nailing patterns used in diaphragm construction.

The simplest nailing pattern is found in *unblocked diaphragms*. See Fig. 9.5a. An unblocked diaphragm is one that does *not* have two of the panel edges *supported* by lumber framing. These edges may be completely unsupported or there may be some other type of edge support such as T&G   **285** edges or panel clips (Fig. 8.14). For an unblocked diaphragm the standard nail spacing requirements are

1. Supported plywood edges—6 in. o.c.

2. Along intermediate framing members (known also as *field* nailing)—12 in. o.c. for roofs, 10 in. o.c. for floors.

EXAMPLE 9.3   Plywood Diaphragm Nailing

Required nail size is a function of the plywood thickness and shear requirements. Plywood is shown spanning in the strong direction for normal (sheathing) loads. See Figs. 9.5a and 9.5b.

The *direction of the continuous panel joint* and the *direction of the unblocked edge* are used to determine the *load case* for diaphragm design. Different panel layouts and framing arrangements are possible. The continuous panel joint and the unblocked edge are not necessarily along the same line. The diaphragm *load case* (described in Example 9.4) is considered in blocked diaphragms as well as unblocked diaphragms.

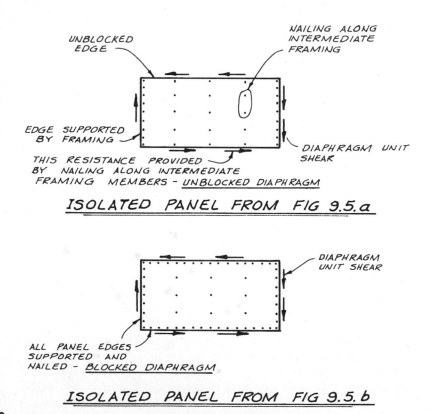

Fig. 9.5c Shear transfer in a plywood diaphragm.

286

From a comparison of the isolated panels it can be seen that the addition of nailing along the edge of a blocked panel will produce a much stronger diaphragm.

In diaphragms using relatively *thin* plywood, there is a tendency for the plywood to buckle. This is caused by the edge bearing of panels in adjacent courses as they rotate slightly under high diaphragm loads.

Fig. 9.5d Nail deformation at ultimate load. *(Courtesy of APA.)*

When loaded to failure, the nails in a plywood diaphragm deform and the *head* of the nail may pull through the face of the plywood. Other possible modes of failure include pulling the nail through the *edge* of the plywood, and *splitting* the framing members to which the plywood is attached. A minimum distance of ⅜ in. from the edge of a panel to the center of a nail is required to develop the design capacity of a nail in plywood.

When a diaphragm has the plywood edges supported with *blocking*, the *minimum* nailing requirements are the same as for unblocked diaphragms. There are, however, more supported edges in a blocked diaphragm. See Fig. 9.5b.

Although the minimum nailing requirements (i.e., maximum nail spacing) are the same for both blocked and unblocked diaphragms, the allowable unit shears are much higher for blocked diaphragms. These higher unit shears **287** are a result of the more positive, direct transfer of stress provided by nailing all four edges of the plywood panel. See Fig. 9.5c.

When loaded to ultimate, there are several possible modes of failure that can occur in a plywood diaphragm. Perhaps the most common type of failure is by the nail head pulling through the plywood. See Fig. 9.5d.

The nail spacing given above is used for all *unblocked* diaphragms. For *blocked* diaphragms, however, these spacing provisions are simply the maximum allowed spacing. Much higher allowable unit shears can be obtained by using a decreased nail spacing. Allowable unit shears for both unblocked and blocked diaphragms are given in UBC Table 25J. This table also gives the required size of nails for different thicknesses of plywood.

When the design unit shears are large, the required nail spacing for blocked diaphragms must be carefully interpreted. The nail spacing along the following lines must be considered

1. Diaphragm boundary

2. Continuous panel joints

3. Other plywood panel edges

4. Intermediate framing members (field nailing)

The field nailing provisions are always the same: 12 in. o.c. for roofs, and 10 in. o.c. for floors. The nail spacing along the other lines may all be the same (6 in. o.c. maximum), or some may be different. Allowable unit shears are tabulated for a nail spacing as small as 2 in. o.c.

There is a great deal of information incorporated into UBC Table 25J. Because of the importance of this table, the remainder of this section deals with a review UBC Table 25J (this and other UBC tables are included in Appendix D).

The table is divided into several main parts. The top part gives allowable unit shears for diaphragms which have the STRUCTURAL I designation included in the plywood grade (Sec. 8.4). The values in the bottom half of the table apply to *all other plywood grades.*

The latter represents the large majority of plywood, and it includes both engineered and appearance grades. STRUCTURAL I is stronger (and more expensive) and should be specified when the added strength is required. The allowable shear values for STRUCTURAL I are 10 percent larger than the values for other plywood grades. Although the table indicates that the values also apply to STRUCTURAL II, it should be remembered that this grade is not generally available.

In working across the table it will be noted that the left side is basically descriptive, and it requires little explanation. The two right-hand columns give the allowable unit shears for *unblocked* diaphragms. The next four columns give allowable shears for *blocked* diaphragms.

Throughout the allowable shear tables, reference is made to the *load cases.* Six load cases are defined and some examples of these are given in UBC

Table 25J. Although somewhat different plywood arrangements can be used, all layouts can be classified into one of the six load cases. Because the proper use of Table 25J depends on the load case, the designer must be able to determine the load case for any layout under consideration.

The *load case* essentially depends on two factors. The *direction of the lateral load* on the diaphragm is compared to the direction of the

1. Continuous panel joint
2. Unblocked edge (if blocking is not provided)

For example, in load case 1 the applied lateral load is perpendicular to the continuous panel joint, and it is also perpendicular to the unblocked edge. Load cases 1 through 4 consider the various combinations of these alternative arrangements. Load cases 5 and 6 have continuous panel joints running in both directions. Load cases 1 through 4 are more common than cases 5 and 6.

EXAMPLE 9.4   Plywood Diaphragm Load Cases

Determine the *load cases* for two horizontal diaphragms (Figs. 9.6a and 9.6b) in accordance with UBC Table 25J. Consider both the transverse and longitudinal directions.

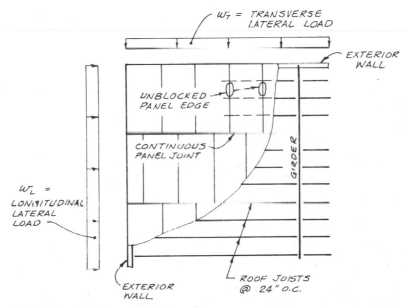

Fig. 9.6a   Partial roof framing plan.

TRANSVERSE DIRECTION

Load is perpendicular to continuous panel joint and parallel to unblocked edge.

∴ Load case 2

LONGITUDINAL DIRECTION

Load is parallel to continuous panel joint and perpendicular to the unblocked edge.

$\therefore$ Load case 4

Practically speaking, for an unblocked diaphragm there is load case 1 and *all* others.
NOTE: Plywood is oriented in the strong direction for sheathing loads.

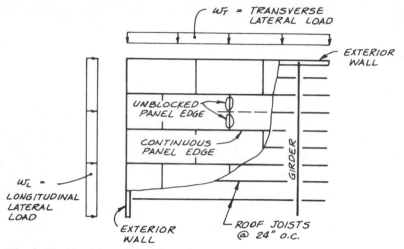

Fig. 9.6*b* Partial roof framing plan.

The building in Fig. 9.6*b* is the same as the one in Fig. 9.6*a* except that the plywood layout has been revised. The plywood here is oriented in the weak direction for sheathing loads. When the criteria defining the load case are considered, it is seen that the load cases are the same as in the previous example:

Transverse—case 2

Longitudinal—case 4

The panel layout must be shown on roof and floor framing plans along with the nailing and blocking requirements. It should be noted that the load case is defined by the criteria given above, and it does not depend on the direction of the plywood panel. See Example 9.4.

For a given horizontal diaphragm problem, two different load cases apply: one for the transverse load and a second for the longitudinal load. The lateral load in the transverse direction normally produces the larger unit shears in a horizontal diaphragm (this may not be the case for shearwalls), but the allowable unit shears may be different for the two load cases. Thus, the diaphragm shears should theoretically be checked in both directions.

There are other factors regarding the makeup and use of UBC Table 25J that should be noted. For example, the width of the framing members must be considered. Allowable shears for diaphragms using 2-in.-nominal framing members are 11 percent less than for diaphragms using wider framing.

The *nail spacing* used for the diaphragm boundary on blocked diaphragms is also the nail spacing required along the continuous panel joint for load cases 3 and 4. This nail spacing is also to be used at all panel edges for load cases 5 and 6. If load case 1 or 2 is involved, the diaphragm boundary nail spacing is not required at the continuous panel joints, and the somewhat larger nail spacing for "other panel edges" may be used. The selection of the proper nailing specifications is illustrated in several examples later in the chapter.

A point should be made about the combination of nail sizes and plywood thicknesses in UBC Table 25J. There are times when the nail size and plywood thickness for a given design will not agree with the combinations listed in the table. If the allowable shears in the tables are to be used without further justification, the following procedure should be followed: if a thicker plywood than given in the table is used, the allowable shear from the table should be based on the nailing that is used. No increase in shear is permitted because of the increased thickness. On the other hand, if a larger nail size is used for a plywood thickness given in the table, the allowable shear from the table should be based on the plywood thickness. No increase in shear is permitted because of the increased nail size. Thus, without further justification, the combination of plywood thickness and nail size given in UBC Table 25J are "compatible." Increasing one item only does not provide an increase in allowable unit shear.

Finally, it should be noted that the other "approved" types of plywood edge support (such as T&G edges or panel clips—Fig. 8.14) are substitutes for lumber blocking for sheathing loads only. They do not qualify as substitutes for blocking for diaphragm design (except 1⅛-in.-thick 2-4-1 plywood with properly stapled T&G edges).

9.4   Diaphragm Chords

Once the diaphragm web has been designed (plywood thickness and nailing), the flanges or chord members must be considered. The determination of the axial forces in the chords is described in Fig. 9.3. The axial force at any point in the chords can be determined by resolving the moment in the diaphragm at that point into a couple (equal and opposite forces separated by a level arm—the level arm is the distance between chords):

$$T = C = \frac{M}{b}$$

The tension chord is often the critical member. There are several reasons for this. One is that the allowable stress in compression is often larger than the allowable stress in tension. This assumes that the chord is laterally supported and that column buckling is not a factor. This is usually the case, **291** but the possible effects of column instability in the compression chord must be evaluated.

Another reason that the tension chord may be critical is that the chords are usually not continuous single members for the full length of the building. In order to develop the chord force, the members must be made effective by splicing separate members together. This is less of a problem in compression because the ends of chord members can transmit loads across a splice in end bearing. Tension splices, on the other hand, must be designed.

Because the magnitude of the chord force is calculated from the diaphragm moment, it should be recognized that the *magnitude of the chord force follows the shape of the moment diagram.* The design force for any connection splice can be calculated by dividing the moment in the diaphragm at the location of the splice by the distance separating the chords. A simpler, more conservative approach would be to design all diaphragm chord splices for the maximum chord force.

It should also be noted that each chord member must be capable of functioning either in tension or compression. The applied lateral load can change direction and cause tension or compression in either chord.

Some consideration should be given at this point as to what elements in a building can serve as the chords for a diaphragm. In a wood-frame building with stud walls, the doubled top plate is usually designed as the chord. See Example 9.5. This type of construction is accepted by contractors and carpenters as standard practice. Although it may have developed through tradition, the concept behind its use is structurally sound.

The top plate members are not continuous in ordinary buildings unless the plan dimensions of the building are very small. Two plate members are used so that the splice in one plate member can be staggered with respect to the splice in the other member. This creates a continuous chord with at least one member being effective at any given point. When chord forces are large, more than two plate members may be required.

In order for the top plate members to act as a chord, they must be adequately connected together. If the chord forces are small, this connection can be made with nails, but if the forces are large the connection will require the use of bolts. These are connection design problems and the procedures given in Chap. 11 and 12 can be used for the design of these splices. It should be noted that the chord forces are the result of wind or seismic loads, and an LDF of 1.33 applies to both the member design and the connections.

Plywood diaphragms are often used in buildings that have concrete tilt-up walls or masonry (concrete block or brick) walls. In these buildings the chord is made up of continuous horizontal reinforcing steel in the masonry or concrete wall. See Example 9.6. If the masonry or concrete is assumed to function in compression only (the usual assumption), the tension chord is critical. The stress in the steel is calculated by dividing the chord force by the cross-sectional area of the horizontal wall steel that is placed at the diaphragm level.

EXAMPLE 9.5   Horizontal Diaphragm Chord—
Double Top Plate

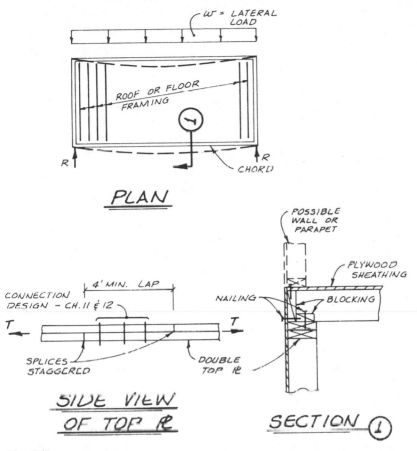

Fig. 9.7

The double top plate in a wood-frame wall is often used as the chord for the horizontal diaphragm. Splices are offset so that one member is effective in tension at a splice. Connections for anchoring the horizontal and vertical diaphragms together are covered in Chap. 14.

The double plate in wood walls and the horizontal steel in concrete and masonry walls are probably the two most common elements that are used as diaphragm chords. However, other building elements can be designed to serve as the chord. As an example of another type of chord member, consider a large window in the front longitudinal wall of a building. See

293

EXAMPLE 9.6 **Horizontal Diaphragm Chord—**
Wall Steel

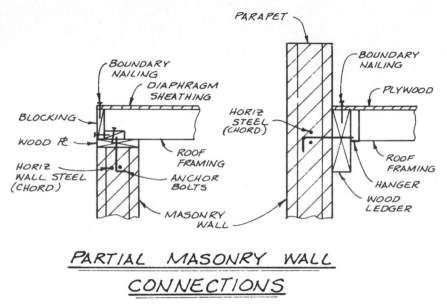

Fig. 9.8 Typical connections of horizontal diaphragms to masonry walls.

The chord for the horizontal diaphragm usually consists of horizontal reinforcing steel in the wall at or near the level of the diaphragm. Attempts to design the wood top plate or ledger to function as the chord are usually considered inappropriate because of the larger stiffness of the masonry or concrete walls.

Development of the chord requires that the horizontal diaphragm be adequately attached (anchored) to the shearwalls. In this phase of the design, the spacing of anchor bolts, blocking, and nailing necessary to transfer the forces between these elements are considered. The *details in this sketch are not complete,* and these sketches (Fig. 9.8) are included here to illustrate the diaphragm chords. Anchorage connections are covered in detail in Chap. 14.

Example 9.7. The top plate in this longitudinal wall is not continuous. Here the window header supports the roof or floor framing directly, and the header may be designed to function as the chord. The header must be designed for both vertical loads and the appropriate combination of vertical and lateral loads. The connection of the header to the shearwall also must be designed. (NOTE: If a cripple stud wall occurs over the header and below the roof framing, a double top plate will be required over the header. In this case the plate can be designed as the chord throughout the length of the wall.)

EXAMPLE 9.7   Header Acting as a Chord

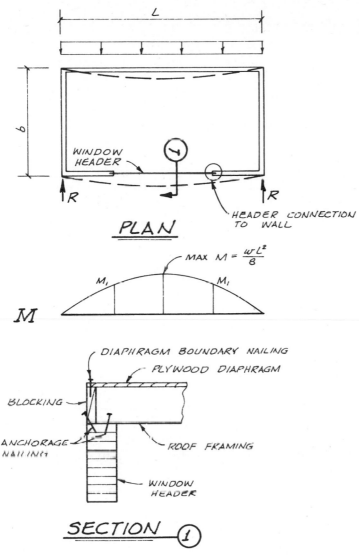

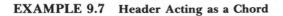

Fig. 9.9

Over the window the header serves as the chord. It must be capable of resisting the maximum chord force in addition to gravity loads. The maximum chord force is

$$T = C = \frac{max. \ M}{b}$$

The connection of the header to the wall must be designed for the chord force at that point:

$$T_1 = C_1 = \frac{M_1}{b}$$

The proper functioning of the chord requires that the horizontal diaphragm be effectively anchored to the chords and the supporting shearwalls. *Anchorage* can be provided by systematically designing for the transfer of gravity and lateral forces. This approach is introduced in Sec. 10.8, and the design for anchorage is covered in detail in Chap. 14.

9.5 Design Problem: Horizontal Roof Diaphragm

In Example 9.8 a plywood roof diaphragm is designed for a one-story building. The design loads, diaphragm unit shears, and plywood thickness requirements for this building are all obtained from previous examples.

The maximum unit shear in the transverse direction is the basis for determining the blocking and *maximum* nailing requirements for the diaphragm. However, the shear in the diaphragm is not constant, and it is possible for the nail spacing to be increased in areas of reduced shear. Likewise, it is possible to omit the blocking where the actual shear in the diaphragm is less than the allowable shear for an *unblocked* diaphragm. The locations where these changes in diaphragm construction can take place are easily determined from the unit shear diagram.

The use of changes in nailing and blocking is fairly common, but these changes can introduce problems in construction and inspection. If changes in diaphragm construction are used, they should be clearly shown on the framing plan. In addition, if blocking is omitted in the center portion of the diaphragm, the requirements for some other type of edge support must also be considered.

In this building the chord members are the horizontal reinforcing bars in the masonry walls, and a check of the chord stress in these bars is shown. The horizontal diaphragm is also checked for the lateral load in the longitudinal direction, and the effect of the change in the diaphragm load case on allowable unit shears is demonstrated.

9.6 Distribution of Lateral Forces in a Shearwall

If a shearwall has door or window openings, the total lateral force in the wall is carried by the effective segments in the wall. In wood shearwalls these

EXAMPLE 9.8   Plywood Roof Diaphragm

This example considers a building that was analyzed for diaphragm unit shears in the examples in Sec. 3.5. Roof framing consists of light-frame wood trusses at 24 in. o.c. The plywood layout is given (Fig. 9.10a) and the shear and moment in the horizontal diaphragm is summarized below the framing plan. In this example, only the exterior walls are assumed to be shearwalls.

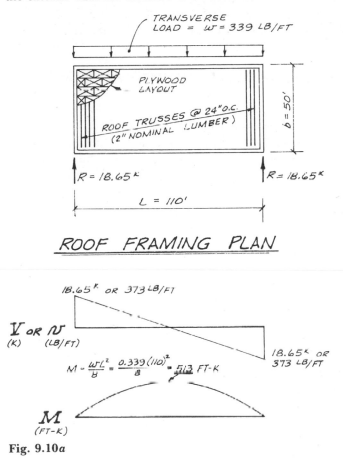

Fig. 9.10a

1. *Plywood for Sheathing Loads*

$$\text{Roof LL} = 20 \text{ psf}$$
$$\text{Roof DL} = 8 \text{ psf}$$

Although the roof framing is different, the plywood sheathing spans the same distance (24 in.) as the sheathing in Fig. 8.16. In fact, since the roof is sloping in this example, the loads along the roof are even less.

The plywood choices given in Example 8.11, part 1 *f* apply to the example at hand. These are:

a. $\frac{3}{8}$-in. C-DX (24/0) with edge support
b. $\frac{1}{2}$-in. C-DX (32/16), edge support not required *except* for bonded roofs

297

For this problem assume that a bonded built-up roof will be used, and the second alternative *with* edge support will be selected.

2. *Diaphragm Nailing.* Although ½-in. C-DX is available in STRUCTURAL I, it will not be used in this problem. If a substantially higher unit shear was involved *or* if the nailing could be considerably reduced, then STRUCTURAL I should be considered. Plywood *load case* (transverse lateral load):

> The lateral load is *perpendicular to the continuous panel joint* and *perpendicular to the unblocked edge.*
> ∴ Load case 1

If blocking is not used to provide the plywood edge support, the maximum allowable unit shear is

$$\text{Allow. } v = 255 \text{ lb/ft}$$

This comes from UBC Table 25J for ½-in. plywood (not STRUCTURAL I) in an unblocked diaphragm, load case 1, 2-in.-nominal framing, and 10d nails. The actual shear is greater than this allowable:

$$v = 373 \text{ lb/ft} > 255 \qquad NG$$

An increased allowable shear can be obtained by increasing the plywood thickness or using STRUCTURAL I plywood. A more efficient method for this design is to provide blocking and increased nailing (i.e., reduced nail spacing). From the blocked diaphragm portion of UBC Table 25J, the following design is chosen:

> Use ½-in. C-DX with 10d nails
> at 4 in. o.c. boundary*
> 6 in. o.c. all other plywood edges
> 12 in. o.c. field

$$\text{Allow. } v = 385 \text{ lb/ft} > 373 \qquad OK$$

The above nailing and blocking requirements can safely be used throughout the entire diaphragm because these were determined for the maximum unit shear. The

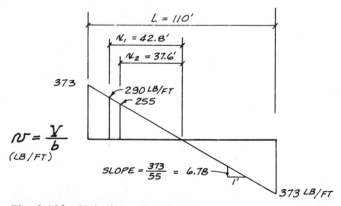

Fig. 9.10*b* Unit shear diagram.

* Because this diaphragm is load case 1, the 4-in. nail spacing is required at the diaphragm boundary only. The 6-in. nail spacing is used at continuous panel joints as well as other plywood edges.

variation of diaphragm shear can be shown on a sketch. The *unit shear diagram* is the total shear diagram divided by the diaphragm width (Fig. 9.10*b*).

Some designers prefer to take into account the reduced unit shears toward the center of the diaphragm (away from the shearwall reactions). If this is done, the nail spacing can be increased to the Code maximum allowed spacing (i.e., 6 in. o.c. at all edges including the diaphragm boundary) where the shear drops off to the corresponding allowable shear for this nailing. This location can be determined by similar triangles. For example,

Allow. $v = 290$ lb/ft   when the nail spacing at the diaphragm boundary is 6 in. o.c. and blocking is provided

Then

$$x_1 = \frac{290}{6.78} = 42.8 \text{ ft}$$

This refinement can be carried one step further, and the location can be determined where blocking can be omitted.

Allow. $v = 255$ lb/ft   for an unblocked diaphragm

Then

$$x_2 = \frac{255}{6.78} = 37.6 \text{ ft}$$

If these changes in diaphragm construction are used, the locations calculated above should be rounded off to some convenient dimensions. These locations and the required types of construction in the various segments of the diaphragm must be clearly shown on the roof framing plan. In this regard, the designer must weigh the savings in labor and materials gained by the changes described above, against the possibility that the diaphragm may be constructed improperly in the field. The more variations in nailing and blocking, the greater the chance for error in the field. Increased nail spacing and omission of blocking can represent substantial savings in labor and materials, but inspection becomes increasingly important.

In addition, it should be remembered that for this problem some type of edge support is required. Therefore, if blocking is only provided in the areas where the shear exceeds 255 lb/ft, some other type of edge support must be provided in the center portion of the diaphragm. Thus, if blocking is omitted the plywood must have T&G edges or panel clips must be used (Fig. 8.14).

3. *Chord.* Axial force in chord is obtained by resolving the diaphragm moment into a couple

$$T = C = \frac{M}{b} = \frac{513}{50} = 10.3 \text{ k}$$

Depending on the type of wall system, various items can be designed to function as the *chord* member.

In this example, the walls are made from 8-in. grouted concrete block units, and the horizontal wall steel (two #5 bars) will be checked for chord stresses. (Examples of wood chords are covered elsewhere.)

The tension chord is critical, and the stress in the horizontal wall steel can be calculated as the chord force divided by the area of steel. The allowable stress in the reinforcing steel is 20 ksi under normal loads, and this can be increased by a factor of 1.33 for short-term loads (wind or seismic).

299

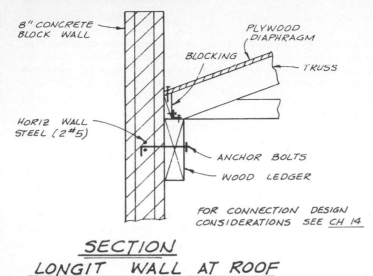

8" CONCRETE BLOCK WALL

PLYWOOD DIAPHRAGM

BLOCKING

TRUSS

HORIZ WALL STEEL (2 #5)

ANCHOR BOLTS

WOOD LEDGER

FOR CONNECTION DESIGN CONSIDERATIONS SEE CH 14

SECTION
LONGIT WALL AT ROOF

Fig. 9.10c

Stress in steel:

$$f_s = \frac{T}{A_s} = \frac{10.3}{2 \times 0.31} = 16.6 \text{ ksi}$$

Allow. $f_s = 20 \times 1.33 = 26.6 \text{ ksi} > 16.6$ *OK*

The design of other wall-reinforcing steel is a problem in masonry design and is beyond the scope of this book. Steps must be taken to ensure that cross-grain bending (Example 6.1) in the ledger does not occur. The design of anchor bolts and other connections for attaching the horizontal diaphragm to the shearwall is covered in Chap. 14.

4. *Longitudinal Lateral Load.* The calculation of the lateral load in the longitudinal direction was not given in the examples in Sec. 3.5. However, the method used is

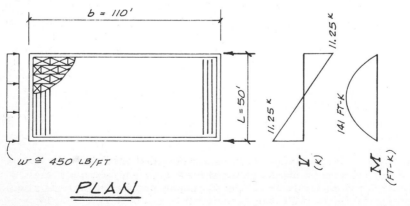

b = 110'

L = 50'

11.25 ᴷ

11.25 ᴷ

141 FT-K

V (K)

M (FT-K)

w ≅ 450 LB/FT

PLAN

Fig. 9.10d

similar to that illustrated for the transverse load. The longitudinal load and the corresponding shear and moment diagrams are shown in Fig. 9.10d.

The unit shear in a horizontal diaphragm is often critical in the transverse direction. However, the longitudinal shear should be checked especially if changes in nailing occur. In any event, the longitudinal direction must be analyzed for anchorage forces for connecting the diaphragm and shearwall.

Unit shear:

$$v = \frac{V}{b} = \frac{11250}{110} = 102 \text{ lb/ft}$$

The load in the longitudinal direction is *parallel* to the *continuous panel joint* and *parallel* to the *unblocked edge.*

$$\therefore \text{ Load case 3}$$

If blocking is omitted in the center portion of the diaphragm (up to 37.6 ft on either side of the center line) the allowable unit shear for an unblocked diaphragm must be used. From UBC Table 25J,

Allow. $v = 190$ lb/ft > 102 OK

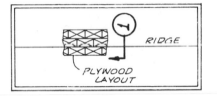

PLAN

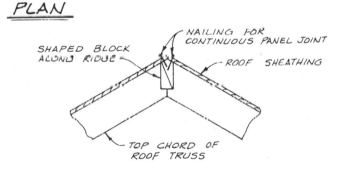

SECTION

Fig. 9.10e

If the unit shear in the longitudinal direction had exceeded 190 lb/ft, then blocking would have been required for the longitudinal loading. When blocking is required for load cases other than cases 1 and 2, the continuous panel joints must be nailed with the same spacing as the diaphragm boundary (UBC Table 25J). The chord force in the transverse walls is less than the chord force in the longitudinal walls.

$$T = C = \frac{M}{b} = \frac{141}{110} = 1.28 \text{ k} < 10.3 \text{ k}$$

$\therefore$ Two #5 bars   *OK* for chord around entire building

5. *Nailing Along Ridge.* The plywood skin in this building is attached directly to the top chord of the light-frame wood roof trusses (see Fig. 9.10*e*). These trusses probably have sufficient strength and stiffness to maintain diaphragm continuity at the ridge. However, it is desirable to provide a positive attachment of the sheathing through doubled 2× blocking or a shaped 3× or 4× block.

---

effective segments are known as *shear panels* (or shearwalls), and in concrete and masonry walls they are referred to as *wall piers*. The designer must understand how to distribute the total force to these resisting elements so that the unit wall shears can be calculated.

The procedures for designing wood shearwalls are given in Chap. 10, but the method for distributing the lateral load is covered at this time. The distribution needs to be understood at this point because it is used to determine the magnitude of the load for the horizontal diaphragm *strut*. A different procedure is used to distribute the horizontal diaphragm reaction to wood shearwalls and concrete or masonry shearwalls.

In a wood-frame shearwall, the most common approach is to assume that the unit shear is uniform throughout the total length of all shear panels. Thus the force in a given panel is in direct proportion to its length. See Example 9.9. The unit shear in one panel is the same as the unit shear in all other panels.

In order for this distribution to be reasonably correct, the wall must be constructed so that the panels function as separate shear resisting elements. The height-to-width ratio h/b should be less than or equal to 3½ for blocked plywood shear panels. In addition, the shear panels are assumed to all have the same height. This is obtained by using continuous full-height studs at each end of the shear panels. The panels must all deflect laterally the same amount. This is accomplished by tying the panels together with the strut.

There are several types of load resistance developed in a wall subjected to a lateral load. These include bending resistance and shear resistance. For wood-frame walls constructed as described above, the shear resistance is the significant form, and the unit shear in this case approaches a uniform distribution.

In buildings with reinforced concrete or masonry shearwalls, the distribution of forces to the various elements in the wall differs from the uniform distribution assumed for wood shearwalls. Concrete and masonry walls may have significant combined bending and shear resistance. This combined resistance is measured by the *relative rigidity* of the various piers in the wall. Under a lateral load the piers are forced to undergo the same lateral deformation.

302

EXAMPLE 9.9   **Distribution of Shear in a Wood Shearwall**

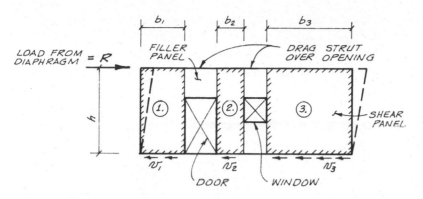

ELEVATION
WOOD SHEARWALL

Fig. 9.11

In a wood shearwall, the wall is usually assumed to be made up of *separate* shear panels. The shear panels are connected together so that they deflect laterally the same amount at the horizontal diaphragm level. Drag strut or tie members provide the connection between the shear panels.

The three shear panels in the wall shown in Fig. 9.11 all have the same height h. This behavior results from the use of double full-height studs at both ends of the shear panels and appropriate shearwall nailing (Chap. 10). Filler panels (wall elements above and below openings) may be nailed less heavily (i.e., with the Code minimum nailing) to further isolate the shear panels (Ref. 14.12).

For the typical wood shearwall, the resistance to the applied lateral load is assumed to be uniform throughout the combined length of the shear panels. Thus

$$v_1 = v_2 = v_3 = \frac{R}{\Sigma b}$$

The force required to deflect a long pier is larger than the force that would be calculated on the basis of its length only.

Thus, in a concrete or masonry wall, the *unit shear* in a longer pier is greater than the unit shear in a shorter wall segment. The relative rigidity is a function of the height-to-width ratio h/b of the pier and not just its length. The relative rigidity and the question of force distribution is related to the *stiffness* of a structural element: the stiffer (more rigid) the element, the greater the force it "attracts." See Example 9.10. The calculation of relative

303

EXAMPLE 9.10 **Distribution of Shear in a Concrete or Masonry Shearwall**

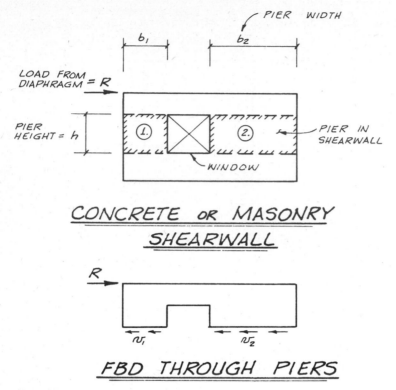

CONCRETE OR MASONRY SHEARWALL

FBD THROUGH PIERS

Fig. 9.12

In a concrete or masonry wall the "piers" are analyzed to determine their *relative rigidities*. The smaller the height-to-width ratio h/b of a pier, the larger the relative rigidity. The larger the relative rigidity, the greater the percentage of total force R carried by the pier. Pier 2 in the wall shown in Fig. 9.12 has a greater rigidity and a greater unit shear (not just total shear) than pier 1; that is,

$$v_2 > v_1$$

rigidities for these wall elements is a concrete or masonry design problem and is beyond the scope of this text.

In buildings with concrete or masonry walls, the loads over openings in the walls can be supported in two different ways. In the first system, there is a concrete or masonry wall element over the openings in the wall. See Fig. 9.13a in Example 9.11. This element can be designed as a beam to carry the gravity loads across the opening. Concrete and masonry beams of

304

this type are known as *lintels*. In the second case a wood *header* may be used to span the opening (Fig. 9. 13*b*).

A distinction is made between these systems because in the first case the load from the horizontal diaphragm is transmitted to the shearwalls through

---

EXAMPLE 9.11

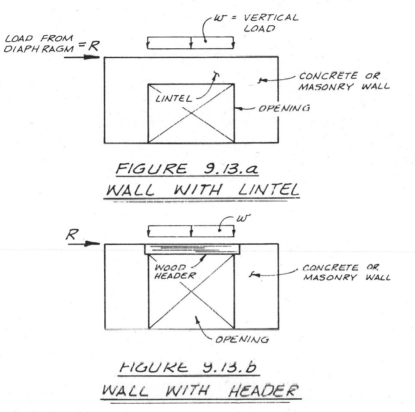

FIGURE 9.13.a

WALL WITH LINTEL

FIGURE 9.13.b

WALL WITH HEADER

Fig. 9.13

A concrete or masonry lintel, or a wood (or steel) header can be used to span over an opening in a wall (see Fig. 9.13). The member over the opening must be designed for both vertical and lateral loads. Load combinations are

1. Vertical loads DL + RLL (or DL + snow).
2. Combined vertical and lateral (RLL can be omitted but all or part of the snow load is required).
a. For lateral loads parallel to the wall, the load in the header is the strut force.
b. For lateral loads perpendicular to the wall, the load in the header is the diaphragm chord force.

the lintel, and the diaphragm design is unaffected. In the case of the wood header, however, the unit shear from the horizontal diaphragm must be transmitted through the header/strut into the shearwalls. The load combinations required for the design of this member are given in Example 9.11.

In summary, the distribution of the *unit shear* in a wall subjected to a lateral load will be as follows:

1. Wood shearwalls—uniform distribution to all shear panels
2. Concrete and masonry shearwalls—distribution is a function of the relative rigidities of the wall piers

For problems in this book involving concrete or masonry walls, it is assumed that the distribution of wall forces and the unit shears are known by previous analysis.

9.7 Drag Strut Forces

The perimeter members of a horizontal diaphragm that are parallel to the applied lateral load are the struts. These members are also known as collectors, drag members, or ties, and their function is illustrated in Fig. 9.4. Essentially the strut pulls or drags the unit shear in an unsupported segment of a horizontal diaphragm (the portion over an opening in a shearwall, for example) into the supporting elements of the shearwall.

The members in a building that serve as the *struts* are often the same members that are used as the *chords* for the lateral load in the other direction. Thus, the design of the chord and the strut for a given wall may simply involve the design of the same member for different forces. In fact, for a given perimeter member, the chord force may be compared with the strut force, and the design can be based on the critical load.

The wood design principles for struts and chords are covered in Chap. 7. These members are either tension or compression members, and they may or may not have bending. Bending may result from vertical loads as in the header in Example 9.11. Connections are also an important part of the collector design. The member itself may be sufficiently strong, but little is gained unless the strut is adequately connected to the supporting shearwall elements. The design of connections is covered in later chapters, but the calculation of the force in the drag member will be covered at this time.

The building in Fig. 9.4 has an end shearwall with an opening located in the *center* of the wall. This presents a simple problem for the determination of the drag strut forces because the total load is shared *equally* by both shear panels. This would be true for equal-length piers in a concrete or masonry wall also. The force in the drag strut is maximum at each end of the wall opening. This maximum force is easily calculated as the unit shear in the horizontal diaphragm times one-half the length of the wall opening.

The force in the strut varies linearly throughout the length of the strut. At one end of the opening the strut is in tension, and at the other end it is in compression. Since the lateral load can come from either direction, the ends may be stressed either in tension or compression. The magnitude of the drag force in the two shear panels is not shown in Fig. 9.4. However, the force in a panel decreases from a maximum at the wall opening to zero at the outside end of the shearwall.

A somewhat more involved problem occurs when an opening is not symmetrically located in the length of the wall. See Example 9.12. Here a greater portion of the total wall load is carried by the longer shear panel. A correspondingly larger percentage of the unsupported diaphragm shear over the opening must be transmitted by the drag strut to the longer panel.

The magnitude of the force at some point in the strut can be determined in several ways. Two approaches are shown in Example 9.12. The first method

---

EXAMPLE 9.12   Calculation of Force in Strut

Determine the magnitude of the drag strut force throughout the length of the front wall in the building shown in Fig. 9.14a. The walls are wood-frame construction.

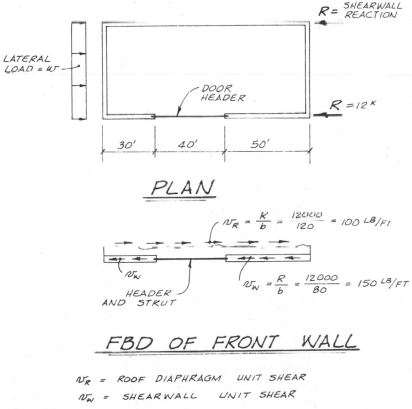

PLAN

$$\nu_R = \frac{R'}{b} = \frac{12000}{120} = 100 \ ^{LB}/_{FT}$$

$$\nu_W = \frac{R}{b} = \frac{12000}{80} = 150 \ ^{LB}/_{FT}$$

FBD OF FRONT WALL

ν_R = ROOF DIAPHRAGM UNIT SHEAR

ν_W = SHEARWALL UNIT SHEAR

Fig. 9.14a

DESIGN OF WOOD STRUCTURES

Method 1: FBD Approach

The force in the strut at any point can be determined by cutting the member and summing forces. Two examples are shown (Figs. 9.14*b* and 9.14*c*). One cut is taken through the header and the other is taken at a point in the wall.

a. Strut force at point *A* (40 ft from left wall):

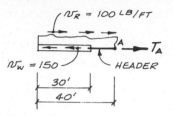

Fig. 9.14*b*

$$\Sigma F_x = 0$$
$$T_A + 100(40) - 150(30) = 0$$
$$T_A = +500 \text{ lb (tension)}$$

NOTE: The lateral load can act toward the right or left. If the direction is reversed, the strut force at *A* will be in compression.

b. Strut force at point *B* (100 ft from left wall)

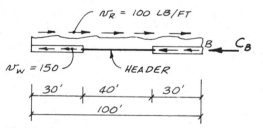

Fig. 9.14*c*

$$\Sigma F_x = 0$$
$$C_B + 150(30 + 30) - 100(100) = 0$$
$$C_B = 1000 \text{ lb (compression)}$$

NOTE: This force would be tension if the direction of the lateral load is reversed.

Method 2: Strut Force Diagram

Construction of the strut force diagram (see Fig. 9.14*d*) involves:

a. Drawing the unit shear diagrams for the roof diaphragm and shearwalls.
b. Drawing the *net* unit shear diagram by subtracting the diagrams in *a.*
c. Constructing the strut force diagram using the *areas* of the net unit shear diagram as *changes* in the magnitude of the strut force. The strut force is zero at the outside ends of the building.

The relationship between the net unit shear diagram and the strut force diagram is similar to the relationship between the shear and moment diagrams for a beam.

308

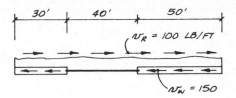

PLAN

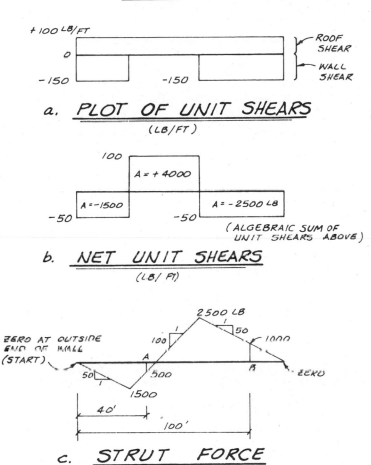

a. PLOT OF UNIT SHEARS
(LB/FT)

b. NET UNIT SHEARS
(LB/FT)

c. STRUT FORCE
(LB)

Fig. 9.14d

NOTE: Signs of forces are not important because loads can reverse direction.
The values obtained by taking FBDs at points *A* and *B* can be verified by similar triangles on the strut force diagram.

**EXAMPLE 9.13   Drag Strut Force in Building with
                  Masonry Walls**

Draw the distribution of the collector force for the building of Example 9.12, assuming
that the walls are of reinforced masonry construction. The wall unit shears have been
determined by a previous analysis using the relative rigidities of the pier elements.

Wall shears:

$$50\text{-ft pier} \qquad v_{w1} = 155 \text{ lb/ft}$$
$$30\text{-ft pier} \qquad v_{w2} = 141.7 \text{ lb/ft}$$

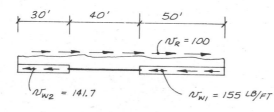

PLAN

UNIT SHEARS

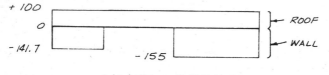

NET UNIT SHEAR

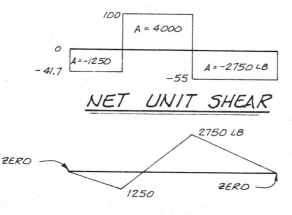

STRUT FORCE

Fig. 9.15

310

simply involves summing forces on a free-body diagram (FBD) of the wall. The cut is taken at the point where the strut force is required. The other forces acting on the FBD are the unit shears in the roof diaphragm and the unit shears in the shearwall.

The second method provides a plot of the force in the strut along the length of the building. With this diagram the strut force at any point can readily be determined.

A third approach is easy to apply, but it may be overly conservative. *The maximum possible strut force is the roof diaphragm unit shear times the length of the opening in the wall.*

Once the strut forces have been determined, they can be compared with the chord forces at corresponding points along the wall.

The procedure illustrated in Example 9.12 for plotting the strut load can be used for either wood or concrete or masonry shearwalls. The only difference is that for masonry or concrete walls, the unit shear in the walls will probably be different.

For example, the unit shear in the wood shearwall in Example 9.12 is 150 lb/ft in both the 30-ft and 50-ft segments. In a masonry or concrete wall with the same dimensions, the unit shear might be 155 lb/ft in the 50-ft segment, and 141.7 lb/ft in the 30-ft portion. The actual values depend on relative rigidities of the piers. The force at any point in the strut is somewhat different than the strut force for wood shearwalls. See Example 9.13.

Notice that if the strut force is approximated as the roof unit shear times the length of the opening, the same value ($T = 100 \times 40 = 4000$ lb) is obtained for the building in Examples 9.12 and 9.13.

9.8 Diaphragm Deflections

The deflection of a horizontal diaphragm has been illustrated in a number of sketches in Chap. 9. This topic is the object of some concern because the walls that are attached to (actually supported by) the horizontal diaphragm are forced to deflect along with the diaphragm. See Example 9.14. If the walls are constructed so they can undergo or "accommodate" these deformations without failure, there is little need for concern. However, a potential problem exists when the deflection is large, or when the walls are constructed so rigidly that they can tolerate little deflection. In this case, the diaphragm imposes a deflection on the wall that may cause the wall to be overstressed.

There are at this time two methods used to account for diaphragm deflections. The method that is the most widely used is basically a rule of thumb, in that no attempt is made to calculate the magnitude of the actual deflection. This method makes use of the span-to-width ratio of the diaphragm. In this simple approach the span-to-width ratio L/b is checked to be less than some acceptable limit. For example, the Code sets an upper limit on the span-to-width ratio for horizontal plywood diaphragms of 4.

EXAMPLE 9.14   **Diaphragm Deflections and Span-to-Width Ratios**

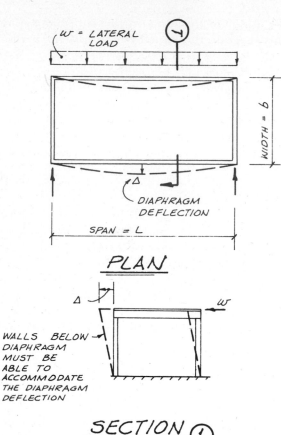

Fig. 9.16

Wood-frame walls are capable of accommodating larger deflections than are more rigid wall systems such as concrete or masonry.

Diaphragm Proportions

One method of controlling deflection is to limit the span-to-width ratio L/b for the diaphragm. For horizontal plywood diaphragms, $L/b \leq 4.0$. For other types of wood diaphragms see UBC Table 25I.

Thus, this approach says that if the *proportions* of the diaphragm are *reasonable*, deflection will not be excessive. Although no numerical value for

312

the deflection of the diaphragm is calculated, this method has traditionally been applied. It is probably an acceptable approach for most wood-frame buildings, and many designers consider it to be sufficient for buildings with concrete and masonry walls.

The second technique of accounting for diaphragm deflections is to calculate a numerical value for the deflection. The total deflection that occurs in an ordinary beam is made up of two deflection components: bending and shear. However, for most practical beams the shear deflection is negligible, and only the bending deflection is calculated.

In a horizontal diaphragm, the total deflection is the sum of the deflection caused by a number of factors in addition to bending and shear. All these factors may contribute a significant amount to the total deflection. In the normally accepted formula, the deflection of a horizontal diaphragm is made up of four parts:

$$\text{Total } \Delta = \Delta_1 + \Delta_2 + \Delta_3 + \Delta_4$$

where $\Delta_1 =$ bending deflection
$\quad \Delta_2 =$ shear deflection
$\quad \Delta_3 =$ deflection due to nail slip (deformation)
$\quad \Delta_4 =$ deflection due to slip in chord connection splices

The methods used to evaluate the various deflection components are described in Ref. 14.12 and are not included here. It should be noted that this formula applies only to *blocked* diaphragms. Attempts are being made to develop appropriate deflection calculation procedures for all types of diaphragms.

Reference 16 suggests that the basic elements in the existing formula should be generalized to handle a variety of diaphragm loading conditions. This reference also suggest an alternate method for calculating the deflection due to nail slip.

One problem with applying the deflection calculation is that there are no definite criteria against which the calculated deflection must be checked. The simple deflection limits used for ordinary beams (such as $L/180$ and $L/160$) are not applied to diaphragm deflections. These types of limits do not evaluate the effects that the deformations impose on the walls supported by the diaphragm.

Reference 10 suggests a possible deflection criterion for diaphragms that support concrete or masonry walls, but generally the evaluation of the calculated deflection is left to the judgment of the designer. It should be noted that the limits on the span-to-width ratio given in UBC Table 25I are a requirement of the Code, and they must be satisfied.

In this introduction to diaphragm design, the limits on the span-to-width ratio will be relied upon for controlling deflections. As noted, this is fairly common practice in ordinary building design. For those interested in examining the problem of diaphragm deflections in greater detail, information is available from APA and Ref. 16.

9.9 Diaphragms with Interior Shearwalls

All of the diaphragms considered up to this point have been supported by shearwalls that are the exterior walls of the building. Although it is convenient to introduce horizontal diaphragms by using this basic system, many buildings make use of interior shearwalls in addition to exterior shearwalls. In this case the roof or floor assemblage is assumed to act as a number of separate horizontal diaphragms. See Example 9.15.

Because the diaphragms are assumed to be separate elements, they are treated as simply supported beams that span between the respective shearwall supports. Thus the *shear in the diaphragm* can be determined using the methods previously covered for buildings with exterior shearwalls. The difference is that the span of the diaphragm is now measured between adjacent parallel walls and not simply between the exterior walls of the building. With these separate diaphragms there will be different unit shears in the diaphragms unless the spans and loads happen to be the same for the various diaphragms.

The *loads to the shearwalls* are calculated as the sum of the reactions from the horizontal diaphragms that are supported by the shearwall. Thus, for an exterior wall the shearwall load is the reaction of one horizontal diaphragm. For an interior shearwall, however, the load to the wall is the sum of the diaphragm reactions on either side of the wall.

The chord force for the diaphragms is determined in the same manner as for a building with exterior shearwalls. The moments in the respective diaphragms are calculated as simple beam moments. Once the chord forces are determined, the chords are designed in accordance with appropriate procedures for the type of member and materials used.

9.10 Interior Shearwalls with Drag Struts

If a horizontal diaphragm is supported by an interior shearwall that is not the full width of the building, or if there are openings in the wall, a drag tie will be required. As with the collectors in exterior walls (Sec. 9.7), the strut will transfer the unit shear from the unsupported portion of the horizontal diaphragm into the shearwall. In the case of an interior shearwall, the drag force will be the sum of the forces from the horizontal diaphragms on both sides of the shearwall. See Example 9.16.

The distribution of the force in the strut throughout its length can be plotted from the areas of the net unit shear diagram. From this plot it can be seen that the critical axial force in the strut is at the point where it connects to the shearwall. The member and the connection must be designed for this force. In addition, the combined effects of gravity loads must also be considered.

All interior walls are not necessarily shearwalls. For an interior wood-frame wall to function as a shearwall, it must have sheathing of adequate strength, and it must be nailed to the framing so that the shear resistance is developed.

EXAMPLE 9.15 **Horizontal Diaphragm with Interior Shearwall**

Determine the unit shear in the horizontal diaphragm and shearwalls in the building shown in Fig. 9.17a for the transverse lateral load of 300 lb/ft. Also calculate the diaphragm chord forces.

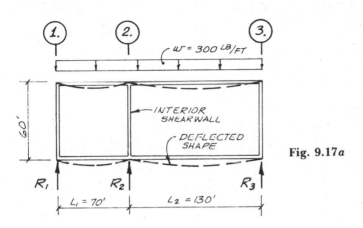

Fig. 9.17a

Horizontal diaphragm

Flexible horizontal diaphragms are assumed to span between the exterior and interior shearwalls. The continuity at the interior wall is disregarded, and two simply supported diaphragms are analyzed for internal unit shears. See Fig. 9.17b.

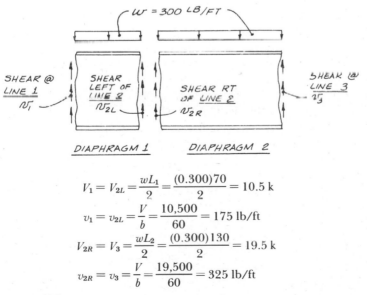

$$V_1 = V_{2L} = \frac{wL_1}{2} = \frac{(0.300)70}{2} = 10.5 \text{ k}$$

$$v_1 = v_{2L} = \frac{V}{b} = \frac{10,500}{60} = 175 \text{ lb/ft}$$

$$V_{2R} = V_3 = \frac{wL_2}{2} = \frac{(0.300)130}{2} = 19.5 \text{ k}$$

$$v_{2R} = v_3 = \frac{V}{b} = \frac{19,500}{60} = 325 \text{ lb/ft}$$

Fig. 9.17b

315

DESIGN OF WOOD STRUCTURES

The nailing requirements for each diaphragm can be determined separately on the basis of the respective unit shears. Note that both diaphragms 1 and 2 will have their own nailing requirements, including boundary nailing along line 2. These nailing provisions must be clearly shown on the design plans.

Shearwalls

The unit shears in the shearwalls can be determined by calculating the reactions carried by the shearwalls. Shearwall reactions must balance the loads (shears) from the horizontal diaphragm. See Fig. 9.17c.

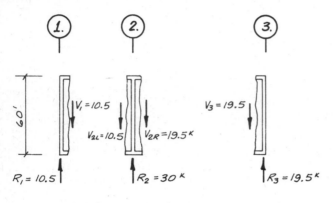

SHEARWALL REACTIONS

Fig. 9.17c

Unit wall shears (no openings in walls):

$$v_1 = \frac{R}{b} = \frac{10,500}{60} = 175 \text{ lb/ft}$$

$$v_2 = \frac{30,000}{60} = 500 \text{ lb/ft}$$

$$v_3 = \frac{19,500}{60} = 325 \text{ lb/ft}$$

ALTERNATE SHEARWALL REACTION CALCULATION:

Because the diaphragms are assumed to be simply supported, the shearwall reactions can also be determined using the tributary widths to the shearwalls. See Fig. 9.17d.

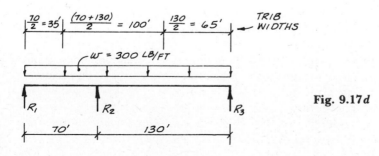

Fig. 9.17d

$$R_1 = (0.300)35 = 10.5 \text{ k}$$
$$R_2 = (0.300)100 = 30 \text{ k}$$
$$R_3 = (0.300)65 = 19.5 \text{ k}$$

NOTE: The calculated unit shears for the shearwalls are from the horizontal diaphragm reactions. If the lateral load is a seismic force, there will be an additional wall shear generated by the dead load of the shearwall itself (Chap. 3).

Diaphragm Chord Forces

The distribution of force in the chord follows the shape of the simple beam moment diagram for the diaphragms. See Fig. 9.17e.

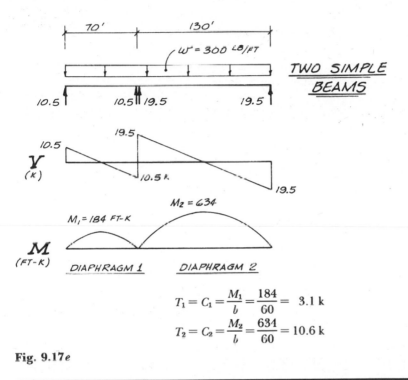

$$T_1 = C_1 = \frac{M_1}{b} = \frac{184}{60} = 3.1 \text{ k}$$

$$T_2 = C_2 = \frac{M_2}{b} = \frac{634}{60} = 10.6 \text{ k}$$

Fig. 9.17e

In addition the height-to-width ratio h/b of the shear panel must be less than the limits given in UBC Table 25I. (Procedures for designing shearwalls are covered in Chap. 10.) The diaphragm must also be connected to the shearwall so that the shears are transferred.

In addition to these strength requirements, some consideration should be given to the relative lengths and locations of the various walls. To illustrate this point, consider an interior wall with a length that is very small in comparison with an adjacent parallel wall. See Example 9.17.

Because of the large length of the right exterior wall, it will be a more rigid support for the horizontal diaphragm than the short interior wall. The

317

EXAMPLE 9.16 Building with Interior Drag Strut

The building of Example 9.15 is modified in this example so that the interior shearwall is not the full width of the building. Determine the strut load for the interior shearwall. See Fig. 9.18.

Unit shear in shearwall:

$$v_2 = \frac{R}{b} = \frac{30,000}{40} = 750 \text{ lb/ft}$$

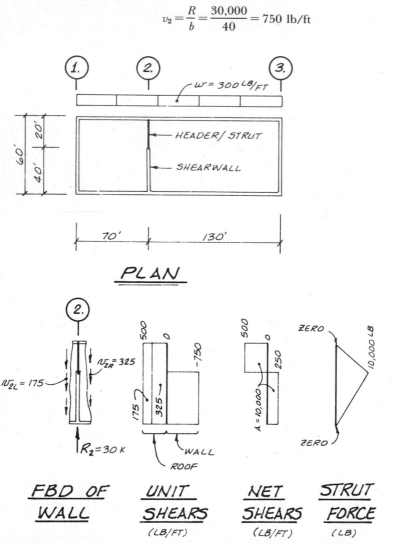

FBD OF WALL UNIT SHEARS (LB/FT) NET SHEARS (LB/FT) STRUT FORCE (LB)

Fig. 9.18

The interior shearwall carries the reactions of the horizontal diaphragms. Part of this load is transferred to the shearwall by the drag strut (strut force = 10k).

EXAMPLE 9.17   **Effectiveness of Interior Shear-walls**

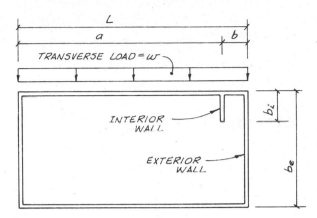

PLAN

Fig. 9.19

Using tributary widths the *calculated* load to the interior shearwall is

$$R_i = \left(\frac{a}{2} + \frac{b}{2}\right)W$$

The load to the right exterior wall would be much smaller:

$$R_e = \left(\frac{b}{2}\right)w$$

Under *actual* loading, however, the load carried by the right exterior wall will approach

$$R_e = \left(\frac{L}{2}\right)w$$

The reason for this is that the large exterior wall is more rigid than the interior wall. Thus, in order to distribute the load from the horizontal diaphragm to the walls using tributary widths, the shearwalls must be *effective*.

The effectiveness of a wall in providing true shearwall action depends on several factors including:

1. Relative lengths of the walls. If b_i and b_e were more nearly equal, the interior wall would be more effective.

2. Relative location of the walls. If b was larger in comparison to a, the interior shearwall would also begin to function as a separate support.

The evaluation of shearwall effectiveness is, to a large extent, a matter of judgment. Walls which are judged to be ineffective (i.e., ignored as shearwalls) should be constructed so that they do not interfere with the assumed action of the diaphragm.

large rigidity of the exterior wall will "attract" and carry a large portion of the lateral load tributary to *both* walls.

Because of this, it would be better practice to ignore the short interior wall and design the nearby exterior wall for the entire tributary lateral load. The short interior wall should then be sheathed and nailed lightly to ensure that is does not interfere with the diaphragm span.

Attempts can be made to compare the relative rigidities of the various shearwalls. This is similar to the rigidity analysis of the various piers in a single concrete or masonry wall (Sec. 9.6). However, in wood-frame construction the question of whether to consider a wall as an "effective" shearwall is normally a matter of judgment. The relative lengths of the shear walls and their proximity to other effective walls are used in making this determination.

9.11 Diaphragm Flexibility

Horizontal diaphragms can be classified according to their tendency to deflect under load. Concrete floor or roof slabs deflect very little under lateral loads and are defined as *rigid* diaphragms. Wood diaphragms, on the other hand, deflect considerably more and are classified as *flexible* diaphragms. There are other systems which classify diaphragms into additional categories such as semirigid, semiflexible, and very flexible (Ref. 10). For purposes of this text, only the basic classification of rigid and flexible diaphragms is considered.

The purpose of introducing these terms is to describe the differences between the distribution of lateral forces for the two types of diaphragms.

With a rigid diaphragm there is a torsional moment that must be considered. A *torsional moment* is developed if the centroid of the applied lateral load does not coincide with the center of resistance (center of rigidity) of the supporting shearwalls. See Example 9.18. Even if there is no theoretical eccentricity, the Code requires that some minimum eccentricity be used in design to account for "accidental" torsion.

The effect of the torsional moment is to cause wall shears in addition to those which would be developed if no eccentricity existed. This is essentially a combined stress problem in that shears are the result of direct loading and an eccentric moment.

It should be noted that the Code does not permit wood framing to support the dead load of concrete roof or floor slabs (except nonstructural coverings less than 4 in. thick). Thus, rigid diaphragms of reinforced concrete will not be supported by wood wall systems.

Wood diaphragms are, in most cases, assumed to be flexible, and rotation is usually not considered. Here the lateral load to the supporting shearwalls is simply determined by using the respective tributary widths to the walls (Sec. 9.9). The load is then distributed to the various resisting elements in the wall by the methods covered in Sec. 9.6.

In certain wood buildings rotation may be used to resist lateral loads. This occurs in buildings with a side that is essentially open. In other words, there is no segment of a wall of sufficient length to provide an "effective" shear panel. See Example 9.19. The only way that a building of this nature can be designed is to take the torsional moment in rotation. The torsional moment is resolved into a couple with the forces acting in the transverse walls.

EXAMPLE 9.18   Comparison of Flexible and Rigid Diaphragms

Basic classification of horizontal diaphragms:

1. Flexible—such as a wood roof or floor system
2. Rigid—such as a concrete roof or floor slab

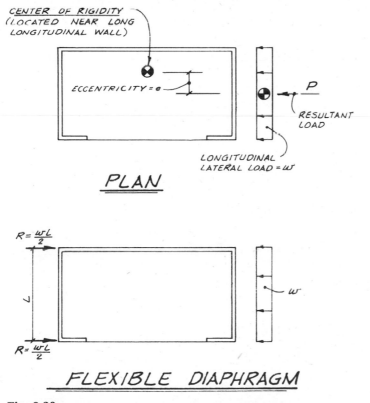

PLAN

FLEXIBLE DIAPHRAGM

Fig. 9.20a

Forces to supporting walls are simple beam (or tributary load) reactions for a *flexible* diaphragm.

321

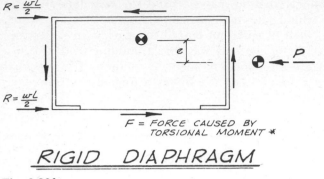

$F = FORCE\ CAUSED\ BY$
$TORSIONAL\ MOMENT\ *$

RIGID   DIAPHRAGM

Fig. 9.20b

Forces to walls supporting a *rigid* diaphragm are the sum of the forces due to direct shear (as for a flexible diaphragm) and those due to the torsional moment *Pe*.

  *Note that the force due to rotation adds to the direct shear force in the open front wall and subtracts from the force in the rear wall. Only the increases are considered in design. Calculation of torsional forces in buildings with rigid diaphragms is beyond the scope of this book.

EXAMPLE 9.19   Rotation in a Flexible Diaphragm

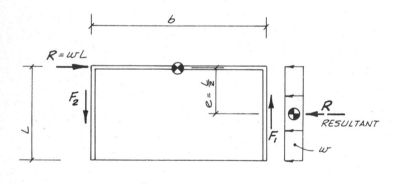

PLAN

Fig. 9.21

In buildings with no effective shearwall on one side, rotation of a *flexible* diaphragm can be used to carry the torsional moment into the walls perpendicular to the applied lateral load (not allowed for buildings with concrete or masonry walls). Even in entirely wood-framed buildings, the Code limits the depth of the diaphragm normal to the open side to 25 ft or two-thirds the diaphragm width, whichever is smaller. (For one-story buildings the width is limited to 25 ft or the full diaphragm width, whichever is smaller.)

Rear wall must carry entire longitudinal load:

$$R = wL$$

Force to transverse walls due to rotation:

$$\text{Torsion} = T = Re = \frac{wL^2}{2}$$

Resolve torsion into a couple separated by the distance between transverse walls:

$$F_1 = F_2 = \frac{T}{b} = \frac{wL^2}{2b}$$

It should be noted that this is not necessarily a desirable design technique. In fact, the *Code does not allow the consideration of rotation of a wood diaphragm in a building with concrete or masonry walls.* In these types of buildings, and the desirable approach in all cases, is to provide a shearwall along the open side of the building.

The question of rotation brings up an important point. The designer, insofar as practical, should strive for structural *symmetry*. In Fig. 9.21 it is quite unlikely that the design could be modified to have walls of exactly the same length at the front and back of the building. However, the design of a shearwall with an effective length sufficient to develop unit shears of a reasonable magnitude should be possible. The addition of this amount of symmetry will aid in reducing damage due to rotation.

Problems involving rotation were responsible for a number of the failures in residential buildings in the 1971 San Fernando earthquake. Reference 11 analyzes these failures and recommends construction details to avoid these problems.

9.12 Problems

9.1 *Given:* The single-story commercial building in Fig. 9.A. Plywood is ½-in. C-DX. No blocking. Nails are 8d. Critical lateral loads (wind or seismic) are

$$w_T = 140 \text{ lb/ft}$$
$$w_L = 320 \text{ lb/ft}$$

Vertical loads are

$$\text{Roof DL} = 10 \text{ psf}$$
$$\text{Snow} = 30 \text{ lb/ft}$$

Find: a. For the transverse direction
 1. Roof diaphragm unit shear
 2. Plywood load case (state criteria) and allowable unit shear
  b. For the longitudinal direction

323

      1. Roof diaphragm unit shear
      2. Plywood load case (state criteria) and allowable unit shear
   c. What is the required nailing?
   d. If blocking is not required, do the edges of the plywood need some
      other type of edge support?

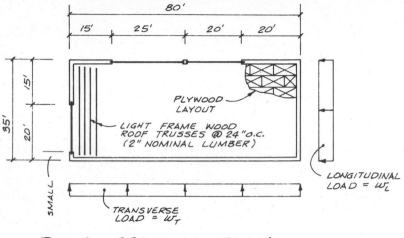

ROOF FRAMING PLAN

Fig. 9.A

9.2 *Given:* The single-story commercial building in Fig. 9.A. Plywood is ½-in. C-DX STR I. Critical lateral loads (wind or seismic) are

$$w_T = 320 \text{ lb/ft}$$
$$w_L = 320 \text{ lb/ft}$$

  Find: *a.* Roof diaphragm unit shear.
        b. Is the UBC span-to-width ratio satisfied? Show criteria.
        c. Is blocking required for lateral loads? If it is required, calculate at what point it can be omitted. Show all criteria such as diaphragm load case, allowable unit shears, and nailing requirements.
        d. The maximum chord force.

9.3 *Given:* The single-story wood-frame warehouse in Fig. 9.B. Roof sheathing of ⅜-in. C-DX STR I is adequate for vertical "sheathing" loads. Lateral loads are

$$w_T = 210 \text{ lb/ft}$$
$$w_L = 360 \text{ lb/ft}$$

  Find: *a.* Design the roof diaphragm, considering lateral loads in both directions. Show all criteria including design shears, load cases, allowable shears, and nailing requirements. If blocking is required, determine at what points it can be omitted.
        b. Calculate the maximum chord forces for both lateral loads. Also calculate the chord forces at the ends of all headers in the exterior walls.
        c. Plot the distribution of the strut force for each of the walls with openings. Compare the strut forces and chord forces to determine the critical loading.

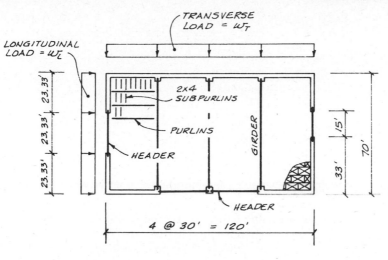

ROOF FRAMING PLAN

Fig. 9.B

9.4 Problem 9.4 is the same as Prob. 9.3 except that the plywood is ½-in. C-C STR I and the lateral loads are

$$w_T = 400 \text{ lb/ft}$$
$$w_L = 500 \text{ lb/ft}$$

9.5 *Given:* The building in Fig. 9.A and the lateral loads

$$w_T = 300 \text{ lb/ft}$$
$$w_L = 500 \text{ lb/ft}$$

Find: a. The maximum chord forces for the above loads. Also determine the chord forces at each end of the wall openings.
b. Plot the distribution of the strut forces for the two walls that have openings. Compare the magnitude of the strut forces with the chord forces.

9.6 *Given:* Assume that the framing plan in Fig. 9.A is to be used for the second-floor framing of a two-story retail sales building. Lateral loads are

$$w_T = 450 \text{ lb/ft}$$
$$w_L = 450 \text{ lb/ft}$$

Floor LL is to be in accordance with UBC. Plywood is C-C STR I.

Find: a. Required plywood thickness.
b. Required plywood nailing. Check unit shears in both the transverse and longitudinal directions. Show all criteria in design calculations.
c. Calculate the maximum chord force and drag force in the building. Assume that three 2 × 6s of No. 1 DF-larch are used as the top plate in the building. Check the maximum stress in this member if it serves as the chord and strut. At any point in the plate, two 2 × 6s are continu-

325

ous (i.e., only one member is spliced at a given location). A single row of 1-in.-diameter bolts is used for the connections.

9.7 The one-story building in Fig. 9.C has the following loads:

Roof DL= 12 psf
Wall DL = 150 pcf (normal-weight concrete)
Wind = 20 psf
Seismic coefficient = 0.186W
Trib. wall height to roof diaphragm = 10 ft
Roof LL in accordance with UBC Table 23C

Plywood is C-C (four-ply, three-layer). Diaphragm chord consists of two #5 continuous horizontal bars in the wall at the diaphragm level.

Find: a. Required plywood thickness.
b. Critical lateral loads.
c. Required plywood nailing based on the diaphragm unit shears. Show all design criteria. Consider loading in both directions.
d. Check the stress in the chord reinforcing steel. Allow. $f_s = 20$ ksi × 1.33.

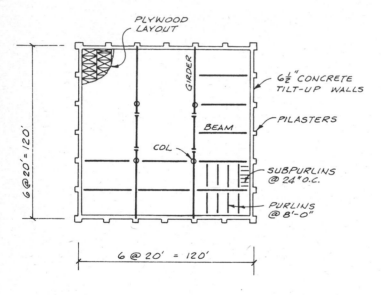

ROOF FRAMING PLAN

Fig. 9.C

9.8 *Given:* The one-story building in Fig. 9.D. Roof sheathing is ½-in. C-DX plywood spanning between roof joists spaced 24 in. o.c. Nails are 10d. For the transverse lateral load, the plywood layout is load case 1.

Find: a. Design the roof diaphragm assuming that only the exterior walls are effective in resisting the lateral load. Show all design criteria including span/width ratio, nailing, and blocking requirements.

b. Redesign the roof diaphragm assuming that the interior *and* exterior walls are effective shearwalls. Show all criteria.

c. Compare the maximum chord forces for *a* and *b*.

d. Plot the strut force diagram for the interior shearwall.

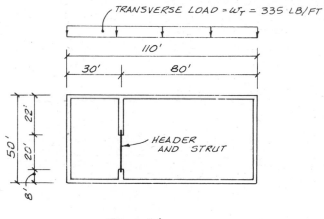

PLAN

Fig. 9.D

9.9 The header/strut over the opening in the interior shearwall in Fig. 9.D is a 4×14 No. 1 DF-larch. The vertical load is

$$w_{DL} = 100 \text{ lb/ft}$$
$$w_{RLL} = 160 \text{ lb/ft}$$

The axial force is to be determined from a plot of the drag strut force.

Find: Analyze the member for combined stresses. Full lateral support is provided for the weak axis by the roof framing.

CHAPTER TEN

Shearwalls

10.1 Introduction

Shearwalls comprise the vertical elements in the lateral-force-resisting system (LFRS). They support the horizontal diaphragm, and transfer the lateral loads down into the foundation. The procedures for calculating the loads to the shearwalls are covered in Chaps. 3 and 9.

Plywood horizontal diaphragms are often used in buildings with masonry or concrete shearwalls as well as buildings with wood-frame shearwalls. However, the design of masonry and concrete walls is beyond the scope of this book, and Chap. 10 deals only with the design of wood-frame shearwalls.

There are a number of sheathing materials that can be used to develop shearwall action in a wood-frame wall. These include

1. Plywood
2. Gypsum wallboard (drywall)
3. Interior and exterior plaster (stucco)
4. Fiberboard
5. Lumber sheathing

There may be other acceptable materials but these are representative. When the design loads (wall shears) are relatively small, the normal wall covering material may be adequate to develop shearwall action. However, when the unit shears become large, it is necessary to design special sheathing and nailing to develop the required load capacity.

10.2 Basic Shearwall Action

Essentially a shearwall cantilevers from the foundation as it is subjected to one or more lateral loads. See Example 10.1. As the name implies, the basic form of resistance is that of a shear element (Fig. 3.5c). The concept of "shear panels" in a wood-frame wall leads to the usual assumption that the lateral load is distributed uniformly throughout the total length of all panels (Sec. 9.6).

EXAMPLE 10.1   Cantilever Action of a Shearwall

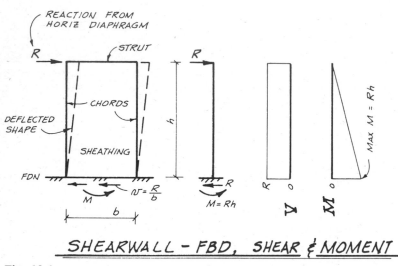

SHEARWALL – FBD, SHEAR & MOMENT

Fig. 10.1

Figure 10.1 shows a typical shear panel for a one-story building. If there are additional stories there will be additional loads applied to the shearwall at the diaphragm levels. If the lateral load is a seismic load there will also be an inertial force generated by the mass of the shearwall (Chap. 3).

A variety of sheathing materials can be used to develop shearwall action for resisting lateral loads. The Code sets limits on the height-to-width ratios h/b for shear panels in order to develop effective shear (deep-beam) resistance. Limits on h/b vary with different sheathing materials.

There are a number of items which should be considered in the design of a shearwall. These include:

1. Sheathing thickness
2. Shearwall nailing

330

3. Chord design

4. Strut design

5. Shear panel proportions

6. Anchorage requirements

These factors are essentially the same as those required for a horizontal diaphragm, but there are some differences in procedures.

The *sheathing thickness* depends on the type of material used in the wall construction. Loads normal to the surface and the spacing of studs in the wall may determine the thickness of the sheathing, but the unit shear often controls the thickness. In other cases the sheathing thickness may be governed by the required fire rating of a wall. *Shearwall nailing* or stapling requirements are a function of the design unit shears and the materials of construction.

As with a horizontal diaphragm, the *chords* are designed to carry the moment, and chords are required at both ends of shear panels. The *strut* for a shearwall is the same member as the strut for the horizontal diaphragm. It is the connecting link between the shearwall and the horizontal diaphragm for loads parallel to the shearwall. Design loads for the strut were covered in detail in Chap. 9.

The *proportions* of a shear panel are measured by its height-to-width ratio h/b. In buildings with two or more stories, the height of the shearwall is the vertical distance between horizontal diaphragms. The Code sets upper limits on the height-to-width ratio for various wall sheathing materials used as shear panels. Specific values of these limits will be covered in the sections dealing with the various construction materials. Panels cannot be considered as effective shearwall elements if these limits are exceeded. Shear panels which satisfy the height-to-width ratio criteria are considered by many designers to also have proportions which will not allow the panel to deflect excessively under load (this is similar to the span-to-width ratios for horizontal diaphragms—Sec. 9.8).

10.3 Plywood-Sheathed Walls

Perhaps the most widely used wood-frame shearwall is the type that uses plywood sheathing or plywood siding. The grades of plywood that are commonly used for sheathing and siding are described in Sec. 8.15. Plywood can be installed rapidly in the field, and with proper nailing it can be used to resist relatively high unit shears.

Plywood sheathing is often provided only on one side of a wall and then finish materials are applied to both sides of the wall. If the design unit shears are higher than can be carried with a single layer, another layer of plywood sheathing can be installed on the other side of the wall. See Fig. 10.2a. The addition of a second layer of plywood sheathing doubles the shear capacity of the wall. The capacity of the wall covering (drywall, plaster, stucco, etc.) is not additive to the shear capacity of the plywood.

331

Plywood siding can also be used to resist shearwall loads. This siding can be nailed directly to the studs, or it may be installed over a layer of ½-in. gypsum sheathing. See Fig. 10.2*b*. A wall construction that gives a 1-hour fire rating uses 2 × 4 studs at 16 in. o.c. with ½-in. gypsum sheathing and ⅜-in. plywood siding on the outside, and ⅝-in. type X gypsum wallboard on the interior. When grooved plywood panels are used, the thickness to be considered in the shearwall design is the net thickness where the plywood is nailed (i.e., the thickness at the groove).

EXAMPLE 10.2   Plywood Shearwalls

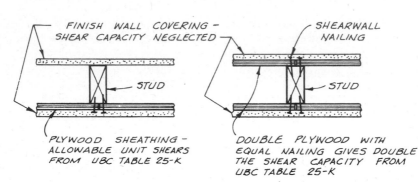

Fig. 10.2*a*   Plywood sheathing.

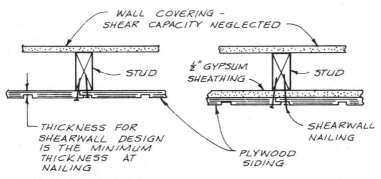

Fig. 10.2*b*   Plywood siding.

Allowable unit shears for plywood sheathing and siding are given in UBC Table 25K. This table covers both siding attached directly to framing, and siding applied over ½-in gypsum sheathing.

Allowable shear values are given for plywood nailed with *common or galvanized box nails* and *galvanized casing nails*. Finishing nails are not allowed for plywood shearwalls.

The allowable unit shears and nailing requirements for plywood shearwalls are given in UBC Table 25K. It will be helpful at this point to review the organization of this table. The shears given in the right side of the table are for plywood installed over the ½-in. gypsum sheathing. The allowable shears to the left of these values apply to plywood that is nailed directly to the wall framing.

Ultimate load tests of plywood shear panels indicate that failure occurs by the nail head pulling through the plywood panel. Common nails and box nails have the same-diameter head, and for this reason, tests of panels built with *common nails and galvanized box nails* develop approximately the same shear resistance. Allowable unit shears in the top portion of UBC Table 25K apply to both of these types of nails. The first section applies to plywood with the STRUCTURAL I addition to the plywood grade. The section below this

---

EXAMPLE 10.3   **Height-to-Width Ratios for Wood-Frame Shearwalls**

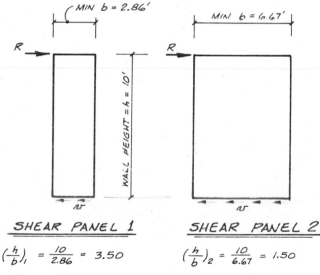

Fig. 10.3

Blocked plywood shearwalls have an upper limit on h/b of 3.5. Thus, for a 10-ft-high shearwall the minimum width is 2.86 ft. (shear panel 1).

Some common wall coverings can be used in shearwall design (Sec. 10.4). Many of these materials have a limit on h/b of 1.5. Here a 10-ft-high wall must have a length of 6.67 ft (shear panel 2). A wall length less than this cannot be used as a shear panel.

applies to all other plywood grades when nailed with common or galvanized box nails.

Although common and box nails can be used to install finish plywood siding, the size of the nail head may be objectionable from an appearance standpoint. Therefore, allowable shears are tabulated at the bottom of Table 25K for plywood siding installed with *galvanized casing nails.* Casing nails have a head diameter that is smaller than common or box nails but larger than finishing nails. Because of the tendency of the nail head to pull through the plywood, the allowable unit shears are considerably smaller when casing nails are used instead of common or box nails. The various types of nails are described more fully in Chap. 11.

UBC Table 25K applies to shear panels with the plywood installed either horizontally or vertically. The table also assumes that all plywood edges are supported either by the wall studs, plates, or blocking. Footnotes give adjustments to the tabulated shears where appropriate.

UBC Table 25I sets an upper limit on h/b of 3.5 for blocked plywood shearwalls. This is the largest height-to-width ratio permitted for wood-frame shear panels. For many other sheathing materials the maximum value for h/b is 1.5 (Sec. 10.4).

These limits can be important in the selection of the materials for a shear panel. See Example 10.3. The larger the allowable height-to-width ratio h/b, the shorter the permitted length of a shear panel for a given wall height. Thus if the length available for a shear panel is restricted, the use of a plywood shearwall may be required in order to satisfy the height-to-width limitations.

10.4   Sheathing Other than Plywood

It was mentioned at the beginning of Chap. 10 that a number of wall materials other than plywood can be used to develop shearwall action. If the design shears are high, plywood sheathing may be required. However, if the design loads are relatively low, these other wall coverings may have sufficient strength to carry the shears directly.

Some of the more common materials are plaster, stucco (exterior plaster), and drywall. UBC Table 47I gives the allowable unit shears and construction requirements for several of these commonly used materials. The allowable unit shears for these materials range from 75 to 250 lb/ft, and the maximum height-to-width ratio h/b is 1.5 (Example 10.3).

Other building materials that are not listed in the body of the Code may be tested and approved as an alternative method of construction. For example, certain forms of stucco construction have been approved for allowable unit shears up to 325 lb/ft. Values for the various approved types of construction are available from the Code research reports and from the manufacturers of the specific construction products.

Fiberboard wall sheathing, often used for insulation, can also be designed for shearwall action when the design shears are low. UBC Table 25O gives the allowable unit shear for fiberboard as either 125 or 175 lb/ft, depending on panel thickness and nailing. The upper limit on h/b fiberboard shearwalls is also 1.5.

In the past, lumber sheathing was used extensively for wood-frame walls. Lumber sheathing can be applied horizontally (known as transverse sheathing), but it is relatively weak and flexible. Diagonally applied sheathing is considerably stronger and stiffer because of its triangulated (truss) action. See Fig. 10.4. Shearwalls using lumber sheathing have largely been replaced

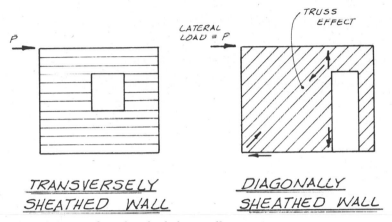

Fig. 10.4 Lumber-sheathed shearwalls.

with plywood shearwalls. Because of their relatively limited use, the design of lumber sheathed walls is not covered in this text. A treatment of lumber shearwalls is given in Ref. 1.

10.5 Bracing in Wood-Frame Walls

The Code requires that most buildings and other structures be "designed" for both vertical and lateral loads. However, light-frame wood construction has a history of satisfactory performance, and the Code generally accepts this type of construction without further justification. Residential structures often fall into this category.

The requirements for this type of construction are given in the *Conventional Construction Provisions* of the Code (UBC Chapter 25). Lateral load resistance is ensured in these types of buildings by requiring some form of *wall bracing*.

Diagonal "let-in" (or light-gauge-metal) braces have often been used in the past for this type of bracing. See Fig. 10.5. These braces are still allowed, but the Code also recognizes alternate forms of acceptable wall bracing for conventional construction. These alternate forms consist of a panel of solid sheathing (in effect, a shearwall). Specific details are given in the Code.

335

Experience in the San Fernando earthquake demonstrated the inadequacy of let-in braces. Many of these braces either failed in tension or they pulled out of the bottom wall plate (Ref. 11). If these braces are used in buildings that require structural calculations, they are essentially ignored for design purposes.

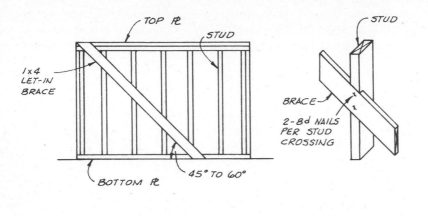

ELEVATION
WALL WITH LET-IN

LET-IN
DETAIL

Fig. 10.5 Let-in bracing.

10.6 Shearwall Chord Members

The vertical members at both ends of a shear panel are the chords of the vertical diaphragm. See Example 10.4. As with a horizontal diaphragm, the chords are designed to carry the entire moment. The moment at the base of the shear panel is resolved into a couple which creates axial forces in the end posts.

Some designers analyze the chords in a shearwall for forces created by the *gross overturning moment* in the wall. This is a conservative approach in designing for *tension* because it neglects the reduction in moment caused by the dead-load framing into the wall. For a definition of gross, net, and design net overturning moments, see Example 2.10.

Even if the dead-load-resisting moment is considered, the critical stress may be the tension at the net section. Reductions in the cross-sectional area often occur at the base of the tension chord for tie-down connections.

In the *compression* chord the gross overturning force alone may not represent the critical design force. *Gravity loads* may be carried by the compression chord *in addition to the force due to overturning.* If the chord members are studs (rather than a post or column), compression perpendicular to the grain of

EXAMPLE 10.4 Shearwall Chord Members

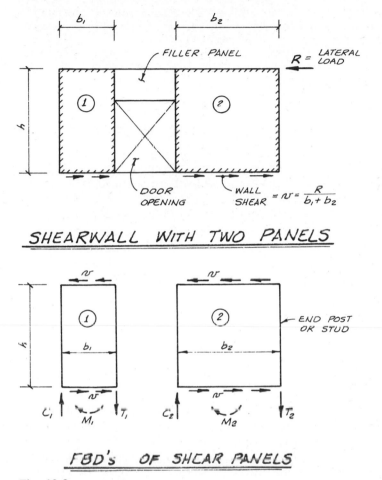

SHEARWALL WITH TWO PANELS

FBD's OF SHEAR PANELS

Fig. 10.6

For a wood wall the unit shear is assumed uniform throughout. For a lateral wind load the unit shear is the same at the top and bottom of the two shear panels. Gross overturning moment is calculated at the base of each shear panel:

$$M_1 = v \times b_1 \times h$$
$$M_2 = v \times b_2 \times h$$

Chord forces are calculated by dividing gross OMs by the length of the shear panels:

$$T_1 = C_1 = \frac{M_1}{b_1} = vh$$

$$T_2 = C_2 = \frac{M_2}{b_2} = vh$$

337

Thus, considering the effects of the *gross OM*:

$$T_1 = C_1 = T_2 = C_2 = vh$$

A somewhat modified (but similar) procedure is used to calculate the gross chord forces when additional lateral forces are involved. These forces can be generated by additional stories (horizontal diaphragms) or, in the case of seismic loads, a lateral force due to the mass of the shearwall itself (Fig. 10.7*d*).

Chord forces are *not equal* if the resisting dead-load moments are considered. When the effects of the dead load are taken into account, large chord forces will occur in the shorter length shear panel.

---

the wall plates can be a problem. With a column serving as chord, a bearing plate may be used to reduce the bearing stress. Lateral loads may occur in either direction, and chords must be designed for both tension and compression.

In order to ensure that the panels in a shearwall function as assumed (i.e., separate panels of equal height), double full-height studs should be used for the chords at the ends of the shear panels (Example 9.9). These will tend to emphasize the vertical continuity of the panels. In addition, the lighter nailing of filler walls can be used to further isolate the shear panels.

10.7   Design Problem: Shearwall

In Example 10.5 the exterior transverse walls of a rectangular building are designed for a seismic load. There is a roof overhang on the left end of the building which develops a smaller lateral seismic load than the main portion of the roof diaphragm. The difference in these two loads is the seismic load generated by the two longitudinal walls. The reaction of the left shearwall is made up of the cantilevered roof overhang plus the tributary load of the main roof diaphragm.

The total shear at the base of the transverse walls is made up of the tributary roof diaphragm reactions plus the seismic force due to the dead load of the shearwall itself. In this example the entire weight of the shearwall was considered in the calculation of the seismic load. Another common practice is to calculate the seismic load at the midheight of the wall.

The unit shear in the left wall can be carried by stucco of special construction. However, in this example a plywood shear panel is used. In the right wall, two shear panels are designed using the same type of plywood as the left wall, but the nailing is increased to meet the higher shear requirements.

The maximum chord force occurs in the walls with the higher unit shears. Calculations indicate that one 2 × 6 stud is sufficient to resist the tension chord force, but two 2 × 6 studs are required to carry the compression chord force.

EXAMPLE 10.5   Shearwall Design

Design the two transverse shearwalls in the building shown in Fig. 10.7a. Studs are 2 × 6 at 16 in. o.c. (No. 1 DF-L) and walls are of stucco construction. Seismic is critical and the seismic coefficient is 0.186W. Wall DL = 20 psf.

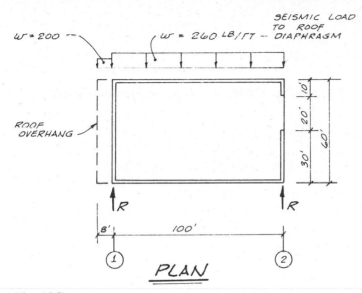

Fig. 10.7a

Diaphragm Reactions

Loads to shearwalls from horizontal diaphragm are calculated on a tributary width basis.

Wall 1: R = reaction from overhang + reaction from main roof

$$- (200 \times 8) + \left(260 \times \frac{100}{2}\right) - 14.6 \text{ k}$$

Wall 2: $R = 260 \times \dfrac{100}{2} = 13.0 \text{ k}$

Shearwall 1

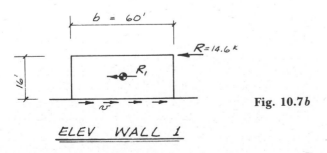

Fig. 10.7b

ELEV   WALL 1

Calculate wall seismic force using total wall dead load (Chap. 3).

$$R_1 = 0.186W = 0.186(20 \text{ psf} \times 16 \times 60) = 3.57 \text{ k}$$

$$V = R + R_1 = 14.6 + 3.57 = 18.17 \text{ k}$$

$$v = \frac{V}{b} = \frac{18,170}{60} = 303 \text{ lb/ft}$$

From UBC Table 47I:

$$\text{Shear capacity of cement plaster} = 180 \text{ lb/ft} > 303$$

Certain types of stucco construction have been approved for unit shears in excess of 300 lb/ft. One of these approved types of stucco can be used, or plywood sheathing may be provided.

Because the research report for these higher-strength types of stucco construction is not included in this text, the design of plywood sheathing will be shown. From UBC Table 25N, $\frac{5}{16}$-in. plywood can be used, but $\frac{3}{8}$-in. C-DX plywood is chosen. From UBC Table 25K and for 8d common nails,

$$\text{Allow. } v = 320 \text{ lb/ft} \times 1.20 \qquad \text{footnote 3}$$
$$= 384 \text{ lb/ft} > 303 \text{ lb/ft} \qquad OK$$

Use $\frac{3}{8}$-in. C-DX plywood (24/0)
with 8d common nails at
4 in. o.c. plywood edges
12 in. o.c. field.
Blocking required

Shearwall 2

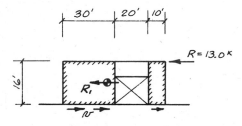

Fig. 10.7c

ELEV WALL 2

Use R_1 from wall 1 (conservative):

$$V = R + R_1 = 13.0 + 3.57 = 16.57 \text{ k}$$

$$v = \frac{V}{b} = \frac{16,570}{40} = 414 \text{ lb/ft}$$

Use plywood similar to wall 1
except 8d common nails at
2½ in. o.c. plywood edges
12 in. o.c. field.
Blocking required

$\text{Allow. } v = 470 \times 1.20 = 564 \text{ lb/ft} > 414 \qquad OK$

Tension Chord

The chord stresses are critical in wall 2 (Fig. 10.7*d*) because of the higher wall shears. The tension chord can be checked conservatively by the gross overturning moment:

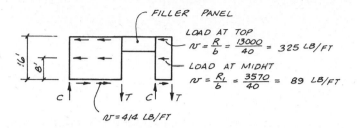

WALL 2 - CHORD FORCES

Fig. 10.7*d*

$$T = C = vh = 325(16) + 89(8) = 5.91 \text{ k} \qquad \text{(using gross OM)}$$

Check the tension stress in one stud

$$\text{Net area} = A_n = 1\tfrac{1}{2}(5\tfrac{1}{2} - 1) = 6.75 \text{ in.}^2$$

$$f_t = \frac{T}{A_n} = \frac{5910}{6.75} = 876 \text{ psi}$$

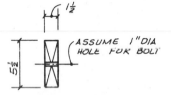

Fig. 10.7*e*

Lumber is No. 1 Douglas Fir.

$$\text{Adj. } F_t = \text{Tab. } F_t \times \text{LDF} = 1000 \times 1.33$$
$$= 1330 > 876 \text{ psi} \qquad OK$$

One 2 × 6 is adequate for the tension chord.

If the tension chord had been overstressed using the gross OM, the resisting DL moment can be considered (Example 2.10). The same result can be obtained by subtracting two-thirds of the DL tributary to the chord from the gross uplift force.* For a given unit shear, the shorter panel in a given shearwall (10 ft in this example) will be critical. Show calculations for a reduced tension chord force using the following dead loads:

Say 12-ft-long roof joists (not shown) frame into the right end wall, and the roof DL = 15 psf. Trib. length to one chord is one-half the wall length (five ft).

---

* The Code requires a FS of 1.5 for overturning in design for lateral *wind* loads. Thus, two-thirds of the RM is subtracted from the gross OM in calculating the design OM (Example 2.10). Although not required by the Code, the ⅔ factor is applied to the *seismic* load calculation in this problem.

$$\text{Roof DL} = 15 \times 6 = 90 \ \text{lb/ft}$$
$$\text{Wall DL} = 20 \times 16 = \underline{320 \ \text{lb/ft}}$$
$$\text{DL} = \overline{410 \ \text{lb/ft}} \ \text{of wall†}$$

$$\text{Design uplift} = T = 5.91 - \tfrac{2}{3}(0.410)(5)$$
$$T = 4.55 \ \text{k} \qquad \text{(considering DL)}$$

Compression Chord

Assume that the loads over the opening in wall 2 are supported by a separate column or trimmer.

Wall studs are 16 in. o.c. and the DL to a chord (stud) in the 10-ft wall is

$$P_{DL} = 410 \ \text{lb/ft} \times 1.33 \ \text{ft} = 545 \ \text{lb}$$

Add the DL of 545 lb to the gross chord force:

$$C = 5.91 + 0.545 = 6.46 \ \text{k}$$

Check the column capacity of one stud. Sheathing provides lateral support about the weak axis.

$$\left(\frac{L}{d}\right)_y = 0$$

$$\left(\frac{L}{d}\right)_x = \frac{16 \times 12 \ \text{in./ft}}{5.5} = 34.9$$

$$K = 0.671 \sqrt{\frac{E}{F_c \ (\text{LDF})}} = 0.671 \sqrt{\frac{1,800,000}{1250(1.33)}}$$

$$= 22.1 < 34.9 \qquad \therefore \ \text{long column}$$

$$F_c' = \frac{0.3E}{(L/d)^2} = \frac{0.3(1,800,000)}{(34.9)^2} = 443 \ \text{psi}$$

Check bearing perpendicular to grain on wall plate:

$$\text{Adj.} \ F_{c\perp} = \text{tab.} \ F_{c\perp} \times \text{LDF} = 385 \times 1.33 = 512 \ \text{psi} > 443$$
$$\therefore \ F_c' \ \text{governs}$$

$$\text{Allow.} \ P = F_c' A = (0.443)(8.25) = 3.66 \ \text{k}$$

For two studs,

$$\text{Allow.} \ P = 2(3.66) = 7.31 \ \text{k} > 6.46 \qquad OK$$

NOTE: These column calculations for the compression chord are conservative. The maximum compressive force at the base of the wall was used for the design load. However, the base of the chord is attached at the floor level, and column buckling is prevented at that point. Some reduced axial column force occurs at the critical location for buckling.

Check the compressive stress at the base of the chord using the net area (Fig. 10.7e). Column buckling does not occur

$$f_c = \frac{C}{A_n} = \frac{6460}{2(6.75)} = 479 \ \text{psi}$$
$$\text{Adj.} \ F_c = \text{tab.} \ F_c \times \text{LDF} = 1250 \times 1.33$$
$$= 1660 \ \text{psi} > 479 \qquad OK$$

---

342 † Roof LL is omitted when checking overturning.

> Use two 2 × 6 studs of No. 1 DF-L for all shearwall chords.

10.8   Anchorage Considerations

Six design considerations for shearwalls were listed in Sec. 10.2. The first five of these items have been covered (loads for designing the drag strut were covered in Chap. 9). The last item (anchorage) will now be considered. Generally speaking, anchorage refers to the tying together of the main elements so that the building will function as a unit in resisting the design loads. Although gravity loads need to be considered, the term anchorage emphasizes the design for lateral loads.

EXAMPLE 10.6   Basic Anchorage Criteria

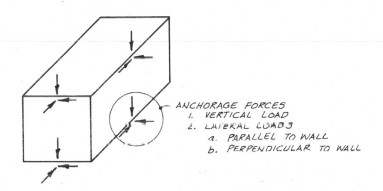

ANCHORAGE FORCES
1. VERTICAL LOAD
2. LATERAL LOADS
   a. PARALLEL TO WALL
   b. PERPENDICULAR TO WALL

VIEW   SHOWING
ANCHORAGE   FORCES

Fig. 10.8

The process of tying the building together can be systematically approached by considering loads in the three principle directions of the building. The critical locations are where the horizontal diaphragms connect to the vertical diaphragms, and where the shearwalls tie into the foundation. See Fig. 10.8.

Anchorage can be systematically provided by considering the transfer of the following loads:

1. Vertical (gravity) loads
2. Lateral loads parallel to a wall
3. Lateral loads perpendicular to a wall

If this systematic design approach is followed where the horizontal diaphragm connects to the various shearwalls, and where the shearwalls are attached to the foundation, the building will naturally be tied together. See Example 10.6.

The remainder of Chap. 10 provides an introduction to anchorage considerations. The transfer of gravity loads is first summarized. For lateral loads, it is convenient to separate the discussion of anchorage into two parts. The transfer of lateral forces at the *base of the shearwall* is fairly direct, and Chap. 10 concludes with a consideration of these anchorage problems. The anchorage of the *horizontal diaphragm to the shearwall* is somewhat more complicated, and this problem is covered in detail in Chap. 14.

The question of anchorage is basically a connection design problem. The scope of the discussion in Chap. 10 is limited to the calculation of the forces involved in the connections. Once the magnitude of the forces is known, the techniques of Chaps. 11 through 13 can be used to complete the problem.

10.9   Vertical (Gravity) Loads

As noted, the term anchorage implies connection design for lateral loads. However, this can be viewed as part of an overall connection design process. In general connection design, the first loads that come to mind are the gravity loads. These start at the roof level and work their way down through the structure. The "flow" of these forces is fairly easy to visualize. See Example 10.7.

The magnitude of the forces required for the design of these connections can usually be obtained from the beam and column design calculations. The hardware used for these connections is reviewed in Chap. 13.

10.10   Lateral Loads Parallel to a Wall

A shearwall supports lateral loads that are parallel to the wall. In doing so, it cantilevers from the foundation, and both a shear and a moment are developed at the base of the wall. Anchorage for the shear and moment are provided by separate connections.

The connections for resisting the *moment* will be considered first. It will be recalled that the moment is carried by the chords, and the chord forces are obtained by resolving the moment into a couple. See Fig. 10.10. The

EXAMPLE 10.7 Connections for Gravity Loads

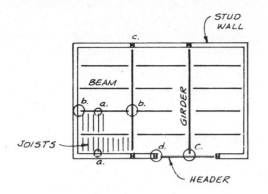

TYP ROOF FRAMING PLAN

Fig. 10.9

Design for gravity loads involves the progressive transfer of loads from their origin at the lightest framing member, through the structure, and eventually into the foundation. Consider the "path" for the transfer of gravity loads in the roof framing plan (Fig. 10.9). This requires the design of the following connections:

a. Joist connections to roof beam and bearing wall

b. Roof beam connections to girder and bearing wall

c. Girder connections to column and header

d. Header connection to column

e. Column connection to footing, and stud connection to wall plate (these connections are not shown on the sketch)

chord forces may be calculated from the gross overturning moment, or the resisting moment provided by the DL may be considered. Both procedures were illustrated in Example 10.5.

For connection design, the tension chord is generally of main concern. The tension chord force and the connection at the base of the tension chord will become important for

1. Tall shear panels (i.e., when h is large)

2. Narrow panels (i.e., when b is small)

3. When the resisting DL is small

When these conditions exist, a large tension uplift force develops at the base of the wall. If these conditions do not exist, there may be no uplift force at the base of the wall.

It was noted in Chap. 2 that a factor of safety of 1.5 is required for the overall moment stability of a structure. Actually, the Code requires this FS for lateral wind loads only. This FS is not required for seismic loads, and previous editions of the Code even had a reduction coefficient J, which was used to calculate a smaller design overturning moment. Although not required by the Code, the FS of $\frac{3}{2}$ is applied in this book to overturning calculations for both *wind* and *seismic* loads.

After the shearwall has been anchored to the foundation for any tension chord force, the overall stability of the wall must be checked. In this case the DL of the foundation is included in the resisting moment.

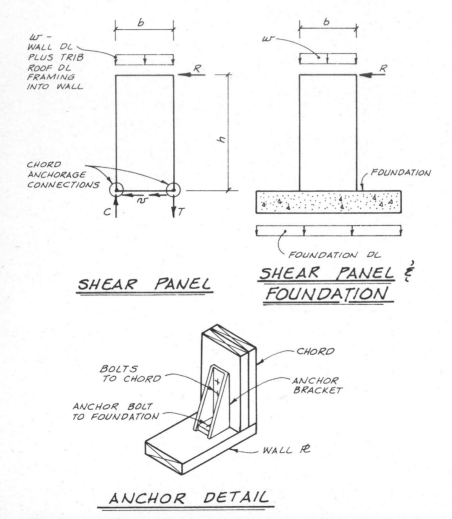

SHEAR PANEL

SHEAR PANEL & FOUNDATION

ANCHOR DETAIL

Fig. 10.10 Anchorage for shearwall moment. The FBD on the left is used to design the anchorage connection of the shear panel to the foundation. The FBD on the right is used to check overall moment stability.

The tension tie for the shearwall anchorage is often made with a prefabricated metal bracket. These brackets are attached to the chord member with bolts and to the foundation with an anchor bolt. The number and size of bolts, and the required embedment length for the anchor bolt, depend on the design load.

Connections of this type can be designed by using the wood design methods of Chap. 12 and the anchor bolt values given in UBC Table 26G. However, prefabricated brackets of this type usually have Code-accepted load values which can be used directly.

Allowable loads for commercially available anchorage hardware range from 1500 lb to 7500 lb for normal duration loads. Most of these values can be increased by 1.33 for short-term wind and seismic loads.

The connection for transferring the *shear* to the foundation will now be considered. This connection is normally made with a series of anchor bolts. See Example 10.8. These bolts are in addition to the anchor bolts used to anchor the chord forces.

It will be recalled that the shear is essentially carried by the wall sheathing. The attachment of the sheathing to the bottom wall plate will transfer the shear to the base of the wall. The anchor bolts, then, are simply designed to transfer the shear from the bottom plate into the foundation.

The design of bolted connections is covered in a later chapter, but the

---

EXAMPLE 10.8   Anchorage for Shearwall Shear

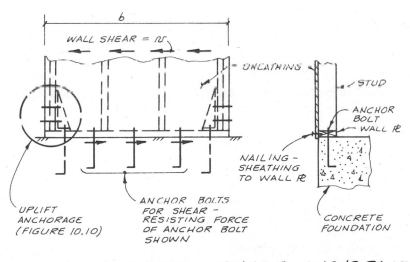

SHEAR PANEL   SHEAR ANCHORAGE

Fig. 10.11

The anchor bolt requirement can be determined by first assuming a size of anchor bolt and determining the allowable load per bolt. If the allowable load per anchor bolt is p (parallel to grain), and the total lateral load parallel to the shear panel is the unit shear times the panel width $(v \times b)$, the required number of anchor bolts is

$$N = \frac{v \times b}{p}$$

The average spacing is approximately

$$\text{Spacing} = \frac{b}{N} = \frac{bp}{vb} = \frac{p}{v}$$

(This spacing is approximate because starting and ending anchor bolts must be set in from the panel ends a sufficient distance to clear the chords and tie-down brackets.)

The minimum anchor bolt requirement for wood-frame walls is

½-in. anchor bolts at 6 ft-0 in. o.c. (UBC Chapter 29)

A minimum of two anchor bolts is required per wall plate, and one bolt is required within 12 in. of the end of each plate piece. Anchor bolts for shearwalls are usually larger and more closely spaced than the Code minimum.

---

basic design procedure is briefly described here. The strength of the anchor bolt is determined by the bearing capacity of the bolt parallel to the grain in the wood plate *or* by the strength of the anchor bolt in the concrete footing. The capacity of the bolt in the wood plate is covered in Chap. 12, and the capacity of the anchor bolt in the concrete footing is given in UBC Table 26G. The smaller of these two values is used to determine the size and number of anchor bolts.

Because the bottom wall plate is supported directly on concrete, there is a potential problem with termites and decay. If this hazard exists, the Code requires that protection be provided. This is accomplished by using lumber for the sill plate which has been pressure-treated or wood that has a natural resistance to decay (e.g., foundation redwood). An introduction to pressure-treated lumber is given in Sec. 4.6.

10.11   Lateral Loads Perpendicular to a Wall

In addition to being designed for the shearwall forces covered in Sec. 10.10, walls must also be anchored to the foundation for lateral loads that are normal to the wall. These include the tributary wind and seismic loads.

The lateral seismic load to be used in the design of this attachment is the force F_p. The seismic load F_p was compared with the wind load for a building with a wood roof and masonry walls in Example 2.15. In this earlier

EXAMPLE 10.9 Load Perpendicular to Wall

Compare the lateral design loads (wind and seismic) for the building in Example 2.15. Substitute wood-frame walls for the masonry walls and calculate the anchorage force at the bottom of the wall. Wall DL = 20 psf. Wind = 15 psf. The seismic coefficients from the previous example apply.

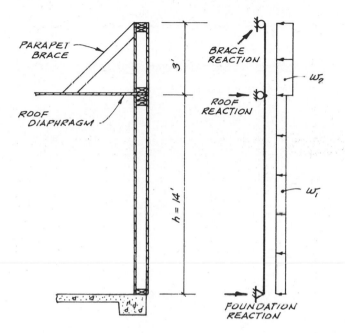

WALL SECTION

Fig. 10.12

Load to main wall:

$$w_1 = 0.3\,Wp$$
$$= 0.3(20)$$
$$= 6.0\,\text{psf} < 15$$

Wind governs main wall design and anchorage to footing

Load to parapet wall:

$$w_2 = 0.8\,Wp$$
$$= 0.8(20)$$
$$= 16\,\text{psf} > 15$$

Seismic governs parapet design and bracing

Anchorage force at bottom of wall calculated using tributary height:

$$R = w_1\left(\frac{h}{2}\right) = (15\,\text{psf})\left(\frac{14}{2}\right) = 105\,\text{lb/ft}$$

example, the seismic load was found to control because of the large dead load of the masonry wall. However, in buildings with wood-frame walls the seismic load is often less than the design wind pressure. See Example 10.9.

In the case of the lateral load *parallel* to the wall, the sheathing was assumed to transfer the load into the bottom plate of the wall. However, for the lateral load *normal* to the wall, the sheathing cannot generally be relied upon to

---

EXAMPLE 10.10   **Anchorage for Loads Perpendicular to Wall**

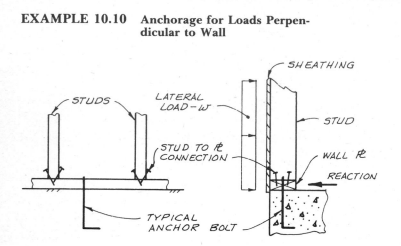

ANCHORAGE FOR NORMAL LOADS

Fig. 10.13

The lateral load normal to the wall is usually transferred from the *stud to the bottom wall plate* by nails. The nailing schedule in UBC Table 25P gives the standard connection for the stud to the bottom (sole) plate as four 8d toenails. An alternate connection is with two 16d end nails. The capacity of these connections can be evaluated by the techniques of Chap. 11. If the capacity of the standard stud-to-plate connections is less than the applied load, some other type of connection (such as a framing anchor—Chap. 13), will be required.

The connection of the *plate to the foundation* is accomplished with anchor bolts. The anchor bolt requirement can be determined by first assuming a size of anchor bolt and determining the allowable load per bolt.

If the allowable load per anchor bolt is q (perpendicular to grain), and the total foundation reaction (Example 10.9) is the reaction per foot R times the length of the wall b, the required number of anchor bolts is

$$N = \frac{R \times b}{q}$$

This number must be compared to the required number determined for anchoring the parallel to wall forces (Example 10.8). The larger number of bolts controls.

transfer these forces. Connections should be designed to transfer the reaction from the stud to the bottom plate, and then from the plate to the foundation. See Example 10.10.

The standard method of connecting wall studs to the bottom plate is with toenails or nails driven through the bottom plate into the end of the stud. The latter connection is often used when a wall panel is preframed and then lifted into place. Methods for evaluating the capacity of nailed connections are given in Chap. 11.

Experience during the 1971 San Fernando earthquake with conventional wood-frame houses proved the standard connection between wall studs and the bottom plate to be inadequate in many cases. Reference 11 was developed as a guide to home designers in an attempt to reduce these and other types of failures. This reference recommends the following anchorage for conventional wood-frame walls:

1. Framing anchors (Chap. 13) are to be used to fasten studs to the bottom wall plate at exterior corners. A ½-in.-diameter anchor bolt is to be installed within 2¾ in. of the corner studs (in addition to other anchor bolts). This is essentially an uplift connection.

2. In high-risk seismic zones (e.g., zone 4), the following additional anchorage is suggested. Framing anchors are to be used to connect the first two studs to the plate in each wall adjacent to the exterior corners.

3. Studs in the first story of two-story constructions are to be attached to the sole plate with framing anchors at 4-ft. intervals along the wall. This is required for all wall coverings except plywood.

For additional information see Ref. 11.

The load from the bottom plate is transferred to the foundation with anchor bolts. Because lateral loads parallel and perpendicular to the wall are not considered simultaneously, the same anchor bolts can be used for both of these loads.

The procedure for determining the required number of bolts is much the same as for parallel-to-wall loads. Here the strength of the anchor bolt is determined by the bearing capacity of the bolt perpendicular to the grain in the wood plate or by the strength of the anchor bolt in the concrete footing. The smaller of these two values is used as the allowable load per anchor bolt.

10.12 Problems

10.1 *Given:* The single-story wood-frame building in Fig. 10.A. The critical lateral load (maximum wind or seismic) in the longitudinal direction is given. Plywood sheathing is ½-in. STR I.

Find: a. The unit shear in the shear panel along line 1.
b. The required nailing for the shearwall along line 1.

    c. The net maximum uplift force at the base of the wall along line 1. Consider the resisting moment provided by the weight of the wall (assume no roof load frames into longitudinal walls). Wall DL = 20 psf.

    d. Are the proportions of the shear panel in the above wall within the limits given in the UBC?

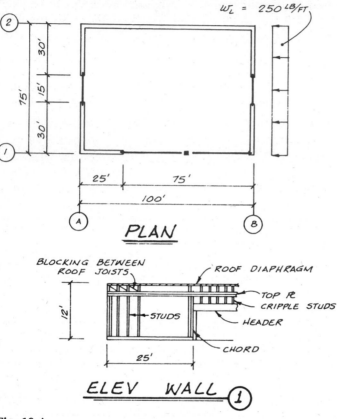

PLAN

ELEV WALL ①

Fig. 10.A

10.2   *Given:* The single-story wood-frame building in Fig. 10.A. The critical lateral load (maximum wind or seismic) in the longitudinal direction is given. Plywood wall sheathing (C-C exterior) will be designed to function as a shearwall. Studs are spaced 16 in. o.c.

    Find: *a.* The minimum required thickness of plywood sheathing, assuming that the face grain of the plywood may run either horizontally or vertically. (Plywood sheathing is used under a nonstructural wall covering.)

    b. The unit shear in the shear panel along line 1.

    c. Using the plywood thickness determined in *a.*, select the required nailing for the shearwall along line 1. Indicate blocking requirements (if any). Check the proportions of the shear panel.

   d. The maximum chord force in the shear panel along line 1, neglecting any resisting moment. Is a doubled 2 × 6 stud adequate to resist this chord force? Lumber is Stud grade DF-larch. A single row of ¾-in.-diameter bolts is used for the anchorage connection at the base of the wall.
   e. The uplift force for the design of the connection at the base of the tension chord, considering the resisting moment provided by the wall and the roof dead load supported by the wall. The wall supports a 5-ft trib. width of roof DL. Roof DL = 10 psf. Wall DL = 15 psf.

10.3 *Given:* The single-story wood-frame building in Fig. 10.A. The *transverse* lateral wind load (not shown in the sketch) is 375 lb/ft. The dead load of the roof and the wall is 250 lb/ft along the wall. Texture 1–11 plywood siding is applied directly to studs spaced 16 in. o.c.

   Find: a. Unit shear in the roof diaphragm and in the shearwalls along lines *A* and *B*.
   b. The required nailing for the plywood siding to function as a shear panel. Use casing nails. Show all criteria. Check the proportions of the shear panel.
   c. The net design uplift force at the base of the shear panel chords.

10.4 *Given:* The wood-frame building in Fig. 10.A. The critical lateral load in the longitudinal direction is given. The exterior wall finish is stucco (⅞-in. portland cement plaster over woven wire mesh).

   Find: Determine if the longitudinal wall along Line 2 is capable of functioning as a shearwall for the given lateral load. Specify nailing. Check diaphragm proportions. Show all criteria.

10.5 *Given:* The one-story wood-frame building in Fig. 10.B. The critical lateral load in the longitudinal direction is the given seismic load. The roof overhang has a higher lateral load because of the larger dead load of the plastered eave. Neglect the seismic load generated by the DL of the wall itself. Plywood wall sheathing is ½-in. C-C STR I.

   Find: a. Design the plywood shear panels in the wall along line 1. Show all criteria including nailing, blocking, and shear panel proportions.
   b. Check shear panel chords using two 2 × 6s of No. 2 DF-larch. Consider the resisting moment provided by a wall DL of 20 psf. Assume no roof DL frames into wall. A single row of ¾-in.-diameter bolts is used to anchor the chord to the foundation.
   c. Plot the distribution of the strut force along line 1. Check the stresses caused by the maximum strut force if a double 2 × 6 wall plate serves as the strut. A single row of ⅞-in.-diameter bolts is used for the splices. Lumber is No. 2 DF-larch.

10.6 *Given:* The one-story building in Fig. 10.B. The longitudinal seismic load to the roof diaphragm is given. Texture 1–11 plywood siding is applied over ½-in. gypsum sheathing for the wall along line 2. In addition to the load shown, consider the seismic force generated by the DL of the top half of the wall. Assume Wall DL = 15 psf throughout and a seismic coefficient of 0.186.

Find: *a.* Design the shear panels in the wall along line 2 using casing nails. Show all criteria.

      b. Check the shear panel chords using two 2 × 4s of No. 2 DF-larch. Neglect any resisting moment. A single row of ⅝-in.-diameter bolts is used to anchor the chord to the foundation.

      c. Plot the distribution of the strut force along line 2. Check the stresses caused by the maximum strut force if a double 2 × 4 wall plate serves as the strut. A single row of ⅝-in.-diameter bolts is used for the splices. Lumber is No. 2 DF-larch.

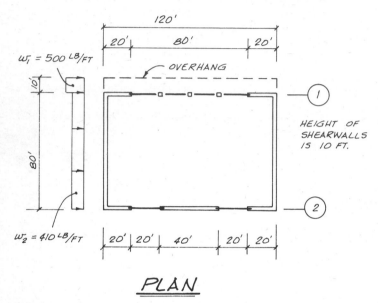

PLAN

Fig. 10.B

10.7   *Given:*   The one-story wood-frame building in Fig. 10.B. Roof DL = 10 psf. Wall DL = 18 psf (assume constant throughout). Wind = 20 psf. Seismic coefficient = 0.186.

    Find: *a* Calculate the critical lateral load in the transverse direction.

      b. The exterior wall covering for the transverse walls is stucco (⅞-in. portland cement plaster on expanded metal lath). Determine if these walls can function adequately as shearwalls. Consider the shear at the midheight of the wall. Show all criteria and give construction requirements.

10.8   *Given:*   The one-story wood-frame building in Chap. 9, Fig. 9.D. The transverse lateral load is $w_T = 355$ lb/ft. The height of the building is 10 ft.

    Find: *a.* The unit shears in the three transverse shearwalls.

      b. Design the three transverse shearwalls using ⅜-in. C-DX plywood wall sheathing. Show all criteria including shearwall proportions.

      c. Design the three transverse shearwalls using ½-in. C-D STR I plywood wall sheathing.

    d. Calculate the chord forces for each of the three transverse shearwalls. Because the layout of the roof framing is not completely given, calculate the chord forces using the gross overturning moments.

10.9   *Given:* The elevation of the two-story wood-frame shearwall in Fig. 10.C. The reactions of the roof diaphragm and second-floor diaphragm are shown in the sketch. Vertical loads carried by the wall are also shown in the sketch (DL values include the weight of the wall). Plywood wall sheathing is ½-in. C-C STR I.

      Find: *a.* The required nailing for the shear panel. Give full specifications.
           b. The required tie (drag strut) force for the connections at *A*, *B*, *C*, and *D*.
           c. The net design uplift force for the tension chord at the first- and second-floor levels.
           d. The maximum force in the compression chord at the first- and second-floor levels. Consider the effects of both gravity and lateral loads. Assume that the header is supported by the shear panel chord.
           e. If the allowable load on one anchor bolt is 1500 lb, determine the number and approximate spacing of anchor bolts necessary to transfer the shear from the bottom wall plate to the foundation. These bolts are in addition to those required for uplift.

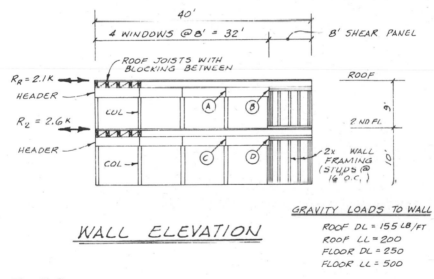

WALL ELEVATION

ROOF JOISTS WITH BLOCKING BETWEEN

$R_R = 2.1 K$

HEADER

$R_2 = 2.6 K$

HEADER

COL

COL

40'

4 WINDOWS @ 8' = 32'

8' SHEAR PANEL

ROOF

2 ND FL.

2x WALL FRAMING (STUDS @ 16' O.C.)

GRAVITY LOADS TO WALL

ROOF DL = 155 LB/FT
ROOF LL = 200
FLOOR DL = 250
FLOOR LL = 500

Fig. 10.C

10.10   *Given:* The wood-frame shearwall in Fig. 10.D. The lateral load from the roof diaphragm to the end shear panel is given. Gravity loads are also shown on the sketch. The DL includes both roof and wall DL. The shear panel has ½-in. C-D plywood sheathing.

      Find: *a.* Determine the required nailing for the plywood.
           b. Calculate the net design uplift for the shear panel chord. Does this occur at column *A* or *B*?

c. Determine the maximum compressive force in the shear panel chord. Consider the effects of gravity loads. Assume that the header is supported by the shear panel chord. What is the critical load combination? Specify which chord is critical.

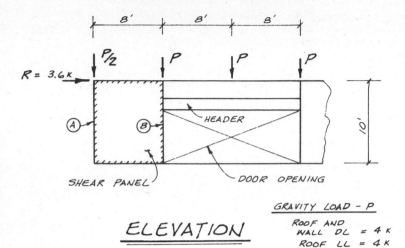

GRAVITY LOAD – P

ROOF AND
WALL DL = 4 K
ROOF LL = 4 K

ELEVATION

Fig. 10.D

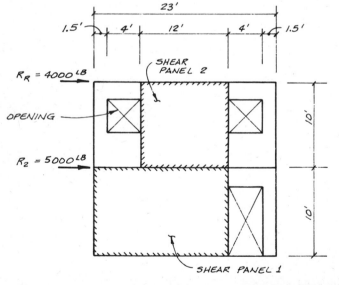

GRAVITY LOADS TO WALL

ROOF DL = 140 LB/FT
ROOF LL = 200
FLOOR DL = 200
FLOOR LL = 500
WALL DL = 16 PSF

ELEVATION

     d. Assume that the allowable load per anchor bolt is 1000 lb. Determine the required number and approximate spacing of anchor bolts necessary to transfer the wall shear from the bottom plate to the foundation. These bolts are in addition to those required for uplift.

10.11   *Given:* The elevation of the two-story wood-frame shearwall in Fig. 10.E. Vertical and lateral loads are shown in the sketch. Wall framing is 2 in. nominal. Studs are 16 in. o.c. Wall sheathing is ½-in. C-C STR I.

    Find: *a.* The required nailing for the first- and second-story shear panels.
           b. The net design uplift force at the base of shear panels 1 and 2.
           c. The maximum compressive force at the end of the first-floor shear panel. State the critical load combination.
           d. Assume that the capacity of anchor bolts parallel to the wall is 1500 lb. Determine the required number and approximate spacing of anchor bolts necessary to transfer the shear from the wall into the foundation. These bolts are in addition to those required for uplift.

10.12   Problem 10.12 is the same as Prob. 10.11 except that shear panel 1 is 12 ft long and lies directly below shear panel 2.

Nailed and Stapled Connections

11.1 Introduction

The use of nails in the construction of horizontal diaphragms and shearwalls was described in previous chapters. In addition, nails are used in a number of other types of structural connections. Nails are used when the loads are relatively small, and other types of fasteners (e.g., bolts) are used for larger loads.

The basic concept of a nail attaching one member to another member has undergone little change, but many developments have occurred in the configuration of nails and in the methods of installation. For example, Ref. 20 lists over 235 definitions and descriptions relating to the use of nails. The majority of these nails are fasteners for use in specialized applications, and the structural designer deals with a relatively limited number of nail types. Power driving equipment has contributed substantially to the advances in this basic type of wood connection.

The different types of nails are distinguished by the following characteristics:

1. Nail head
2. Shank
3. Nail point
4. Material type
5. Surface condition

359

The first three items have to do with the configuration of the nail. See Fig. 11.1, in which four typical nail heads are shown along with the widely used diamond-shaped nail point. Many other types of nails are described in ASTM Standard D2478 (included in Ref. 29) and in Refs. 20 and 30.

Chapter 11 introduces the main types of nails used in structural applications. The factors affecting the strength of nailed connections are reviewed, and the design procedures for the various types of nailed connections are outlined.

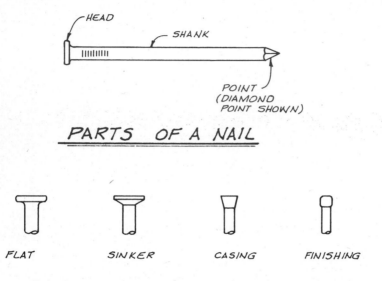

Fig. 11.1 Definitions of nail terms.

11.2 Types of Nails

The NDS provides structural design values for the following basic types of nails:

1. Common nails
2. Box nails
3. Common spikes
4. Threaded hardened steel nails

The size of nails is specified by the pennyweight of the nail, abbreviated d. The pennyweight for different types of nails specifies the length, shank diameter, and head size of the nail. The sizes of the nails listed above are tabulated in NDS Sec. 8.8, *Nailed and Spiked Joints*.

For a given pennyweight, all four basic types of nails have the same length.

The first three types have the same basic form, but the diameters are different. They are fabricated from low-carbon-steel wire and have a flat head, plain shank, and diamond point (like the nail shown in Fig. 11.1). The fourth type of nail is made from high-carbon steel and has an annularly or helically threaded shank, a flat head, and a diamond point (some low-carbon threaded steel nails have sinker heads). See Fig. 11.2.

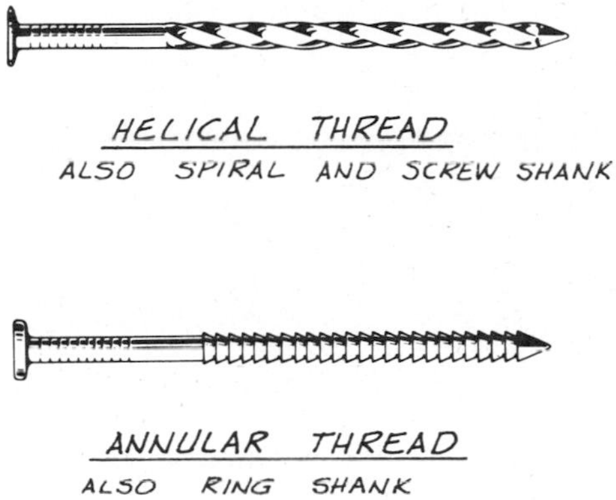

HELICAL THREAD
ALSO SPIRAL AND SCREW SHANK

Fig. 11.2 Threaded hardened steel nails.

ANNULAR THREAD
ALSO RING SHANK

Common nails represent the basic structural nail. Tabulated design values for common wire nails are for a nail surface condition that is described as *bright.* This term is used to describe nails with a natural bare metal finish, and load values for this surface condition can be considered conservative for other surface conditions. The diameter of common wire nails is larger than the diameter of box nails, and allowable loads are correspondingly higher. Because of their large diameter, common nails have less tendency to bend when driven manually.

Box nails are also widely used. They are available in several different surface conditions including bright, galvanized (zinc-coated), and cement-coated. Tabulated values for nails in the bright condition may conservatively be used for other surface conditions. Coated nails are discussed below. The wire diameter of box nails is smaller than for common nails, and the allowable loads are correspondingly smaller. Box nails also have a greater tendency to bend during driving.

Common spikes are similar in form to common wire nails except that they have a larger diameter. Because of their larger diameter, spikes have higher allowable loads, but the spacing requirements between fasteners are greater in order to avoid splitting of the wood. Predrilling of holes may also be necessary to avoid splitting.

Threaded hardened steel nails are made from high-carbon steel and are heat treated and tempered to provide greater strength. These nails are especially useful when the lumber may have varying moisture contents. The threaded shank provides better load resistance in these circumstances.

361

The structural designer may encounter some types of *nails not covered in the NDS*. These may include:

1. Finishing nails
2. Casing nails
3. Cooler nails
4. Zinc-coated nails
5. Cement-coated nails

Finishing nails are slender, bright, low-carbon-steel nails with a finishing head and a diamond point. These nails are used for finish or trim work, and they may be used for the installation of nonstructural panelling. However, because of the small nail head, finishing nails are not allowed for the attachment of structural sheathing or siding (Sec. 10.3).

Casing nails are slender, low-carbon nails with a casing head and a diamond point. These are available with a zinc coating to reduce staining of wood exposed to the weather. Casing nails can be used for the structural attachment of plywood sidings when the appearance of a flat nail head is considered objectionable. Because of the smaller diameter of the nail head, allowable plywood shearwall values for casing nails are smaller than for common or galvanized box nails (Sec. 10.3).

Cooler nails are slender, low-carbon-steel nails with a flat head and diamond point, and are usually cement coated. The head diameter is the same or smaller than that of a common wire nail of the same length. When gypsum wallboard (drywall) of the required thickness is properly installed with the Code-required size and number of cooler nails, a wall may be designed as a shear panel using the allowable loads from UBC Table 47I (Sec. 10.4).

It was noted earlier that nails can be obtained with a number of different surface conditions. The bright condition was previously described, and several others are introduced in the remainder of this section.

Zinc-coated (galvanized) nails are intended primarily for use where corrosion and staining resistance are important factors in performance and appearance. In addition, resistance to withdrawal may be increased if the zinc coating is evenly distributed. Extreme irregularities of the coating may actually reduce withdrawal resistance. It should also be pointed out that the use of galvanized nails to avoid staining is not fully effective. Where staining should be completely avoided for architectural purposes, stainless steel nails can be used. The expense of stainless steel nails is much greater than galvanized nails.

Cement-coated nails are coated by tumbling or emergence in a resin or shellac (cement is not used). Resistance to withdrawal is increased because of the larger friction between the nail and the wood. However, there are substantial variations in the uniformity of cement coatings, and much of the coating may be removed in the driving of the nail. These variables cause large differences in the relative resistance to withdrawal for cement-coated nails. The increased resistance to withdrawal for these types of nails is, in most cases, only temporary.

11.3   Power-Driven Nails and Staples

Major advances in the installation of driven fasteners such as nails and staples have been brought about by the development of pneumatic, electric, and mechanical guns. The use of such power equipment greatly increases the speed with which these fastening devices can be installed. The installation of a nail or other fastener has been reduced to the pull of a trigger and can be accomplished in a fraction of a second. See Fig. 11.3.

Fig. 11.3   Typical power fastening equipment. (*a*) Nailing through bottom wall plate with Senco SN-IV nailer. (*b*) Stapling plywood roof sheathing with Senco M-II stapler. (*Courtesy of Senco Products, Inc.*)

Conventional round flathead nails can be installed with power equipment. In order to accomplish this, the nails are assembled in clips or coils that are fed automatically into the driving gun. The round head on these types of fasteners prevents the shanks of the nails from coming in close contact with adjacent nails, and therefore, a relatively large clip is required for a given number of nails.

The size of the clip can be reduced by modifying the shape of the conventional nail. A number of different configurations have been used in order to make more compact fastener clips. *Wire staples, T nails,* and *modified roundhead nails* are all in current use. See Fig. 11.4. Fasteners such as staples are often

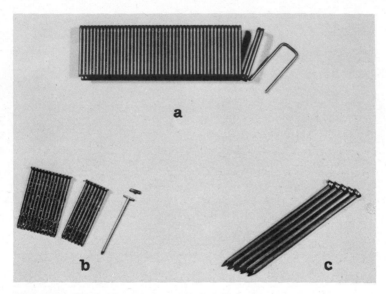

Fig. 11.4 Typical power-driven fasteners. *(a)* Wire staples; *(b)* T nails; *(c)* Modified round-head nails. *(Photograph by Mike Hausmann.)*

made from relatively thin wire. The tendency of this type of fastener to split the wood is reduced. In addition, standard power-driven nails are often thinner and shorter than corresponding pennyweight sizes of hand-driven nails.

Testing programs have resulted in Code acceptance of a number of power-driven fasteners for use in structural design. Information on fasteners and design load values is given in Ref. 18 and in Code research reports. In addition to these sources, information and literature can be obtained from distributors of power fastening equipment. The literature giving design information is extensive and is not reproduced in this text. The remainder of Chap. 11 deals specifically with the four basic types of nails covered in the NDS. The designer, however, should be able to locate *equivalent* power-driven fasteners in the references.

It should be pointed out that the setting of power driving guns is an important factor in the performance of a fastener. Care should be taken that the driving gun does not cause the head of a fastener to penetrate below the surface of the wood. Fasteners improperly installed will have a substantially lower load-carrying capacity. Precautions should also be taken in the manual driving of nails, and the head of a nail should not be "set" deeply below the surface of the wood. Because of the rapidity and consistency obtained with power driving equipment, the installation setting is of particular importance with power-driven fasteners. The type of member will affect the setting of the gun. For example, when plywood roof sheathing is installed on a panelized roof, the gun should be adjusted when nailing into the different supporting members (2 X's, 4 X's, and glulams).

11.4 Factors Affecting Strength of Nailed Connections

There are a number of factors that affect the capacity of a nailed connection. These include:

1. Number, size, and type of nail
2. Species of lumber
3. Type of connection (lateral or withdrawal)
4. Direction of nailing (side grain, end grain, or toenail)
5. Condition of use (moisture content at fabrication and in service)
6. Duration of load
7. Type of side plate
8. Spacing of nails

The NDS provides tabulated allowable loads for the four basic types of nails described in Sec. 11.2 for the various commercial species of wood. Several tables in UBC Chap. 25 give similar values for common and box nails in Douglas fir-larch and Southern pine. Numerical values for design can be obtained directly from the NDS (or the Code), and only the general effects of the strength factors are discussed in Chap. 11. For power-driven nails and staples, Ref. 18, or other appropriate source, should be consulted.

The capacity of a nailed connection is calculated as the sum of the nail capacities of the individual fasteners. The tabulated allowable loads apply to lumber seasoned to a moisture content of 19 percent or less at the time of fabrication of the connection *and* that will remain dry in service. NDS Table 8.1B, *Fastener Load Modification Factors for Moisture Content*, gives multiplying factors (CUFs) for moisture-content conditions other than those described. It should be noted that no reductions of allowable load are required (i.e., CUF = 1.0) for connections using threaded hardened steel nails. CUFs for other types of nails range from 0.25 to 1.0.

365

The LDF (load duration factor) was defined in Sec. 4.11 and is used to adjust allowable loads on nailed connections as well as the allowable stresses in wood members.

Before the remaining criteria can be discussed, it is necessary to define some basic terminology. The member that the nail passes through is referred to as the side member or side plate. See Example 11.1. The nail then penetrates into the holding member. The penetration of the nail is the length that extends into the holding member.

EXAMPLE 11.1 Members in a Nailed Connection

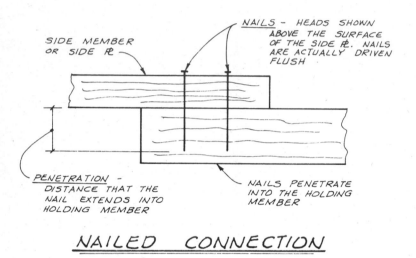

NAILS – HEADS SHOWN ABOVE THE SURFACE OF THE SIDE ₤. NAILS ARE ACTUALLY DRIVEN FLUSH

SIDE MEMBER OR SIDE ₤

PENETRATION – DISTANCE THAT THE NAIL EXTENDS INTO HOLDING MEMBER

NAILS PENETRATE INTO THE HOLDING MEMBER

NAILED   CONNECTION

Fig. 11.5

The connection shown in Fig. 11.5 has a wood side plate. Steel side plates can also be used and are generally made of sheet metal.

Penetration is calculated as the length of the nail minus the thickness of the side plate. If the nail extends completely through the holding member, the penetration is taken as the thickness of the holding member.

11.5 Lateral Load Connections

In a laterally loaded connection the load is applied *perpendicular* to the length of the nail. These are the most common nailed connections.

Two types of lateral load connections are distinguished. In one connection the nail is driven into the side grain of the holding member. See Fig. 11.6a in Example 11.2. This is the strongest type of nailed connection, and it is recommended over the other types. The tabulated allowable values for lateral load connections are for this basic connection.

Allowable lateral loads depend on the species of the wood used in the connection. NDS Table 8.1A, *Grouping of Species for Fastening Design,* divides the commercial lumber species into four species groups (Groups I to IV). Allowable lateral loads P_L for these species groups are given in NDS Table 8.8C, *Nails and Spikes—Lateral Load Design Values.* The remainder of this section should be accompanied by a review of the NDS tables.

Tabulated allowable loads are based on an assumption that a minimum penetration is provided. The penetration requirements differ for the different

---

EXAMPLE 11.2   Lateral Load Connections

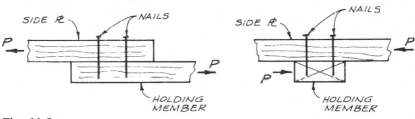

Fig. 11.6a

Lateral Resistance in Side Grain
The *basic* lateral load connection (Fig. 11.6a) has the nail driven into the side grain (i.e., perpendicular to the grain) of the holding member, and the load is applied perpendicular to the length of the nail.

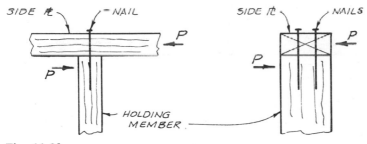

Fig. 11.6b

Lateral Resistance in End Grain
The second type of lateral load connection (Fig. 11.6b) has the nail driven into the end grain (i.e., parallel to the grain) of the holding member, and the load is applied perpendicular to the length of the nail.

species groups. The required length of penetration is given in the NDS load table. These values are calculated as a multiple of the nail diameter. As an example, Southern pine is in species group II, and this group requires a penetration of 11 nail diameters. The strength of nail connections which do not provide the specified penetration can be evaluated by straight-line interpolation between zero and the tabulated load. In any event, the actual penetration must not be less than one-third the specified penetration.

With the other type of lateral load connection the nail is driven into the *end grain* of the holding member. See Fig. 11.6*b*. This is a weaker connection, and allowable load is taken as two-thirds of the value for the lateral load connection with the nail driven into the side grain ($\frac{2}{3} \times P_L$). For all connections, nails are driven through the side grain of the side plate.

The basis for the allowable lateral loads should be understood. A plot of load vs. deflection for a typical lateral load connection with side grain resistance gives a comparison between the allowable load and the ultimate load capacity. See Example 11.3. Allowable loads are based on limiting the deformation that occurs in the joint to some reasonable value. This results in a large factor of safety against ultimate. For softwood lumber, Ref. 4 indicates that this factor of safety is about 5.

It is important to understand that allowable nail values are based on deformation rather than strength. For this reason the tabulated allowable loads

---

EXAMPLE 11.3   Deformation in Lateral Load Connections

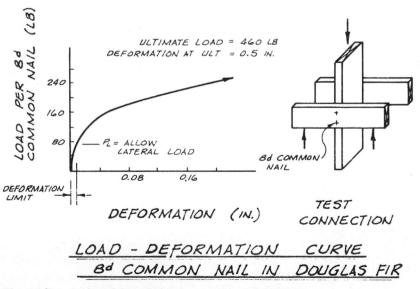

LOAD - DEFORMATION CURVE
8ᵈ COMMON NAIL IN DOUGLAS FIR

368   **Fig. 11.7a**   *(Adapted from Ref. 4 by permission of WWPA.)*

The allowable lateral load on a nailed connection is based on limiting the deformation in the connection.

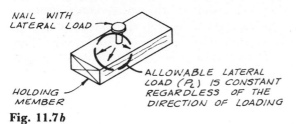

NAIL WITH LATERAL LOAD

HOLDING MEMBER

ALLOWABLE LATERAL LOAD (P_L) IS CONSTANT REGARDLESS OF THE DIRECTION OF LOADING

Fig. 11.7*b*

Because design values are based on deformation, the allowable load per nail is the same regardless of the direction of the lateral load.

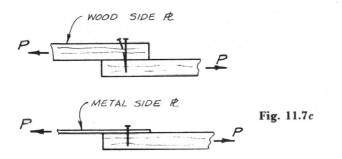

WOOD SIDE P̸

METAL SIDE P̸

Fig. 11.7*c*

Metal side plates limit deformation, and allowable lateral loads can be increased 25 percent.

---

do not depend on the direction of the load with respect to the direction of the grain. (Allowable loads on bolts and other fasteners depend on the direction of the load with respect to the direction of the grain of the wood.)

The deformation in a nailed joint explains why an increased allowable load is permitted when metal side plates are used instead of wood side members. Metal side plates are usually thin steel plates (usually sheet metal), and consequently there is less length in which joint deformation can occur. Even for thicker metal side plates, the nail is held more rigidly and deformation is limited. When metal side plates are used, the allowable lateral load for wood side plates can be multiplied by 1.25. In this book this adjustment is referred to as the Metal Plate Factor (MPF).

11.6   Design Problem: Nailed Connection for Knee Brace

The type of brace used in this example is typical for a number of unenclosed structures including carports, sheds, and patios. See Example 11.4. The total lateral load is assumed to be shared equally by all of the columns, and the reaction at the base of the column is determined first. The horizontal component of force in the brace is then calculated using the free-body diagram (FBD) of the column. From the horizontal component the axial force in the brace can be determined.

In this problem the required nail penetration is 1.78 in. This is less than the actual or furnished penetration, and no reduction of allowable load is required.

EXAMPLE 11.4   Knee Brace Connection

The carport shown below uses 2 × 6 knee braces to resist the longitudinal seismic load. Determine the number of 16d common nails required for the connection of the brace to the 4 × 4 post. Lumber is "dry" at the time of construction. Consider Southern pine or Douglas fir-larch.

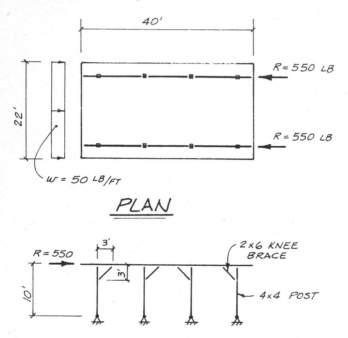

Fig. 11.8a

Load to one row of braces:

$$R = \frac{wL}{2} = 50\left(\frac{22}{2}\right) = 550 \text{ lb}$$

Assume load is shared equally by all braces.

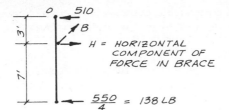

Fig. 11.8*b*

FBD OF COLUMN

$$\Sigma M_o = 0$$
$$3H \quad 138(10) = 0$$
$$H = 458 \text{ lb}$$
$$B = \sqrt{2}\, H = \sqrt{2}\,(458)$$
$$= 648\text{-lb axial force in knee brace}$$

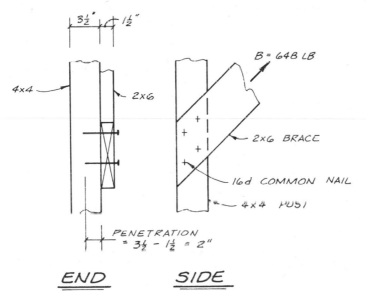

END   SIDE

ELEVATIONS

Fig. 11.8*c*

Southern pine and Douglas fir-larch are both in nail species group II (NDS Table 8.1A, *Grouping of Species for Fastening Design*).

The lateral load capacity of a 16d common nail is

$$P_L = 108 \text{ lb}$$

and the required penetration is

$$11 \times \text{dia.} = 1.78 \text{ in.} < 2.00 \qquad OK$$

371

Because the building is "unenclosed" the brace connection may be exposed to moisture. The severity of this exposure must be judged by the designer. Assume that the reduction of allowable load given in NDS Table 8.1B, *Fastener Load Modification Factors for Moisture Content,* is judged appropriate.

$$\text{Adj. } P_L = \text{LDF} \times \text{CUF} \times P_L = 1.33 \times 0.75 \times 108 = 108 \text{ lb/nail}$$

$$\text{Req'd number} = \frac{B}{\text{Adj. } P_L} = \frac{648}{108} = \boxed{6 \text{ nails}}$$

NOTE: If the reduction for moisture is not required, or if threaded hardened nails are used, the CUF = 1.0.

Then

$$\text{Adj. } P_L = \text{LDF} \times P_L = 1.33 \times 108 = 144 \text{ lb/nail}$$

(Tab. $P_L = 108$ is the same for both common and threaded nails).

$$N = \frac{648}{144} = 4.5$$

$$\boxed{\text{Say 5 nails}}$$

For an unenclosed building of this nature, the designer must evaluate the exposure and moisture-content condition of the members and their connections. This example illustrates the effects of different assumed conditions of use. The number of nails that can be accommodated in a connection may be limited by the spacing requirements (joint details). The layout of this connection is considered in Sec. 11.10.

11.7   Design Problem: Top Plate Splice

In Example 11.5, a splice in the double top plate in a wood-frame wall is designed. The loads in the plate are caused by the horizontal diaphragm action of the roof. The longitudinal wall plate is designed for the *chord force* (produced by the transverse lateral load) or the *drag strut force* (produced by the longitudinal lateral load). The larger of the two loads is used for the connection design. In this building the chord force controls, but if the opening in the wall had been somewhat longer than 14 ft, the drag force could have been the critical load.

In larger buildings with larger lateral loads, the magnitude of the chord or drag force may be too large for a nailed connection. In this case, bolts or some other connection hardware can be used for the plate splice.

At any splice, either the top or bottom chord member will be continuous. The thickness of the 2× plates is such that the required penetration of the nails is not provided. This requires that the tabulated allowable load be reduced by the ratio of the actual penetration to the specified penetration. Because the load is a short-term lateral load, an LDF of 1.33 is used.

Although it was not done in this example, the designer can specify the location of the splices in the top plate. If these locations are clearly shown on the plans, the design forces (chord and strut loads) that actually occur at these points can be used for the connection design. In this design the maximum forces were used, and consequently the top plate splices can occur at any point along the length of the wall.

EXAMPLE 11.5 Top Plate Splice

The building shown in Fig. 11.9a has stud walls with a double 2× top plate around the entire building. Design the splice in the double plate for the *longitudinal wall* for the horizontal diaphragm chord force or drag force (whichever is larger). Lumber is dry Douglas fir that will remain dry in service. Use 16d common nails.

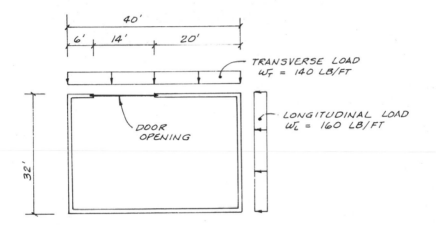

PLAN

Fig. 11.9a

Transverse load

Wall plate serves as chord.
Moment in diaphragm:

$$M = \frac{wL^2}{8} = \frac{(140)(40)^2}{8}$$
$$= 28{,}000 \text{ ft/lb}$$

Chord force:

$$T = C = \frac{M}{b} = \frac{28{,}000}{32}$$
$$= 875 \text{ lb}$$

For a discussion of horizontal diaphragm chords and drag struts see Chap. 9.

DESIGN OF WOOD STRUCTURES

Longitudinal load

Wall plate serves as strut.
Shearwall reaction:

$$R = \frac{wL}{2} = \frac{(160)(32)}{2}$$

$$= 2560 \text{ lb}$$

Wall with opening has the critical drag strut force.

Roof diaphragm unit shear:

$$v_R = \frac{R}{b} = \frac{2560}{40} = 64 \text{ lb/ft}$$

Unit shear in shearwall:

$$v_w = \frac{R}{b} = \frac{2560}{(6+20)} = 98.5 \text{ lb/ft}$$

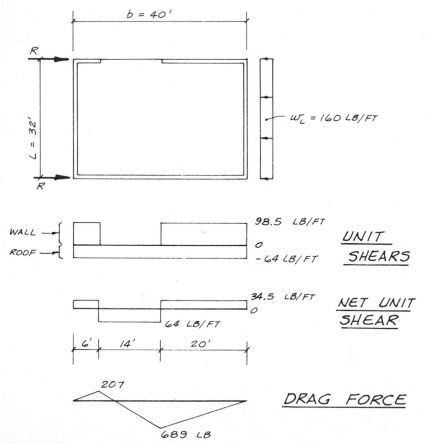

Fig. 11.9*b*

Splice Connection

$$\text{Max. chord force} = 875 \text{ lb} \qquad \text{governs}$$
$$\text{Max. drag force} = 689 \text{ lb}$$

If the code minimum lap of 4 ft is used for the top plate splice, the connection within the lap must transmit the full chord force.

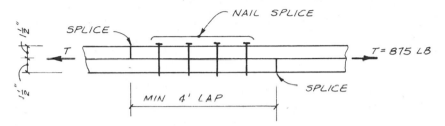

Fig. 11.9c

For 16d nail:

$$\text{Douglas fir} = \text{species group II}$$
$$\text{Req'd penetration} = 1.78 \text{ in.}$$

Penetration = nail length − thickness of side plate
$$= 3\tfrac{1}{2} - 1\tfrac{1}{2} = 2 > 1\tfrac{1}{2} \text{ in. (the thickness of the holding member)}$$
∴ Penetration = 1½ in.

$1\tfrac{1}{2} < 1.78$ ∴ reduce tabulated nail value by interpolation

$$\text{Tab. } P_L = 108 \text{ lb/nail} \qquad \text{(NDS Table 8.8C)}$$

$$\text{Adj. } P_L = \text{LDF} \times \text{CUF} \times P_L \times \frac{1.5}{1.78} = 1.33 \times 1.0 \times 108 \times \frac{1.5}{1.78}$$

$$= 121 \text{ lb/nail}$$

$$\text{Req'd number} = \frac{875}{121} = 7.23$$

> Use eight 16d common nails

11.8 Design Problem: Shearwall Chord Tie

The requirements for anchoring a shearwall panel to the foundation were discussed in Chap. 10. A similar problem occurs in the attachment of a second-story shearwall to the supporting first-story shearwall. The same basic loads must be considered.

In Example 11.6 the tension chord of the second-story wall is spliced to the chord of the first-floor wall. This splice is required because in typical *platform* construction the studs are not continuous from the roof level to the foundation. The first-floor studs stop at the double wall plate that supports

375

EXAMPLE 11.6   Shearwall Chord Tie

Design the connection tie between the first- and second-floor shearwall chords for the right end wall (Fig. 11.10a). A 12-gauge steel strap with 10d common nails is to be used. Lumber is dry Douglas fir that will remain dry in service. Reactions from the horizontal roof and second-floor diaphragms are given.

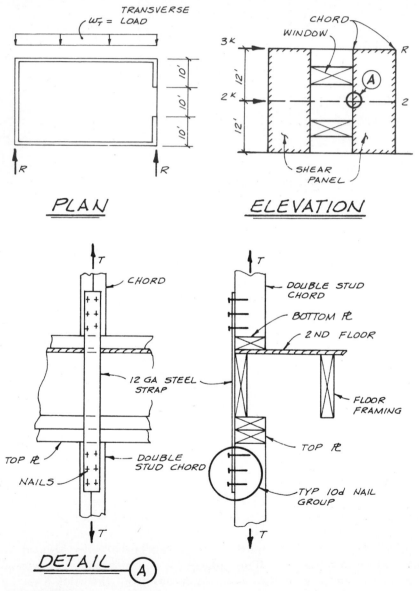

Chord Force

The chord forces at the second-story level are caused by the loads at the roof level.
See Fig. 11.10b.

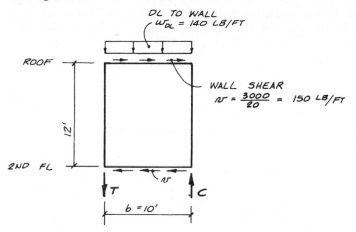

<u>FBD OF SHEAR PANEL</u>

Fig. 11.10b

From the gross OM,

$$T = C = vh = (150)(12) = 1800 \text{ lb (conservative)}$$

From the design net OM,

$$\text{Design net OM} = \text{gross OM} - \tfrac{2}{3} \text{ dead load RM}$$
$$= (150 \times 10)12 - \tfrac{2}{3}(140 \times 10)5$$
$$= 13.3 \text{ ft-k}$$

$$T = \frac{M}{b} = \frac{13,300}{10} = 1330 \text{ lb}$$

$$T = \text{gross } T - \tfrac{2}{3}(\text{trib. DL})$$
$$= 1800 - \tfrac{2}{3}(140 \times 5) = 1330 \text{ lb}$$

Nail Connection

Determine the number of 10d common nails required.

$$\text{Adj. } P_L = \text{tab. } P_L \times \text{LDF} \times \text{CUF} \times \text{MPF}$$
$$= 94 \times 1.33 \times 1.0 \times 1.25 = 156 \text{ lb/nail}$$

For the gross OM:

$$N = \frac{1800}{156} = 11.5 \qquad \boxed{\text{Use twelve 10d common nails}}$$

For the net OM:

$$N = \frac{1330}{156} = 8.5 \qquad \boxed{\text{Use nine 10d common nails}}$$

the second-floor framing. Separate second-floor studs are then constructed on top of the second-floor platform.

The purpose of the splice, then, is to develop a continuous shearwall chord from the roof level to the foundation. Splices of this type are referred to as *continuity ties* or splices. Various types of connections can be used for these ties. In this example the connection consists of a steel strap connected to the first- and second-floor chords with 10d common nails. For larger loads, bolts or lag screws can be used in place of nails. Calculations to verify the tension capacity of the strap are not shown in this example, but they would normally be required in design. In addition, the layout and spacing requirements of the nails in the connection should also be considered.

Chord forces can be calculated using either the gross overturning moment or the net overturning moment. Both are illustrated, and the choice of which method to use in a particular situation is left to the judgment of the designer.

11.9   Withdrawal-Type Connections

The second main use of nails is in withdrawal connections. These can be defined as connections in which the load is applied *parallel* to the length of the nail. The load is attempting to pull the nail out of the holding member. Generally speaking, connections loaded in withdrawal or "pullout" are weaker and less desirable than similar connections subjected to a lateral load.

The basic withdrawal connection has the nail passing through the side plate into the *side grain* of the main or holding member. See Fig. 11.11a. The other possible arrangement is to have the nail penetrate into the *end grain* of the holding member. See Fig. 11.11b. Nails that are driven into the end grain of the holding member have very low capacities and exhibit considerable variation in withdrawal values. Consequently no allowable load values are assigned to these nails, and the discussion in the remainder of this section applies to nails driven into the side grain of the holding member only.

The allowable loads for nails in withdrawal connections depend on the specific gravity of the lumber used. NDS Table 8.1A, *Grouping of Species for Fastening Design*, lists the specific gravities of the commercial lumber species. NDS Table 8.8B, *Nails and Spikes—Withdrawal Design Values*, gives the allowable withdrawal loads P_W for the four basic types of nails described in Sec. 11.2. The values represent the allowable *load per inch* of penetration into the holding member.

Tabulated allowable values can be adjusted for duration of load, but the adjustment for metal side plates applies only to lateral load connections. The adjustments for conditions of use (moisture-content conditions) are given in NDS Table 8.1B, *Fastener Load Modification Factors for Moisture Content*. A substantial reduction is specified when green lumber used in the fabrication of a withdrawal connection later seasons in service. The volume changes that occur cause the wood fibers to pull away from the nail. This reduced withdrawal capacity is taken into account with a CUF of 0.25. A similar reduc-

EXAMPLE 11.7   Withdrawal Load Connections

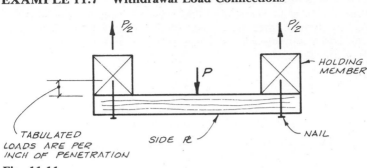

Fig. 11.11a

Withdrawal from Side Grain

The *basic* withdrawal load connection has the nail driven into the side grain (i.e., perpendicular to the grain) of the holding member, and the load is parallel to the length of the nail (Fig. 11.11a).

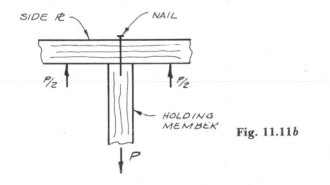

Fig. 11.11b

Withdrawal from End Grain

The second type of withdrawal connection has the nail driven into the end grain (i.e., parallel to the grain) of the holding member, and the load is parallel to the length of the nail (Fig. 11.11b). This is an inherently weak connection, and it is *not permitted* in design (i.e., it has no allowable load).

---

tion applies to initially dry lumber that is subjected to alternate wetting and drying in service. However, with threaded hardened nails the CUF = 1.0 for all conditions.

Because of the variables associated with withdrawal connections, the NDS recommends that, where possible, connections be designed so that nails and spikes are not loaded in withdrawal. For this reason, only a brief example of a withdrawal connection is presented. See Example 11.8.

379

EXAMPLE 11.8   Withdrawal Connection

Assume that the hanger in Fig. 11.11a is used to suspend a dead load P. The side plate is a 2 × 8 and the holding member is a 6 × 6. Lumber is Western hemlock that is initially green, and seasoning will occur in service. A total of twelve 16d box nails is used for all of the connections.

From NDS Table 8.1A,

$$\text{Western hemlock specific gravity} = G = 0.48$$

From NDS Table 8.8B,

$$\text{Allow. withdrawal load} = \text{tab. } P_w = 30 \text{ lb/in.}$$

$$\text{Penetration} = \text{nail length} - \text{side thickness}$$
$$= 3\frac{1}{2} - 1\frac{1}{2} = 2 \text{ in.}$$

Adjusted allow. load:

$$\text{Adj. } P_w = \text{LDF} \times \text{CUF} \times (\text{tab. } P_w \times \text{penetration})$$
$$= 0.9 \times 0.25 \times (30 \times 2)$$
$$= 13.5 \text{ lb/nail}$$
$$\text{Allow. } P = 13.5 \times 12 = 162 \text{ lb}$$

11.10   Spacing Requirements

The spacing requirements for nailed connections given in the NDS are somewhat general. Nail spacing "shall be such as to avoid unusual splitting of the wood." In the past, the layout of nails in a connection has typically been left to the discretion of the field personnel fabricating the connection. However, because there are an increasingly large number of nails used in some structural connections, more specific criteria regarding the spacing of nails are now included in the UBC.

Before reviewing the Code criteria it is necessary to define some of the terms used in the detailing of connections. These definitions apply to nails as well as other types of fasteners used in wood connections. The end distance is the distance from the center of a fastener to the end grain of a piece of lumber. The edge distance is from the center of the fastener to the side grain of a member. Finally, the center-to-center distance is the distance from the center of one fastener to the center of an adjacent fastener.

The UBC has adopted the following nail spacing requirements for wood-to-wood nailed connections. The center-to-center spacing in the direction of stress should not be less than the required penetration of the nail. The minimum end and edge distances in the direction of stress are taken as one-half the required penetration. Spacing requirements perpendicular to the loading are not specified except that splitting of the wood is to be avoided.

In order to avoid splitting of the wood, nail holes can be prebored to a maximum of nine-tenths the nail diameter for Group I species and three-fourths the nail diameter for the other species groups. No reduction in allowable lateral or withdrawal loads is required for nails with prebored holes.

EXAMPLE 11.9   Code Nail-Spacing Criteria

Check the spacing requirements for the connection shown in Fig. 11.12 using the UBC criteria. Can six 16d common nails be used in the connection? Lumber is Douglas fir-larch or Southern pine.

$$\text{Req'd penetration} = 1.78 \approx 1\tfrac{3}{4} \text{ in.}$$

Parallel-to-load spacing requirements:

$$\text{Min. end distance} = \tfrac{1}{2} \times 1\tfrac{3}{4} = \tfrac{7}{8} \text{ in.}$$
$$\text{Min. edge distance} = \tfrac{7}{8} \text{ in.}$$
$$\text{Min. c.-to-c. distance} = 1\tfrac{3}{4} \text{ in.}$$

NOTE: With these minimum requirements only two vertical rows of nails can be accommodated.

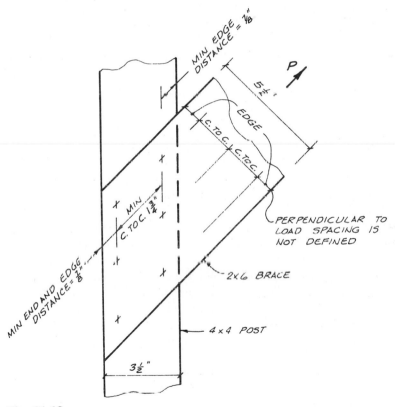

Fig. 11.12

For this connection the layout of six nails meets the *parallel*-to-load spacing criteria in *both* directions. Although this is not a requirement, it is conservative, and the connection is acceptable.

As an example, consider the layout of the nails for the knee brace connection designed in Sec. 11.6 (Example 11.4). A scale drawing is helpful in determining the spacing requirements. See Example 11.9. Because the spacing provisions perpendicular to the load are not specified, the total number of nails that can be used in a limited area is not fully defined. However, in this example the required number of nails can accommodate the parallel-to-load spacings in both directions. Although the application of the parallel-to-load spacing is not required in both directions, it would be conservative to take this approach.

11.11 Toenailed Connections

In many cases it may not be possible to nail directly through the side plate into the holding member. In these circumstances, toenails can be used. See Fig. 11.13. Connections of this type are commonly used for stud-to-plate,

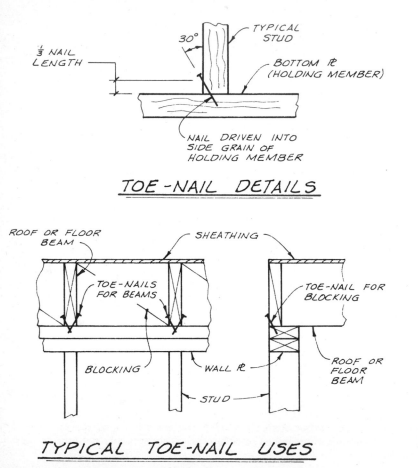

Fig. 11.13 Installation and application of toenails.

beam-to-plate, and blocking-to-plate connections. The dimensions given in the first sketch in Fig. 11.13 are used to calculate the penetration of the toenail into the holding member. The penetration requirements for toenails are the same as for other nailed connections.

Toenailed connections can be loaded laterally or they can be loaded in withdrawal. Withdrawal loads act perpendicular to the surface of the holding member and attempt to pull the nail out of the holding member. Lateral loads are parallel to the surface of the holding member. No distinction is made in this text between lateral loads on toenails and lateral loads on slant-driven nails. See Refs. 3 and 8 for a discussion of slant-driven nails.

Allowable loads on toenails driven into the side grain of the holding member are taken as percentages of the allowable loads for the basic lateral or withdrawal connection values. The allowable *lateral load* on a toenail is taken as five-sixths the design value for a basic lateral load connection. The design allowable is further adjusted by the LDF and CUF.

The allowable *withdrawal load* for a toenail is taken as two-thirds the tabulated allowable value. This is then adjusted for duration of load, but the condition-of-use factor is not applied to toenail connections.

EXAMPLE 11.10   Capacity of a Toenailed Connection

Determine the allowable wind load reaction at the base of a wood-frame shearwall which uses the standard Code toenailing of four 8d common or box nails (Fig. 11.14). Lumber is dry Douglas fir which will remain dry.

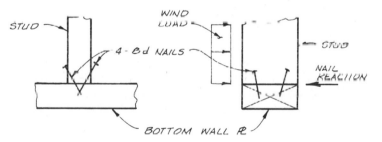

TYPICAL   STUD   TO   R   CONNECTION

Fig. 11.14

Douglas fir is species group II.

$$\text{Tab. } P_L = 78 \text{ lb/nail}$$
$$\text{Adj. } P_L = \text{LDF} \times \text{CUF} \times \tfrac{5}{6} \times P_L$$
$$= 1.33 \times 1.0 \times \tfrac{5}{6} \times 78 = 86.5 \text{ lb/nail}$$
$$\text{Allow reaction} = R = 4 \times 86.5 = 345 \text{ lb/stud}$$

The connection given for a toenail design example is for the Code attachment of a stud to a bottom wall plate. This connection uses four 8d toenails. See Example 11.10. Design loads and special considerations for this attachment in high-seismic-risk areas are covered in Sec. 10.11.

11.12 Nailing Schedule

The UBC Table 25P is a nailing schedule which gives the minimum nailing that is to be used for a number of common connections in wood-frame construction. The table covers such items as the nailing of beams (joists and rafters) to the top plate of a wall. For this connection three 8d toenails are specified (Fig. 11.13). The required nailing for numerous other connections is covered by this schedule.

The requirements given by the Code should be regarded as a *minimum.* They apply to conventional wood-frame construction as well as construction that has been structurally designed. A larger number of nails (or some other type of connection hardware) may be required when structural calculations are performed. Special consideration should be given to the attachment of the various elements in the lateral-force-resisting system.

11.13 Nail Capacity in Plywood

The lateral load capacity of nails driven through *lumber* side plates into a holding member was covered in Sec. 11.5. At times it may be necessary to perform lateral load calculations for nails driven through *plywood* into a holding member (sawn or glulam). These calculations are not required for general plywood diaphragm design because of appropriate design tables. However, in certain anchorage problems (e.g., item *f* in Example 14.4) these calculations are necessary.

In general, the strength of a nail driven through plywood is governed by the *nail capacity in the holding* member. However, the capacity can be governed by the plywood if the panel is too thin or if the nailing is very heavy.

Reference 36 indicates that the following plywood thicknesses develop approximately the bearing strength of common nails. The holding member is assumed to be of adequate thickness and of a similar density to that of the plywood.

Plywood Thickness	Common Nail Size
$\frac{5}{16}$	6d
$\frac{3}{8}$	8d
$\frac{1}{2}$	10d
$\frac{5}{8}$	16d

For these minimum plywood thicknesses, the allowable nail values can be taken from the NDS.

11.14   Problems

11.1   The shed in Fig. 11.A has double 2 × 6 knee braces on each column to resist the lateral load. Lumber is Southern pine. The braces are to be connected to the columns with 16d common nails. Lateral load w_L is 67 lb/ft. Heights: $h_1 = 9$ ft, $h_2 = 2$ ft.

> *Find:*   *a.* The required number of nails for the brace connection assuming that the CUF = 1.0. Site the references where the allowable loads and adjustment factors are found.
>
> *b.* The required number of nails for the brace connection, assuming that the connection is exposed to the weather. Site the references where the allowable loads and adjustment factors are found.
>
> *c.* How would *a.* and *b.* change if threaded hardened nails are used?

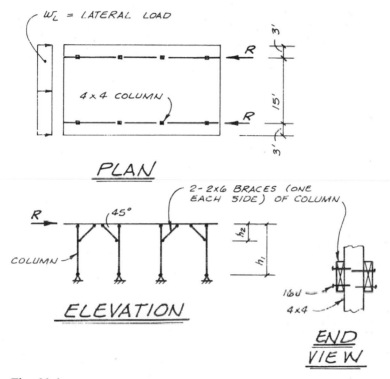

Fig. 11.A

11.2   The shed in Fig. 11.A has double 2 × 6 knee braces on each column to resist the lateral load. Lumber is California redwood (close grain). The braces are to be connected to the columns with 16d common nails. Lateral load $w_L = 46$ lb/ft. Heights: $h_1 = 12$ ft, $h_2 = 4$ ft.

> *Find:*   *a.* The required number of nails for the brace connection assuming that the CUF = 1.0. Site the references where the allowable loads and adjustment factors are found.
>
> *b.* The required number of nails for the brace connection, assuming

385

that the connection is exposed to the weather. Site the references where the allowable loads and adjustment factors are found.

    c. How would *a.* and *b.* change if threaded hardened nails are used?

11.3   Problem 11.3 is the same as Prob. 11.2 except that 16d box nails are to be used.

11.4   The single-story wood-frame building in Fig. 11.B has a double 2 × 4 top wall plate of Southern pine. This top plate serves as the chord and the drag strut along line 1. Loads are $w_T = 134$ lb/ft and $w_L = 223$ lb/ft. CUF = 1.0.

    Find: *a.* The maximum chord force along line 1.
              b. The maximum drag force along line 1.
              c. Determine the number of 16d common nails necessary to splice the top plate for the maximum force determined in *a* and *b.*

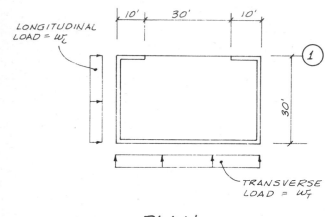

PLAN

Fig. 11.B

11.5   Problem 11.5 is the same as Prob. 11.4 except that 16d box nails are to be used.

11.6   The single-story wood-frame building in Fig. 11.B has a double 2 × 4 top wall plate of Douglas fir-larch. The top plate serves as the chord and drag strut along line 1. Loads are $w_T = 100$ lb/ft and $w_L = 250$ lb/ft. CUF = 1.0.

    Find: *a.* The maximum chord force along line 1.
              b. The maximum drag force along line 1.
               c. Determine the number of 16d common nails necessary to splice the top plate for the maximum force determined in *a* and *b.*

11.7   Problem 11.7 is the same as Prob. 11.6 except that 16d box nails are to be used.

11.8   *Given:* The elevation of the shearwall in Fig. 11.C. The lateral load and the resisting dead load of the shearwall are shown on the sketch. The uplift force at the bottom of the wall is anchored to the foundation by a metal strap embedded in the concrete foundation. The steel strap is attached to the chord (two 2 × 4s) with 10d common wire nails. Lumber is Hem-Fir. CUF = 1.0. Assume that the steel strap is adequate.

Find: Determine the number of nails required to attach the anchor to the chord. Consider the resisting DL when calculating the anchorage force.

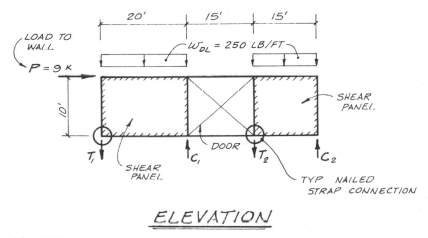

ELEVATION

Fig. 11.C

11.9 Problem 11.9 is the same as Prob. 11.8 except that nails are 16d common wire nails.

11.10 The two-story wood-frame building in Fig. 11.D has plywood shear panels along line 1. The lateral loads are shown in the sketch. Diaphragm span length L = 30 ft. Resisting dead loads to line 1 are roof DL = 100 lb/ft, floor DL = 120 lb/ft, and wall DL = 10 psf. Tie-down anchorage is to be with metal straps and 16d common nails. Studs are 2 × 6 of Idaho white pine. Assume that the metal strap is adequate.

Find: The number of nails required for the anchorage connections at *A* and *B*.

11.11 Problem 11.11 is the same as Prob. 11.10 except that the diaphragm span length is L = 40 ft.

11.12 The attachment of a roof diaphragm to the supporting shearwall is shown in Fig. 11.E. The shear transferred to the wall is v = 230 lb/ft. Toenails in the blocking are used to transfer the shear from the roof diaphragm into the double top plate. Lumber is DF-larch. CUF = 1.0.

Find: The required number of 8d common toenails per block if roof beams are 16 in. o.c.

11.13 Problem 11.13 is the same as Problem 11.12 except v = 180 lb/ft and nails are 8d box toenails.

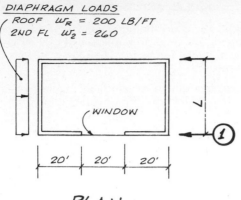

DIAPHRAGM LOADS
ROOF W_R = 200 LB/FT
2ND FL W_2 = 260

WINDOW

20' 20' 20'

L

1

PLAN

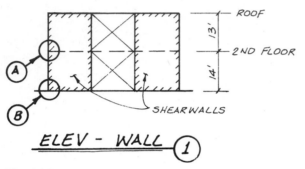

ROOF

13'

2ND FLOOR

14'

A

B

SHEARWALLS

ELEV - WALL 1

Fig. 11.D

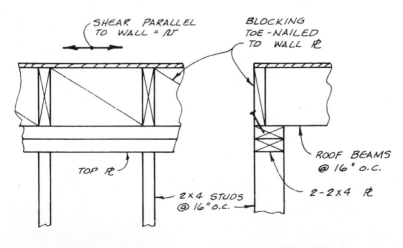

SHEAR PARALLEL
TO WALL = N

BLOCKING
TOE-NAILED
TO WALL R

TOP R

2×4 STUDS
@ 16" O.C.

ROOF BEAMS
@ 16" O.C.

2-2×4 R

ELEVATION

SECTION

11.14 The studs in the wood-frame wall in Fig. 11.F span between the foundation and the horizontal diaphragm. Roof beams are anchored to the wall top plate with 8d box toenails. Lumber is Hem-Fir. Roof beams are spaced 24 in. o.c. Wind = 20 psf. Wall height = 10 ft. CUF = 1.0.

Find: The required number of toenails per roof beam.

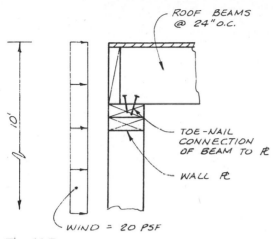

ROOF BEAMS @ 24" O.C.

TOE-NAIL CONNECTION OF BEAM TO ℞

WALL ℞

10'

WIND = 20 PSF

Fig. 11.F

11.15 The support shown in Fig. 11.G carries a DL of 500 lb. Lumber is DF-larch. Nails are 20d common nails driven into the side grain of the holding member.

Find: The required number of nails assuming that
 a. CUF = 1.0
 b. The connection is exposed to the weather.
 c. The connection is exposed to the weather and threaded hardened nails are used in place of the common nails.

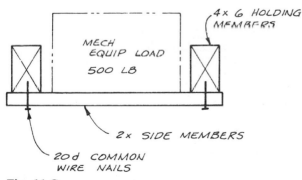

4 x 6 HOLDING MEMBERS

MECH EQUIP LOAD 500 LB

2 x SIDE MEMBERS

20d COMMON WIRE NAILS

Fig. 11.G

11.16 Determine the UBC minimum required nailing for the following connections. Give nail size, type, and required number.
 a. Joist to sill or girder, toenail
 b. Bridging to joist, toenail each end

     c. Sole plate to joist or blocking, face nail
     d. Top plate to stud, end nail
     e. Stud to sole plate, toenail
     f. Doubled studs, face nail
     g. Doubled top plates, face nail
     h. Top plates, laps and intersections, face nail
     i. Ceiling joists to plate, toenail
     j. Continuous header to stud, toenail
     k. Ceiling joists to parallel rafters, face nail
     l. Rafter to plate, toenail

CHAPTER

Bolts,
Lag Screws,
and Timber
Connectors

TWELVE

12.1 Introduction

Once the vertical-load-supporting system and lateral-force-resisting system for a building have been designed, attention is turned to the design of the connections. The importance of connection design has been emphasized in previous chapters, and the methods used to calculate the forces on several types of connections have been illustrated.

When the design forces on a connection are relatively small, the connections may often be made with nails (Chap. 11). However, with larger loads the use of bolts, lag screws, and timber connectors will be required. These connections often involve the use of some form of structural steel connection hardware. For many common connections this hardware may be available prefabricated from manufacturers or suppliers. Other connections, however, will require the fabrication of special hardware. Prefabricated hardware, when available, is often more economical than made-to-order hardware.

Chapter 12 deals with the design procedures given in the NDS for the fasteners (other than nails) that are used in typical wood connections. Because bolts are the most common of these fasteners, Chap. 12 emphasizes bolted connections. The other types of fasteners are also covered but in less detail.

Chapter 13 continues the subject of connection design by giving examples of common types of connection hardware. These examples are accompanied with comments about both good and poor connection layout practices. Chapter 13 concludes with a review of some prefabricated connection hardware.

12.2 Types of Fasteners

The main types of fasteners in wood construction (in addition to nails) are

1. Bolts
2. Lag screws
3. Timber connectors
    a. Split rings
    b. Shear plates

Before going into design specifics, the configuration of some typical connections that use these fasteners will be given for comparison.

The most common use of bolts is in the basic shear connection. See Example 12.1. Bolts are usually stressed in single shear or double shear, but more shear planes can be involved. The side members can be either wood or steel. This basic type of bolted connection is considered in detail in Chap. 12. In addition to shear connections, bolts or threaded rods can be stressed in tension as hangers or anchors, or for bracing.

Lag screws are large wood screws that typically have a square or hex bolt head. Lag screws are used when an excessively long bolt would be required to make a connection, or when access to one side of a through-bolted connection may be prevented. See Example 12.2. Lag screws can be loaded either in shear or in withdrawal, and side plates can be either wood or steel.

Timber connectors (split rings and shear plates) are devices that are installed into precut grooves in wood members. These connectors provide a shearing-type resistance, and a bolt through the center of the connector is required to assemble the connection. Allowable loads are higher for timber connectors than for bolts or lag screws. Sketches of split rings and shear plates can be found in the NDS Sec. 8.4, *Timber Connectors*.

Split rings are for wood-to-wood connections only. The steel ring is "split" to provide simultaneous bearing on the inner core and the outer surface of the groove. Shear plates can be used for wood-to-steel connections because the shear plate is flush with the surface of the wood. Two shear plates can also be used for wood-to-wood connections. Shear plate wood-to-wood connections are easier to assemble than split-ring connections, and may be desirable where ease of erection is important.

Although the allowable loads for split rings and shear plates are much higher than for bolts, the use of timber connectors is limited by the fabrication costs of the grooving. Timber connectors may be used in ordinary building design where relatively large loads must be transferred in a fairly small amount of space, but generally speaking, they are less common than simple bolted

EXAMPLE 12.1   Bolted Connections

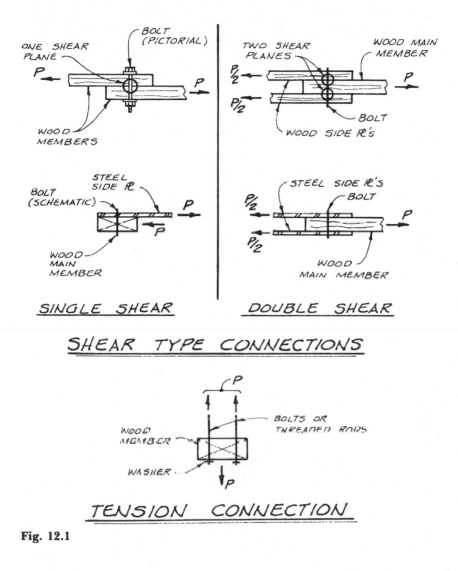

Fig. 12.1

Shear connections are the basic type of bolted connections covered in the NDS. Bolts or threaded rods (such as anchor bolts or cross braces) can also be used in tension.

Washers are required under the bolt head and under the nut in both types of connections. Oversized washers may be required to reduce bearing perpendicular to the grain in the tension connections.

EXAMPLE 12.2   Lag Screw Connections

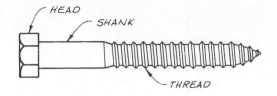

TYPICAL LAG SCREW

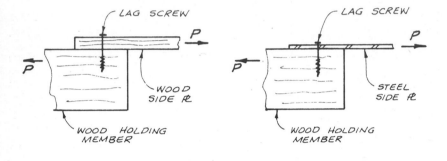

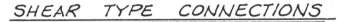

SHEAR   TYPE   CONNECTIONS

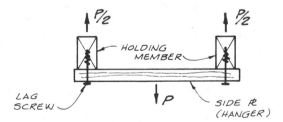

WITHDRAWAL   CONNECTION

Fig. 12.2

Lag screws are installed in special prebored holes to accommodate the shank and provide holding material for the thread (Sec. 12.8). Washers are required under the head, as for the bolts in Fig. 12.1.

connections. Timber connectors are more widely used in glulam arches and heavy timber trusses than in shearwall-type buildings.

In addition to the above types of fasteners, there are other devices such as spiked grids and sheet-metal nail plates that can be used to connect wood members. Spiked grids are designed for use in wood pile and pole construction. Nail plates are usually made from galvanized sheet metal and have barbs or teeth formed by punching holes in the plate. These plates are typically used in the fabrication of light-frame wood truss connections. They are installed by pressing the nail plate into the two or more members framing into a joint in a truss. The design of these fasteners is a specialty item and is beyond the scope of this book. For information see Ref. 35.

12.3 Factors Affecting the Strength of Connections

There are several factors that affect the strength of connections that are common to all of the fasteners described in Sec. 12.2. These include:

1. Direction of loading
2. Duration of load
3. Condition of use
4. Net section calculations
5. Number of fasteners in a row
6. Horizontal shear in members at a connection

The directional strength properties of lumber are introduced in Chap. 4. It is noted there that the allowable compressive stress parallel (0 degree) to the grain is substantially larger than the allowable stress perpendicular (90 degrees) to the grain. In Sec. 6.7 the Hankinson formula was introduced; this formula provides an allowable compressive stress for some angle θ between 0 and 90 degrees to the direction of the grain.

These same principles apply to the design of laterally loaded bolts, lag screws, and timber connectors installed in the side grain of lumber. The parallel-to-grain allowable load is p, and the perpendicular-to-grain allowable load is q. The allowable lateral load at some intermediate angle θ can be determined from the Hankinson formula expressed in terms of p and q instead of F_c and $F_{c\perp}$. See Example 12.3. This formula can be solved directly or the graphical solution in NDS Appendix J, *Hankinson Formula,* can be used.

The capacity of a wood connection, in most cases, is governed by the strength of the wood and not the strength of the fasteners. For these connections the tabulated allowable loads can be adjusted by the load duration factor (LDF). For certain timber connectors the strength of a connection may be governed by the strength of the fastener, and the allowable loads **395**

**EXAMPLE 12.3   Direction of Loading in Wood
Connections**

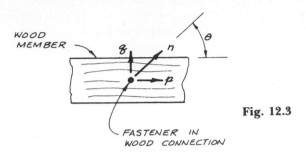

Fig. 12.3

For bolts, lag screws, and timber connectors allowable fastener loads are tabulated
for

a.   Parallel-to-grain loading p

b.   Perpendicular-to-grain loading q

The allowable load parallel to the grain is much larger than the allowable load perpendicular to the grain $(p > q)$. The allowable load at some angle θ is between p and q
and is given by the Hankinson formula

$$n = \frac{pq}{q \sin^2\theta + q \cos^2\theta}$$

are not to be adjusted for duration. Restrictions of this type are noted in
the load tables.

The capacity of any steel connection plates must also be evaluated. The
Code does allow a stress increase of one-third for all building materials when
wind or seismic loads are involved. With this exception, however, adjustments
of allowable stresses based on duration are applied only when the strength
of wood governs.

Tabulated allowable connection loads are for lumber seasoned to a moisture content of 19 percent or less at the time of fabrication. In addition, it
is assumed that the lumber remains continuously dry in service. If the moisture
content *at fabrication* or *in service* is different, the tabulated loads must be
reduced by the appropriate condition-of-use factor (CUF) given in NDS Table
8.1B, *Fastener Load Modification Factors for Moisture Content.*

The calculation of the stresses at the net section of the member was covered
in Secs. 7.2 and 7.4. This basically is a problem in the design of the member,
but the connection is the "cause" of the reduction in cross section. The
reduction of cross-sectional area, resulting from the presence of a fastener,
is to be used in the design of both tension and compression members. The

396

projected area of all material removed from the member by boring, grooving, or dapping should be deducted from the gross area of the member when stresses are checked at the critical section. See Example 12.4. The projected area of a bolt hole can be taken as the hole diameter times the thickness of the member. The actual hole is $\frac{1}{16}$ to $\frac{1}{32}$ in. larger than the bolt. For net area calculations the hole diameter can conservatively be taken as $\frac{1}{16}$ to $\frac{1}{8}$ in. larger than the bolt diameter. The projected areas of timber connectors can be calculated from the dimensions given in NDS Appendix K, *Typical Dimensions for Timber Connectors.*

In the past it was common practice to assume that the total capacity of a connection was the sum (P_s) of the allowable loads of the individual fasteners in the connection. Research has shown, however, that the effectiveness of a

---

EXAMPLE 12.4   Net Areas at Connections

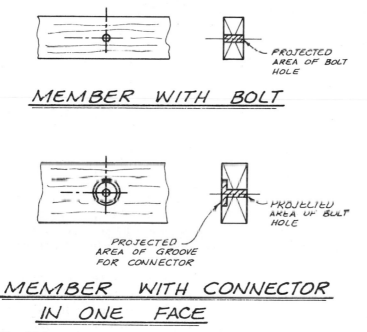

Fig. 12.4

The stress at the net section in an axially loaded member is calculated as the load divided by the net area.

$$\text{Net area} = \text{gross area} - \begin{bmatrix} \text{projected area of material removed} \\ \text{from the member at the connection} \end{bmatrix}$$

fastener is reduced as the number of fasteners in a row increases. To account for this behavior a reduced allowable load for a connection with multiple fasteners in a row is calculated as

$$P_r = KP_s$$

where P_r = reduced allowable load for a row of fasteners in a connection
P_s = sum of the individual allowable loads for the fasteners in a row
K = reduction factor ($K \leqslant 1.0$)

The reduction factor K, for use in connections with wood side plates, is obtained from NDS Table 8.3A, *Wood Side Plate Modification Factors.* For connections with metal side plates, the K factor is obtained from NDS Table 8.3B, *Metal Side Plate Modification Factors.* K depends on the

1. Number of fasteners in a row

2. Gross cross-sectional area of the main member, A_1

3. Ratio A_1/A_2

where A_1 = gross cross-sectional area of the main member
A_2 = sum of the gross cross-sectional areas of the side members

For a given main member and side plates, K becomes smaller as the number of fasteners in a row increases.

The final point about general connection design deals with shear in members at a connection. See Example 12.5. This problem must be considered

---

EXAMPLE 12.5   Shear Stresses at a Connection

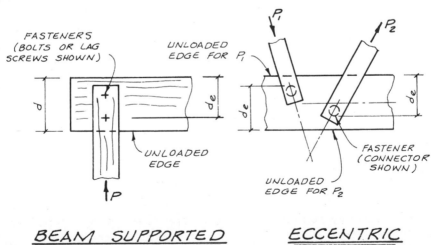

BEAM SUPPORTED
BY FASTENERS

ECCENTRIC
CONNECTION

Fig. 12.5

The horizontal shear in the members at the connections in Fig. 12.5 is checked using the effective depth. For bolts and lag screws d_e is the member depth minus the distance from the unloaded edge to the center of the nearest fastener. For timber connectors d_e is the member depth minus the distance from the unloaded edge to the edge of the connector. The *unloaded edge* is the edge away from the direction of the load.

When the connection is more than $5 \times d$ from the end of the member,

$$f_v = \frac{3V}{2bd_e} \leq (\text{adj. } F_v) \times 1.5$$

Also,

$$f_v = \frac{3V}{2bd} \leq \text{adj. } F_v$$

When the connection is less than $5 \times d$ from the end of the member,

$$f_v = \frac{3V}{2bd_e}\left(\frac{d}{d_e}\right) \leq \text{adj. } F_v$$

For adj. F_v see Chap. 6.

---

in connections with an eccentricity and in beams supported by fasteners. The method of checking the shear stresses in a member is described in the example.

12.4 Bolted Connections

The most common grade of bolt used in the fabrication of wood connections conforms to ASTM Standard A307. In the design of steel structures this grade bolt has largely been replaced by high-strength (A325) bolts. High-strength bolts can be used in wood structures, although, generally speaking, they are not. The allowable load values in the NDS are for A307 bolts, and they can conservatively be applied to high strength bolts.

There are a number of factors that affect the strength of bolted connections in addition to those discussed in Sec. 12.3. These include:

1. Species of lumber
2. Bolt size and slenderness
3. Width of members in the connection
4. Number of shear planes
5. Type of side plate (wood or steel)
6. Spacing requirements
7. Arrangement of side plates with multiple rows of bolts
8. Fabrication techniques

The effects and appropriate adjustments for these items are contained in the NDS, and a review of these criteria should accompany the remainder of **399**

this section. A description of the NDS tables and some background information is provided in the following paragraphs.

Allowable loads for bolts (*p* and *q*) are given in NDS Table 8.5A, *Bolt
Design Values.* This table assigns bolt values for 12 different categories of
commercial lumber species. For example, most stress grades of Douglas fir-
larch and Southern pine are in category 3. However, when the term "dense"
is added to the stress grade (e.g., Dense Select Structural), these species
fall into category 1.

Tabulated allowable loads apply to bolts in *double-shear wood-to-wood connections.* Allowable loads for other conditions (e.g., single shear, metal side plates,
etc.) are obtained by adjusting the values for this basic connection. The NDS
bolt table gives the allowable load directly for a given bolt diameter and a

---

EXAMPLE 12.6   **Slenderness Effect in Bolted Connections**

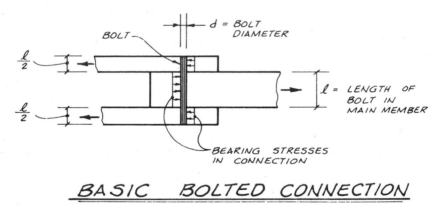

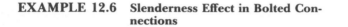

Fig. 12.6

Tabulated bolt load values are for the basic connection shown in Fig. 12.6. The
slenderness ratio of a bolt is defined as the length of bolt in main member divided
by the bolt diameter *(l/d).*

The bearing stresses in the wood members are shown as uniformly distributed.
This is essentially the case for small values of bolt slenderness. However, as the bolt
slenderness ratio *l/d* becomes large, the stress distribution becomes nonuniform.

Therefore, for small *l/d* ratios, the allowable loads are directly proportional to
the bearing area, but for a large bolt slenderness the allowable load may change
little with changes in bearing areas.

When the bolt slenderness reaches 6 or 7, no increase in allowable load occurs
with increased member thickness. In fact, the allowable load on a bolt may actually
decrease with an increase in member thickness. Fortunately, the NDS allowable load
table automatically accounts for bolt slenderness. The designer should simply be aware
of its general effect.

given "length of bolt in main member," l. In general, bolt values are related to the bearing area of the bolt, but the *bolt slenderness* also affects the load capacity. See Example 12.6.

In the NDS load table, the thickness of each side member is assumed to be one-half the thickness of the main member. However, the table can be used for other conditions. Allowable loads are read from the table based on the thickness of the main member. If the side members are greater than one-half the thickness of the main member, the allowable load is determined from the actual thickness of the main member. If the side members are less than one-half the thickness of the main member, the allowable load is determined from a main member thickness which is twice the thickness of the thinner side plate.

As noted, the double-shear bolted connection serves as the basis for determining the allowable load for other types of connections. Single-shear connections occur frequently, and the designer should be familiar with the method used to determine the allowable load for these connections. See Example 12.7. It should be emphasized that the allowable load on a single shear connection is one-half of the double shear allowable load. The application of these procedures will be demonstrated later in an example.

The tabulated allowable loads apply to connections with wood side members. Steel side plates can be used in bolted connections, and this often occurs in the form of connection hardware (Chap. 13). For steel side plates, the allowable load parallel to the grain, p, can be increased by a multiplying factor of 1.25. In this book, metal plate factor (MPF) is the symbol used for this adjustment. No increase is allowed for the perpendicular-to-grain load q (i.e., MPF = 1.0).

Bolts should be installed in properly drilled holes, and the members in a connection should be aligned so that only light tapping is required to insert the bolts. If the bolt holes are too large, nonuniform bearing stresses will occur. On the other hand, if the holes are too small, an excessive amount of driving will be required to install the bolt and splitting of the wood may be induced. Obviously, splitting will greatly reduce the capacity of a bolted connection. Recommended bolt hole diameters are from $\frac{1}{32}$ to $\frac{1}{16}$ in. larger than the bolt diameter.

The manner in which the holes are bored also has an effect on the capacity of a bolted connection. A bolt hole with a smooth surface will develop a higher load with less deformation than one that has a rough surface. Smooth surfaces are obtained by using sharp drill bits and a proper drill speed and rate of feed.

The major factors that affect the strength of bolted connections were listed at the beginning of this section, and most of these have been discussed at least briefly. However, the arrangement of bolts in a connection requires special consideration. Therefore, the spacing requirements for bolts and the detailing of connections are covered separately in Sec. 12.7. Additional considerations are given in Chap. 13.

EXAMPLE 12.7   Single-Shear Bolted Connections

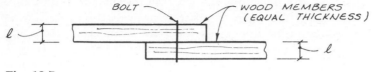

Fig. 12.7a

Allowable load on a bolt in single shear with wood members of equal thickness, *l*, is taken as *one-half* the allowable load on a bolt in double shear. The double shear value is determined using the thickness of *one* of the members, *l.* This represents a change from previous NDS methods, and considerably lower allowable loads are obtained with this new approach.

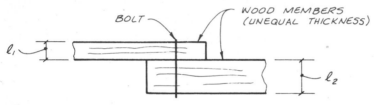

Fig. 12.7b

Allowable load in single shear with members of unequal thickness is taken as *one-half* of the allowable load on a bolt in double shear. The double shear value is taken as the smaller allowable load using the following thicknesses:
a.   The thickness of the thicker member
b.   Twice the thickness of the thinner member

12.5   Design Problem: Bolted Diaphragm Chord Splice

In the building in Example 12.8, the top plate of a 2 × 6 stud wall serves as the chord for the roof diaphragm. The connections for the splice in the top plate are to be designed. The magnitude of the chord force is so large that the splice cannot be made with nails, and a bolted connection is required.

The location of the splices in the chord should be considered. In this building the chord is made up of fairly long members (say 20 ft), and the splices are offset by half the length of the members (10 ft in this example). At some point between the splices, the chord force is shared equally by the two members. Therefore, the connections on either side of the splice can be designed for one-half of the total chord force.

This is in contrast to the nailed chord splice designed in Example 11.5 (Sec. 11.7). If the lap of the members in the chord is kept to a minimum, the entire chord force must be transmitted by the connection within the lap. Either method of splicing the chord can be used in practice.

The connection is designed by first assuming a bolt size. The allowable load on one bolt is determined, and from this the total number of bolts in the connection is estimated. Finally, this number is checked using the reduction based on the number of fasteners in a row.

EXAMPLE 12.8 Bolted Chord Splice

A large wood-frame warehouse uses 2 × 6 stud walls and a double 2 × 6 top plate which functions as the chord of the horizontal roof diaphragm. Determine the number of 3/4-in.-diameter bolts for the splice connection in the top plate. The bolts are in a single row. Lumber is Select Structural Douglas fir-larch. The chord force at the connection has been determined, and the splice is to be designed for one-half of this load. CUF = 1.0. See Fig. 12.8a.

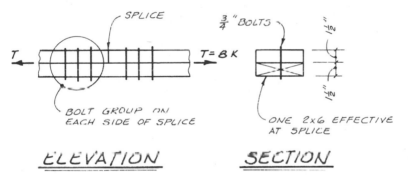

Fig. 12.8a

Bolt Design

Sel. Str. DF L is in bolt species category 3. The allowable load for a single-shear connection is one-half of the double-shear value. Length of bolt in main member is 1½ in. The load is parallel to grain.

From NDS Table 8.5A, *Bolt Design Values,*

$$\text{Adj. } p = \text{tab. } p \times \tfrac{1}{2} \times \text{LDF} \times \text{CUF}$$
$$= 1420 \times \tfrac{1}{2} \times 1.33 \times 1.0$$
$$= 944 \text{ lb/bolt}$$

The splices in double top plate are well spaced (say 10 ft o.c.).

$$\text{Load on one bolt group} = \frac{T}{2} = \frac{8}{2} = 4\text{k}$$

$$\text{Approximate number of bolts} = \frac{4000}{944} = 4.3 \qquad \textit{Try five bolts}$$

403

Check reduction for five bolts in a row. For wood side plate use NDS Table 8.3A, *Wood Side Plate Modification Factors.*

$$A_1 = \text{gross area of main member} (2 \times 6) = 8.25 \text{ in.}^2$$
$$A_2 = \text{gross area of side plate (also } 2 \times 6) = 8.25 \text{ in.}^2$$
$$A_1/A_2 = 1.0 \quad \text{and} \quad A_1 < 12 \text{ in.}^2$$
$$\therefore K = 0.85$$

Reduced allowable load for connection:

$$P_r = P_s K = (5)(944)(0.85) = 4010 \text{ lb} > 4000 \qquad OK$$

> Use five ¾-in. bolts each side of splice

NOTE: The spacing of the bolts must be shown on the design plans.

Net Section

Check the stress in one 2×6 at the bolt hole.

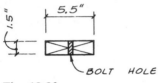

Fig. 12.8*b*

$$A_n = 1.5 [5.5 - (¾ + ⅛)] = 6.94 \text{ in.}^2$$

The ⅛ in. was arbitrarily added to the bolt diameter to allow for clearance and possible damaged material. Actual clearance is less.

$$f_t = \frac{T}{A_n} = \frac{8000}{6.94} = 1153 \text{ psi}$$
$$\text{Adj. } F_t = \text{tab. } F_t \times \text{LDF} \times \text{CUF} \qquad \text{(no reduction for width)}$$
$$= 1200 \times 1.33 \times 1.00$$
$$= 1600 \text{ psi} > 1153 \qquad OK$$

> Use two 2×6 Sel. Str.
> DF-L top plate

12.6 Design Problem: Bolted Diagonal Brace Connection

A 4×10 beam is supported by a double 2×6 diagonal brace. A single ¾-inch-diameter bolt is used to connect these members, and the bolt is located on the center lines of the two members. See Example 12.9. The resultant force on the bolt acts parallel to the grain in the brace and at an angle of 45 degrees with respect to the grain in the beam.

The NDS states that when the thickness of the side plate is less than one-half the thickness of the main member, the allowable load is to be based

404

on twice the thickness of the side member. This, however, only applies when the direction of the load to the grain is the same for both members.

When the direction of loading is different for the two members (as in this problem), it is reasonable to calculate two allowable loads using different thicknesses. The capacity of the connection is governed by the smaller of the two.

Using this approach in the example, the allowable load on the main member at 45 degrees is based on a thickness of 3½ in. The allowable load parallel to the grain in the brace is determined for a thickness of 3 in. (2 × 1½ in.).

EXAMPLE 12.9   Bolted Diagonal Brace

The beam supports a DL of 575 lb at its free end, and is connected to the diagonal brace with a single ¾-in. bolt. Lumber is No. 1 Douglas fir-larch that was dry at the time of fabrication and remains dry in service. Determine if the bolt is adequate, and check the shear stress in the beam at the connection. See Fig. 12.9a.

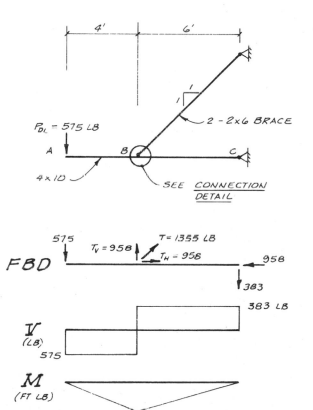

Fig. 12.9a

Solve for force in brace.

$$\Sigma M_c = 0$$
$$6 T_v = 575(10)$$
$$T_v = 958 \text{ lb}$$
$$T = \sqrt{2} \; T_v = \sqrt{2} \; (958) = 1355 \text{ lb}$$

NOTE: Segment BC of the beam is subjected to combined axial and bending stresses (Chap. 7).

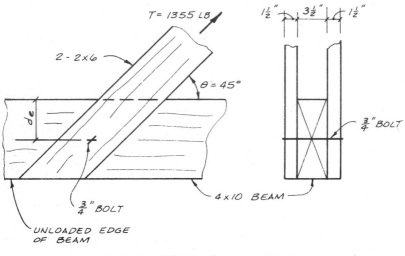

Fig. 12.9*b*

Bolt Capacity
From NDS Table 8.5A, *Bolt Design Values,*
Brace: Load is parallel to grain.

$$l = 2(1\tfrac{1}{2}) = 3 \text{ in.}$$
$$p = 2630 \text{ lb}$$

Main member: Load is at 45 degrees to grain.

$$l = 3\tfrac{1}{2} \text{ in.}$$
$$p = 2800 \text{ lb}$$
$$q = 1260 \text{ lb}$$

Hankinson formula

$$n = \frac{pq}{p(\sin^2 \theta) + q(\cos^2 \theta)}$$
$$= \frac{(2800)(1260)}{2800(\sin^2 45) + 1260(\cos^2 45)}$$
$$= 1740 \text{ lb} < 2630 \qquad \therefore \text{ main member governs}$$

406 Adjusted allowable load:

$$= n(\text{LDF})(\text{CUF})$$
$$= 1740 \, (0.9)(1.0) = 1564 \, \text{lb} > 1355 \qquad OK$$

$$\boxed{\text{¾ in. bolt is } OK}$$

Shear Stress

The shear stress in the 4 × 10 beam is checked by using the criteria given in Example 12.5. Compare (5× beam depth) with the distance of the connection from the end of the beam.

$$5 \times 9.25 = 46.25 \, \text{in.} < 48 \, \text{in.}$$

The joint is more than 5 times the depth from the end.

$$f_v = \frac{3V}{2bd_e} = \frac{3(575)}{2(3.5)\left(\dfrac{9.25}{2}\right)} = 53.3 \, \text{psi}$$

$$\text{Adj. } F_v = \text{tab. } F_v \times \text{LDF} \times \text{CUF}$$
$$= 95 \times 0.9 \times 1.0$$
$$= 85.5 \, \text{psi} > 53.3 \qquad OK$$

When the shear is checked by using the effective depth at the connection, the allowable horizontal shear can be increased by 50 percent if the connection is more than 5 times the beam depth from the end of the member. The increase was not used in this problem because shear is not critical.

$$\boxed{\text{Shear in beam is } OK}$$

It is expected that the NDS criteria will be revised in the future to reflect the interpretation used in this problem.

Because the center of resistance of the bolt coincides with the centroid of the members in the connection, there is no eccentricity. However, the horizontal shear in the beam should be evaluated using the "effective depth" to the fastener. The effective depth d_e is the depth of the beam less the distance from the "unloaded edge" to the fastener.

12.7 Spacing Requirements and Details for Bolted Connections

The dimensions for locating bolts in a connection are given to the center line of the bolts. As with nailed connections, the detailing of a connection requires consideration of:

1. End distance
2. Edge distance
3. Center-to-center spacing

In bolted connections, however, the problem is somewhat more complicated because the direction of loading in relation to the direction of the grain must be considered. Other factors such as bolt slenderness l/d and the sense of force in the member (tension or compression) may also affect the spacing requirements.

The spacing criteria are typically stated as a number of bolt diameters d. The NDS spacing criteria for *parallel-to-grain loading* are summarized in Example 12.10. Spacing for *perpendicular-to-grain loading* is given in Example 12.11.

Specific requirements for the spacing of bolts in a connection where the load is at some angle other than 0 or 90 degrees to the direction of the grain are not given in the NDS. It is stated, however, that "uniform stress in main members and a uniform distribution of load to all bolts . . . require that the gravity axis of all members shall pass through the center of resistance of the bolt groups." In addition to providing this symmetry, the spacing

---

EXAMPLE 12.10   Bolt Spacing Requirements: Parallel-to-Grain Loading

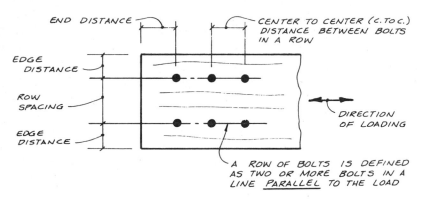

Fig. 12.10

Parallel-to-grain loading spacing requirements (Fig. 12.10) are:

1. *End distance*
    a. Members in tension = $7d$ for softwood species
    b. Members in compression = $4d$
2. *Edge distance*—bolt slenderness = l/d (Sec. 12.4)
    a. For $l/d \leqslant 6.0$: Edge distance = $1\frac{1}{2}d$
    b. For $l/d > 6.0$: Edge distance = $1\frac{1}{2}d$ or $\frac{1}{2} \times$ row spacing, whichever is larger
3. *Center to center* (c. to c.) = $4d$
4. *Row spacing* = $1\frac{1}{2}d$. Row spacing shall not exceed 5 in. unless separate side plates are used for each row.

EXAMPLE 12.11 **Bolt Spacing Requirements: Per-**
pendicular-to-Grain Loading

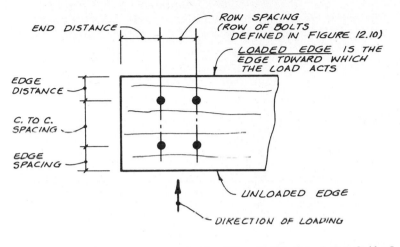

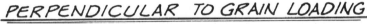

PERPENDICULAR TO GRAIN LOADING

Fig. 12.11

Perpendicular-to-grain loading spacing requirements (Fig. 12.11) are:

1. *End distance* = 4d

2. *Edge distance*
   a. Loaded edge = 4d
   b. Unloaded edge = 1½d

3. *Center to center*—c.-to-c. spacing is governed by the spacing requirements of the member(s) to which a bolt is attached (say steel side plates* or wood member loaded parallel to the grain).

4. *Row spacing*
   a. For $l/d = 2.0$: Row spacing = 2½d
   b. For $l/d \geq 6.0$: Row spacing = 5d
   c. For $2.0 < l/d < 6.0$: interpolate between 2½d and 5d

* The c.-to-c. and edge spacing requirements for steel plates are given in Ref. 12.

provisions for *both* parallel and perpendicular loading can be applied to the connection.

There is an increasing concern about problems in connections that are created by shrinkage (or swelling) caused by changes in moisture content. In Sec. 4.4 it was noted that significant dimensional changes can occur across the grain, but very little change takes place parallel to the grain.

409

Problems involving volume changes develop when bolts, or some other type of fasteners, are held rigidly in position. Perhaps the best illustration of this problem is in a connection that uses a metal side plate and parallel rows of bolts. See Example 12.12. The bolts are essentially fixed in position by the metal side plates (the only movement permitted is that allowed by the clearance between the bolts and the holes). If the lumber is at an initially high moisture content and is allowed to season in place after fabrication of the connection, a serious cracking situation can develop.

The shrinkage that occurs is essentially across the grain, and the result will be the development of cracking parallel to the grain. Obviously this type of cracking greatly reduces the strength of a connection.

The problem of cross-grain shrinkage can often be eliminated by proper detailing of a connection. This can be accomplished by providing for the unrestricted movement of the bolts across the grain. For example, in the case of the steel side plate, separate side members for the two rows of bolts can be used. These separate plates will simply move with the bolts as the moisture content changes, and cross-grain tension and the associated cracking will be avoided. Another approach is to use slotted holes in the steel side plates so that movement is not restricted. Other examples of proper connection detailing to avoid splitting are given in Chap. 13.

EXAMPLE 12.12   Cross-Grain Cracking

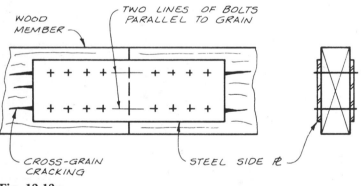

Fig. 12.12a

Figure 12.12a shows a critical connection for cross-grain cracking. As the wood loses moisture, cross-grain shrinkage occurs, and the bolts are rigidly held in position by the steel plates. The lumber becomes stressed in tension across the grain (a weak state of stress for wood).

Condition-of-Use Factors (CUF):
For connections the following definitions apply:

410

"Wet" condition—MC at or above FSP (approximately 30 percent)
"Dry" condition—MC ≤ 19 percent

a. When the lumber is initially dry and remains dry in service, CUF = 1.0.

b. When the lumber used to fabricate a connection is initially wet and seasons in place to a dry condition, the allowable load is reduced by 60 percent (i.e., CUF = 0.40).

c. For lumber that is between the wet and dry condition (partially seasoned) at fabrication and later seasons in place, an intermediate CUF can be used.

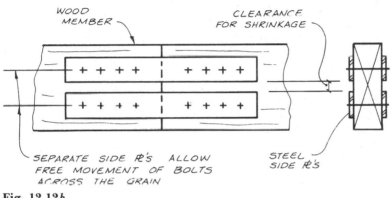

Fig. 12.12*b*

The cross-grain cracking problem can be eliminated by proper joint detailing. The use of separate side plates for each line of bolts allows the free movement of the bolts across the grain. Slotted holes can also be used. With proper detailing the CUF can be taken as 1.0. See Fig. 12.12*b*.

NDS spacing criteria (Example 12.10) limits the spacing between rows of bolts to a maximum of 5 in. unless separate side plates are used. If separate plates are not used, the 5-in. limit on the spacing across the grain provides a limited width across the grain in which shrinkage can occur.

It was noted that very little volume change occurs parallel to the grain. This fact can cause a cross-grain cracking problem when wood members frame together at right angles. See Figure 12.13. The problem develops when there are multiple rows of bolts passing through the two members and significant changes in moisture content occur. Here the problem is caused by cross-grain shrinkage in the mutually perpendicular members with little or no shrinkage parallel to the grain. The use of a single fastener for this connection is desirable. The problem is not critical in connections which are initially dry and remain dry in service.

The importance of proper connection detailing can be seen by the magnitude of the condition-of-use factor required when severe cross-grain cracking can develop. Example 12.12 indicates that a CUF of 0.40 is required when

411

these connections are fabricated with wood members that are initially "wet" and season to a "dry" condition in service. A complete summary of the CUFs applied to wood connections is given in NDS Table 8.1B, *Fastener Load Modification Factors for Moisture Content,* and NDS Table 8.1C, *Modification Factors for Laterally Loaded Bolts and Lag Screws in Timber Seasoned in Place.*

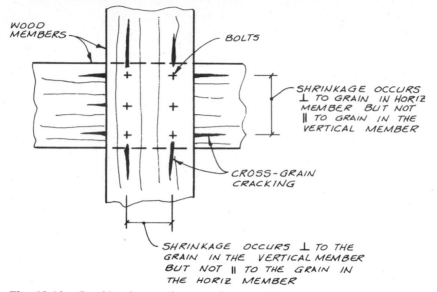

Fig. 12.13   Cracking in wood-to-wood connections.

It should be noted that the CUF of 0.40 does not apply to connections where cross-grain cracking is prevented. The NDS states that the following connections have CUF = 1.0:

1. One fastener only

2. Two or more fasteners in a single line parallel to the grain

3. Fasteners in two or more lines parallel to the grain with separate side plates for each line

According to NDS Table 8.1B the CUF = 1.0 when lumber is initially "dry" and remains dry in service (item *a* in Example 12.12). "Dry" wood is defined as having a moisture content of 19 percent or less. It should be realized that there can be a fairly large difference between the initial and final moisture contents in the *dry* range, and some shrinkage can occur (Example 4.3). This is especially true in dry climate zones. In view of the concern about the effects of cross-grain shrinkage in critical connections, the designer may also want to make reductions of allowable stress when large differences in moisture contents are expected even in the dry range.

As noted earlier, however, the most desirable solution is to detail the connection so that shrinkage can not induce cracking in the first place. Recommended connection detailing practices with this intention are given in Chap. 13.

12.8 Lag Screw Connections

Lag screws and some of their typical connections were introduced in Example 12.2. The general factors that affect the strength of these connections (and wood connections using bolts and timber connectors) are covered in Sec. 12.3.

Other factors that affect the strength of lag screws are

1. Species of lumber
2. Diameter and length of lag screw
3. Thickness of side member
4. Type of connection (lateral or withdrawal)
5. Side-grain or end-grain connection
6. Type of side member (wood or steel)
7. Fabrication techniques
8. Spacing requirements (same as bolts—see Sec. 12.7)
9. Arrangement of side plates with multiple rows of lag screws (same as bolts—see Sec. 12.7)

The allowable lateral load on a lag screw depends on the species group of lumber, and the allowable withdrawal load depends on the specific gravity of the lumber. The species groups and specific gravities are given in NDS Table 8.1A, *Grouping of Species for Fastening Design.*

The configuration of lag screws is given in NDS Appendix L, *Typical Dimensions of Standard Lag Screws for Wood.* Lag screws are installed in prebored holes that accommodate the shank and threaded portion of the screw. See Example 12.13. The hole involves drilling with two different-diameter bits. The larger-diameter hole has the same diameter and length as the shank of the lag screw. The diameter of the lead hole for the threaded portion is given as a percentage of the shank diameter.

It is important that lag screws be properly installed with a wrench. Driving with a hammer is an unacceptable method of installation. Soap or other type of lubricant should be used on the lag screws or in the pilot hole to facilitate installation and to prevent damage to the fastener.

Tabulated allowable lateral loads are for lag screws installed in the *side grain* of the holding member. Allowable loads parallel to the grain, p, and perpendicular to the grain, q, are given for this type of connection. Allowable loads are given in NDS Table 8.6B, *Lag Screws—Lateral Load Design Values with Wood Side Pieces,* and NDS Table 8.6C, *Lag Screws—Lateral Load Design*

EXAMPLE 12.13 Installation of Lag Screws

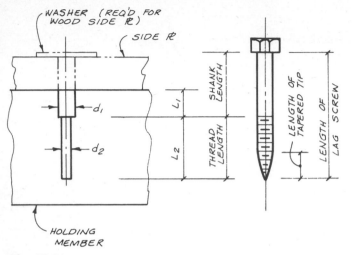

Fig. 12.14

Lag screws are installed in pilot holes. The dimensions of the pilot hole in the holding member are

For the shank:

$$L_1 = \text{shank length} - \text{side } P_L \text{ thickness} - \text{washer thickness}$$
$$d_1 = \text{shank diameter}$$

For the threaded portion:

$$L_2 \geq \text{thread length}$$

$$\text{Group I species---} d_2 = (0.65 \text{ to } 0.85) \times d_1$$
$$\text{Group II species---} d_2 = (0.60 \text{ to } 0.75) \times d_1$$
$$\text{Group III and IV species---} d_2 = (0.40 \text{ to } 0.70) \times d_1$$

The larger fractions of d_1 should be used for larger-diameter lag screws.

Values with ½ in. Metal Side Pieces. These tables apply to single lag screws with loads of normal duration and dry conditions of use. Adjustments for multiple fasteners, duration of loading, and conditions of use are given in Sec. 12.3.

The allowable lateral load for lag screws installed in the *end grain* of the holding member is taken as two-thirds of the allowable load q for the same fastener installed in the side grain of the holding member.

The withdrawal load is similar in concept to a nailed connection. Tabulated allowable values are for lag screws installed in the side grain of the holding

414

member. NDS Table 8.6A, *Lag Screws—Withdrawal Design Values*, gives the allowable load per inch of effective penetration. The *effective penetration* is the length of the threads in the holding member minus the length of the tapered portion of the thread near the point. The effective thread length can be taken directly from NDS Appendix L.

Withdrawal loading of lag screws installed in the end grain of the holding member is not recommended. If this type of loading cannot be avoided, an allowable load of three-fourths of the side grain withdrawal value is used. As with lateral loads, tabulated values are based on normal-duration loading and dry conditions of use. Adjustments for other conditions are given in Sec. 12.3.

12.9 Design Problems: Lag Screw Connections

Two relatively brief examples of lag screw calculations are provided. The first deals with a lateral load connection through a metal side plate. See Example 12.14. A drag strut load is transferred from the glulam beam to the steel splice plate by the lag screws.

The allowable load parallel to the grain for the lag screws can be read directly from the NDS table. (This is in contrast to bolted connections where the allowable load for metal side plates is determined by multiplying the allowable value for wood side plates by the MPF.) The allowable load for the lag screws is subject to the adjustments for duration of load, condition of use, and the number of fasteners in a row.

EXAMPLE 12.14   Splice With Lag Screws

The roof beam in the building in Fig. 12.15 serves as the drag strut for the two horizontal diaphragms. The maximum strut force is 4000 lb at the connection of the beam to the concrete shearwall. Determine the number of $\frac{1}{2}$-in. lag screws, 6 in. long, that are necessary to connect the steel splice plate to the Southern pine glulam header. Assume that the splice plate and the connection to the wall are adequate. A single row of lag screws will be used. CUF — 1.0.

Southern pine is in species group II. From NDS Table 8.6C,

$$\text{Adj. } p = \text{tab. } p \times \text{LDF} \times \text{CUF}$$
$$= 945 \times 1.33 \times 1.0 = 1255 \text{ lb/lag}$$

Approx. number of lag screws:

$$N = \frac{P}{\text{adj. } p} = \frac{4000}{1255} = 3.2 \qquad \textit{Say four lag screws}$$

Check four fasteners in a row:

$$A_1 = 5\frac{1}{8} \times 21 = 108 \text{ in}^2$$
$$A_2 = 1.0 \text{ in}^2 \qquad \text{(given, Fig. 12.15)}$$
$$A_1/A_2 = 108$$

415

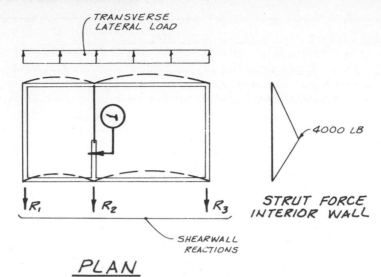

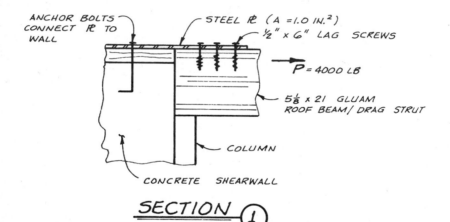

Fig. 12.15

This value is not covered in NDS Table 8.3B. Therefore use values for max A_1/A_2.

$$K = 0.93$$
$$P_r = KP_s = 0.93 \times 4 \times 1255 = 4675 \text{ lb}$$
$$4675 > 4000 \text{ lb} \qquad OK$$

Use four ½ in. × 6 in. lag screws

The spacing requirements for lag screws are the same as for bolts. For parallel-to-grain loading, refer to Example 12.10.

**EXAMPLE 12.15 Lag Screw Withdrawal Connec-
tion**

The load *P* is an equipment DL which is suspended from two glulam supports (Fig.
12.16*a*). Determine the allowable load based on the strength of the lag screws. Lumber
is initially dry and is exposed to the weather in service.

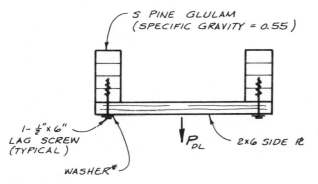

Fig. 12.16*a*

Determine effective penetration of threads (dimensions from NDS Appendix L).

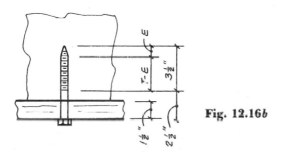

Fig. 12.16*b*

$$\text{Effective thread length} = T - E = 3\tfrac{3}{16}$$

From NDS Table 8.6A,

$$\text{Adj. } P_W = \text{LDF} \times \text{CUF} \times (\text{tab. } P_W \times \text{effective thread length})$$
$$= 0.9 \times 0.75 \times (437 \times 3\tfrac{3}{16})$$
$$= 940 \text{ lb/lag}$$
$$\text{Allow. } P = 2(940) = 1880 \text{ lb}$$

* Washer must be sufficiently large to keep bearing perpendicular to the grain below $F_{c\perp}$.

The second connection example uses lag screws loaded in withdrawal. See Example 12.15. The strength of withdrawal connections is based on the effective thread penetration into the holding member. The allowable load per lag is determined by multiplying the effective length by the tabulated allowable load per linear inch. This value is subject to further adjustments for duration of load and condition of use.

12.10 Timber Connectors

Split rings and shear plates were described in Sec. 12.2. Timber connectors have the advantage of being able to carry high allowable loads on a single fastener. They have the disadvantage of requiring special fabrication equipment for the cutting of the grooves.

Because of the more involved fabrication required with timber connectors, they are used more often in the construction of glulam arches and special truss designs than in ordinary shearwall-type buildings. See Example 12.16. The designer, however, should have some understanding of the procedures for determining allowable loads for timber connectors.

Split rings are available in 2½-in. and 4-in. diameters, and shear plates are available in 2⅝-in. and 4-in. diameters. Other pertinent dimensions are given in NDS Appendix K, *Typical Dimensions for Timber Connectors.*

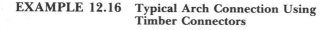

EXAMPLE 12.16   Typical Arch Connection Using Timber Connectors

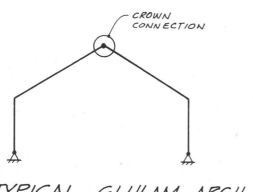

TYPICAL GLULAM ARCH

Fig. 12.17a

Arch connections are shown to illustrate how timber connectors can be used in the end grain of a member. Allowable loads for timber connectors are given in the NDS. Information on the design of arches can be found in Ref. 3.

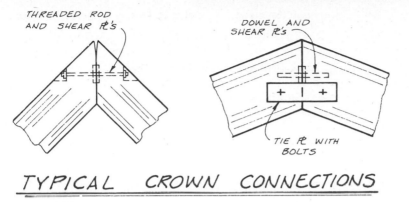

TYPICAL   CROWN   CONNECTIONS

Fig. 12.17*b*

For arches with steep slopes, two shear plates can be used with a threaded rod. Washers are counterbored into the arch.

For flat arches, two shear plates on a dowel can be used for vertical reactions. The tie plate keeps the arch from separating.

EXAMPLE 12.17   Shear Plate Connection

Determine the number of 4-in.-diameter shear plates necessary to develop the chord splice in Fig. 12.18. Lumber is Douglas fir glulam that is dry and will remain dry in service. Assume metal splice plates and glulams are adequate.

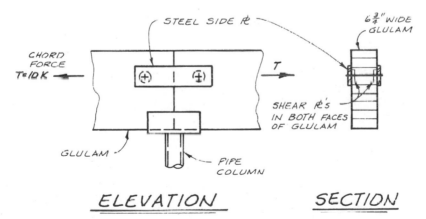

ELEVATION   SECTION

Fig. 12.18

From NDS Table 8.1A Douglas fir-larch is in connector group B.
From NDS Table 8.4B,

$$\text{Adj. } p = \text{tab. } p \times \text{LDF} \times \text{CUF}$$
$$= 4320 \times 1.33 \times 1.0 = 5745 \text{ lb/shear plate}$$

419

NOTE: The 11 percent increase for metal side plates in Group B species is disregarded (NDS Sec. 8.4.9, *Metal Side Plates*).

Allow load for one shear plate in each side of glulam (i.e., two connectors each side of splice).

$$\text{Allow. } T = 2(5745) = 11.5 \text{ k} > 10 \text{ k} \qquad OK$$

> Use total of four 4-in. shear plates
> (as shown in Fig. 12.18)

For timber connector design, the commercial species are divided into four groups (A to D) (NDS Table 8.1A, *Grouping of Species for Fastening Design*).

Tabulated allowable loads are for connectors installed in the side grain of wood members. Both parallel-to-grain p and perpendicular-to-grain q values are given in NDS Table 8.4A, *Split Ring Design Values*, and NDS Table 8.4B, *Shear Plate Design Values*. Allowable loads for timber connectors are subject to the general adjustments given in Sec. 12.3.

The NDS gives the method of obtaining the allowable load for connectors installed in the end grain (as in an arch crown connection). Spacing requirements are also given in the NDS.

Additional information and design aids for obtaining allowable loads and the spacing requirements for timber connectors can be found in Refs. 3 and 24.

An example will illustrate the use of the NDS tables for timber connectors. The connection involves the design of shear plates for a horizontal diaphragm chord splice. See Example 12.17.

12.11   Problems

The allowable loads for fasteners in the following problems are to be in accordance with the NDS.

12.1   The connection in Fig. 12.A uses a single row of five ¾-in. bolts. Lumber is No. 1 Hem-Fir. The load is the result of (DL + wind). CUF = 1.0

Find: a. The allowable load on the bolted connection.
b. The allowable tension in the wood members.
c. Indicate the spacing requirements for the bolts.

12.2   The connection in Fig. 12.A uses a single row of six ¾-in. bolts. Lumber is No. 1 DF-larch. The load is the result of (DL + snow). CUF = 1.0.

Find: a. The allowable load on the bolted connection.
b. The allowable tension in the wood members.
c. Indicate the spacing requirements for the bolts.

12.3   Problem 12.3 is the same as Prob. 12.1 except that the lumber is green at the time of fabrication, and it later seasons in service.

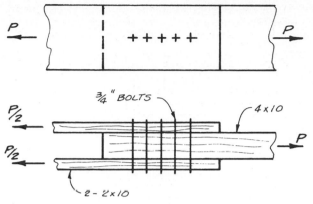

Fig. 12.A

12.4   Problem 12.4 is the same as Prob. 12.2 except that the lumber is green at the time of fabrication and it later seasons in service.

12.5   Problem 12.5 is the same as Prob. 12.1 except that the connection is exposed to the weather.

12.6   Problem 12.6 is the same as Prob. 12.2 except that the connection is exposed to the weather.

12.7   The connection in Fig. 12.B uses steel side plates and four 1-in. bolts on each side of the splice. Lumber is No. 1 DF. The load is seismic.

   Find: The allowable load on the bolts if
      a. CUF = 1.0
      b. Green lumber is used that later seasons in service.
      c. The connection is exposed to the weather.
      d. The lumber is above the FSP in service.

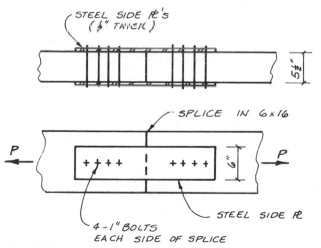

Fig. 12.B

12.8 The beam in Fig. 12.C is suspended from a pair of steel angles. The load is (DL + FLL). Lumber is Select Structural Western hemlock that is initially dry and remains dry in service. Assume that the angles are adequate.

    Find: *a.* Determine if the given bolts are adequate.
               b. Check the shear stress in the beam at the connection.
               c. Are the spacing requirements satisfied?

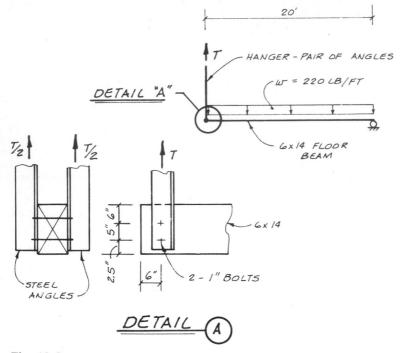

Fig. 12.C

12.9 Problem 12.9 is the same as Prob. 12.8 except that the lumber is initially at the FSP and it seasons in service.

12.10 The connection in Fig. 12.D uses two rows of ⅞-in. bolts with three fasteners in each row. The load carried by the connection is (DL + FLL). The wood member is a 6¾ × 16.5 DF-larch glulam (22 F) that is initially dry (at or near the EMC). The connection remains dry in service.

    Find: *a.* The allowable connection load based on the capacity of the bolts. Show the bolt spacing requirements on a sketch of the connection.
               b. Check the stresses in the main member and side plates for the load in *a.* Assume A36 steel plate (Ref. 12).

12.11 The connection in Fig. 12.D uses two rows of ⅞-in. bolts with *five* fasteners in each row (figure shows three). The load carried by the connection is (DL + snow). The wood member is a 6 × 14 Select Structural DF-larch that is initially wet, and it later seasons in place.

Find: *a.* The allowable load on the connection, based on the strength of the bolts.

      b. The allowable load on the connection if separate side plates are used for the two rows of bolts. Show the spacing requirements for the bolts in the connection.

      c. Check the stresses in the main member and side plates for the load in *b.* Assume A36 steel plate (Ref. 12).

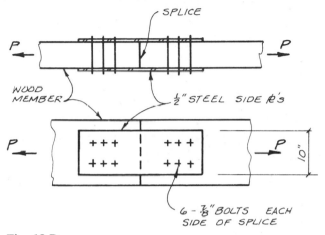

Fig. 12.D

12.12   The connection in Fig. 12.E uses ¾-in. bolts in single shear. Lumber is DF-larch that is initially dry (near the EMC), and it remains dry in service. The load is caused by (DL + FLL). Assume that the members are adequate.

    Find: *a.* The allowable load on the connection, based on the strength of the bolts.

      b. Show the spacing requirements for the bolts on a sketch of the connection.

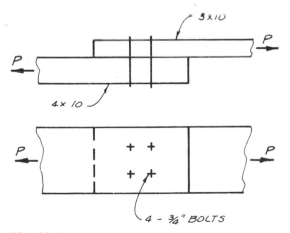

Fig. 12.E

12.13 The connection in Fig. 12.F uses 1-in. bolts and serves as a tie for lateral wind loads. Lumber is DF-larch that is initially dry (near the EMC), and it remains dry in service. Assume that the members are adequate.

Find: a. The allowable load on the connection based on the strength of the bolts. Consider the direction of loading and the respective thicknesses of the main member and side plates.

b. Show the spacing requirements for the bolts on a sketch of the connection.

c. If initially green lumber is used, what would be the effect on the allowable load?

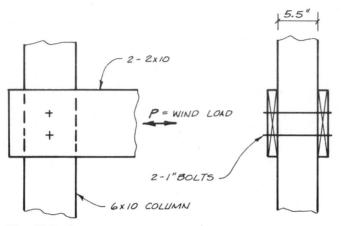

Fig. 12.F

12.14 The connection in Fig. 12.G uses one 1-in. bolt that carries a lateral wind load. Lumber is Southern pine that is dry. EMC ≤ 19 percent. Assume that the members are adequate.

Find: The allowable load on the connection. Consider the direction of loading and the respective thicknesses of the members.

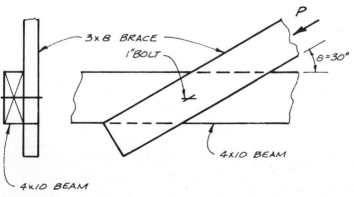

424   **Fig. 12.G**

12.15 The building in Fig. 12.H uses 2 × 6 wall framing of No. 1 Hem-Fir. The double top wall plate serves as the horizontal diaphragm chord and drag strut. The top plate is fabricated from 20-ft.-long lumber and splices occur at 10-ft. intervals. Design lateral loads are given. CUF = 1.0.

Find: Design and detail the connection splice in top plate of wall 1 using ¾-in. bolts.

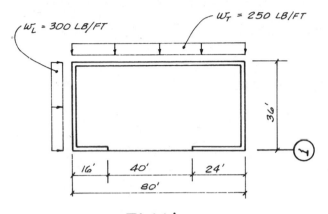

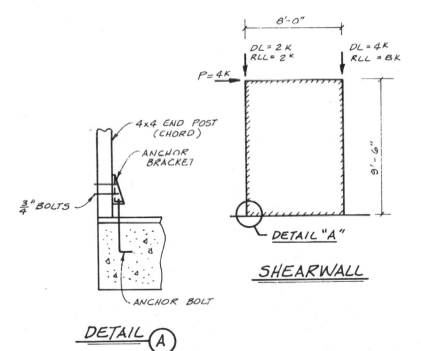

Fig. 12.H

Fig. 12.I

12.16 Problem 12.16 is the same as Prob. 12.15 except that the lumber is No. 1 DF-larch and the bolts are ⅞ in.

12.17 *Given:* The shearwall and anchorage detail in Fig. 12.I. The gravity loads and lateral wind load are shown in the sketch. The chords in the shearwall are 4 × 4 members of No. 2 DF (south). Assume the bracket is adequate. CUF = 1.0.

  Find: *a.* The number of ¾-in. bolts necessary to anchor the chord to the hold down bracket. Sketch the spacing requirements for the bolts.
  b. Check the tension capacity of the chord at the critical section (bolts are in a single row).
  c. Determine the required size of anchor bolt to tie the bracket to the foundation. Use UBC Table 26G.

12.18 The connection of the ledger to the stud wall in Fig. 12.J uses ½-in. × 6-in. lag screws. Patio DL = 6 psf and roof LL = 10 psf (UBC Appendix Chap. 49). Lumber is DF-larch.

  Find: *a.* Determine if the connection is adequate to carry the design loads if the lumber is initially dry and remains dry in service.
  b. Is the connection adequate if the lumber is exposed to the weather?

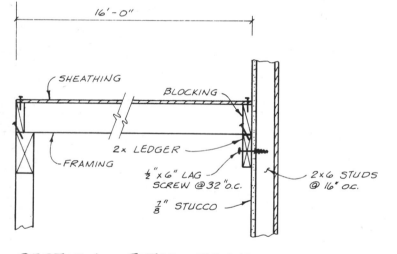

SECTION - PATIO ROOF

Fig. 12.J

12.19 Problem 12.19 is the same as Prob. 12.18 except that the roof LL is 20 psf.

12.20 The building in Fig. 12.K uses the connection shown in section *A* to transfer the lateral load (chord force and strut force) from the wall plate to the header. Lumber is No. 1 DF-larch. CUF = 1.0.

  Find: Determine the required number of ⅝-in. × 6-in. lag screws for the connection.

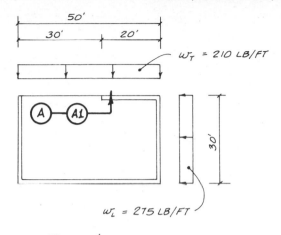

$w_T = 210 \ LB/FT$

$30'$

$w_L = 275 \ LB/FT$

PLAN

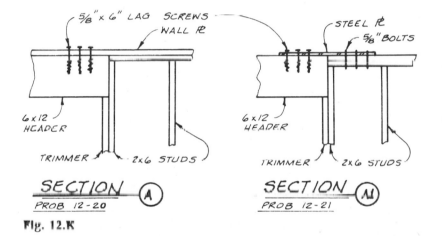

$\tfrac{5}{8}" \times 6"$ LAG SCREWS

WALL ℞

STEEL ℞

$\tfrac{5}{8}"$ BOLTS

6 x 12
HEADER

6 x 12
HEADER

TRIMMER   · 2x6 STUDS

TRIMMER   2x6 STUDS

SECTION (A)

PROB 12-20

SECTION (A1)

PROB 12-21

Fig. 12.K

12.21 Problem 12.21 is the same as Prob. 12.22 except that the lateral load transfer is made with the connection shown in section A1. In addition, determine the requirements for the steel plate.

Connection
Hardware

13.1   Introduction

Chapters 11 and 12 covered the basic fasteners that are used in the construction of wood buildings. In addition to these fasteners most wood connections involve some type of metal connection hardware. This hardware may be a simple tension strap, but considerably more complicated devices are often used.

Chapter 13 introduces some typical connection hardware details. A critique of the connection accompanies a number of the details, and particular attention is given to the effects of cross-grain shrinkage. These principles should be given special attention if it is necessary to design, detail, and fabricate special connection hardware. An example of several connection hardware calculations is given after the general review.

In a number of cases, prefabricated connection hardware may be available for common connections. In any event, the basic principles of connection design should be understood by the designer. Chapter 13 concludes with a review of some typical prefabricated connection hardware.

13.2   Connection Details

The hardware used in a connection must itself be capable of supporting the design loads, and structural steel design principles of Ref. 12 should be used to determine the required plate thickness and

429

weld sizes. Some of these principles are demonstrated in examples, but a comprehensive review of steel design is beyond the scope of this book.

Some suggested details for common wood connections are included in Ref. 3. Reference 25 provides additional information on connection detailing practices with an emphasis on avoiding cross-grain cracking caused by changes in moisture content. These details are aimed at minimizing restraint in connections that will cause splitting. Where restraint does occur, the distance across the grain between the restraining elements should be kept small so that the total shrinkage between these points is minimized. Where possible, the connections should be designed to accommodate the shrinkage (and swelling) of the wood without initiating built-up stresses.

The majority of the details in this section are taken or adapted from Ref. 25 by permission of AITC. The sketches serve as an introduction to the types of connections that are commonly used in wood construction. See Example 13.1. At the same time, both good and poor connection design practices are illustrated. The details were developed specifically for glulam construction, where members and loads are often large. The principles are valid for both glulam and sawn lumber of any size, but the importance of these factors increases with increasing member size.

The connection details in Example 13.1 are not to scale and are to be used as a guide only. Actual designs will require complete drawings with dimensions of the connection and hardware. The designs must take into consideration the spacing requirements for bolts (Sec. 12.7) or the other types of fasteners. Examples of the following typical connections are included in Example 13.1:

a. Typical beam-to-column connections

b. Beam-to-girder saddle connection

c. Beam face hanger connection

d. Beam face clip connections

e. Cantilever beam hinge connection

f. Beam connection for uplift

g. Beam connection to continuous column

h. Beam with notch in tension zone

i. Beam with end notch

j. Inclined beam support details

k. Suspended heavy loads from beams

l. Suspended loads from beams

m. Suspended multiple loads

n. Eccentricity considerations

o. Moisture protection at column base

References 3 and 25 cover a number of connections in addition to these. However, the details given here illustrate the basic principles involved in proper connection design. A numerical example is provided in the following section for the design of a beam-to-column connection. This is followed by some additional considerations for beam hinge connections.

EXAMPLE 13.1   Connection Details*

Beam-to-Column Connections

If the compression perpendicular to the grain at the beam reaction is satisfactory, the beam may be supported directly on a column. A 'T' bracket can be used to tie the various members together (Fig. 13.1a.1).

The bottom face of the beam bears on the column. If cross-grain shrinkage occurs in the length a, cracking or damage to the bracket will result. The problem can be minimized by keeping a small.

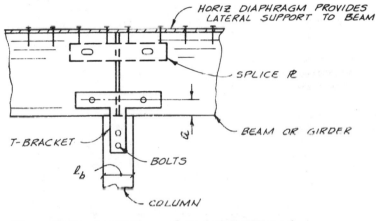

T - BRACKET   CONNECTION

Fig. 13.1a.1

For larger loads the T bracket will be replaced with a U bracket. The U bracket can provide a bearing length l_b which is greater than the width of the column. The spacing requirements for the fasteners must also be considered in determining the length of the bracket.

U brackets (Fig. 13.1a.2) may be fabricated from bent plates when the bracket does not cantilever a long distance beyond the width of the column. For longer U brackets, the vertical plates will probably be welded to a thicker base plate. Calculations for a U-bracket connection are given in Sec. 13.3.

---

* Adapted from Ref. 25, courtesy of AITC.

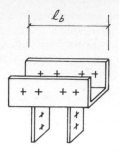

Fig. 13.1a.2

U BRACKET

If a separate splice plate is used near the top of the beam (Fig. 13.1a.1), the effects of the end rotation and beam separation should be considered. Bolts in tightly fitting holes can restrict the movement of the end of the beam and splitting may be induced. Slotted holes can reduce this problem. If the beam carries an axial drag or chord force, the T bracket or U bracket can be designed to transmit the lateral load.

Beam Saddle Connection

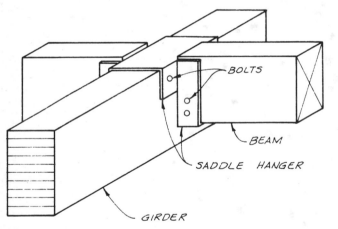

Fig. 13.1b

Cross-grain tension is not developed by shrinkage in this type of connection (Fig. 13.1b). The beam reaction is transferred by bearing perpendicular to the grain on the bottom of the hanger. The load on the hanger is transferred by bearing perpendicular to the grain through the top of the saddle to the girder. This type of connection is recommended for larger loads.

Note that the bolts are located *near* the bearing points to minimize cross-grain effects. Although the bolts should be located near the bearing face of the member, care should be taken to avoid boring holes in the high-quality outer lamination in bending glulams. This is especially important when bolts occur in the tension zone of the girder.

The tops of beams are shown higher than the girder to allow for shrinkage in beams without affecting roof sheathing or flooring. The idea is to have the tops of beams and girders at the same level after shrinkage occurs. For shrinkage calculations see Example 4.3.

Beam Face Hanger Connection

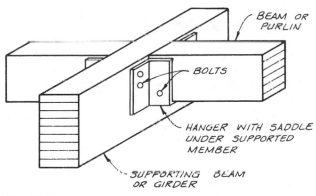

BEAM OR PURLIN

BOLTS

HANGER WITH SADDLE UNDER SUPPORTED MEMBER

SUPPORTING BEAM OR GIRDER

Fig. 13.1c

When the reaction of the beam or purlin is relatively small, the hanger can be bolted to the face of the girder (Fig. 13.1c). The bolts in the main supporting beam or girder should be placed in the upper half of the member, but not too close to the top of the beam where extreme fiber-bending stresses are maximized. This is especially important in the tension zone which occurs on the top side of cantilever girders.

Beam Face Clip Connection

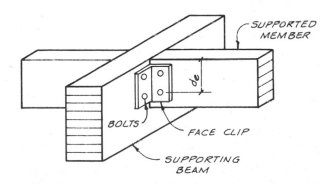

SUPPORTED MEMBER

d_c

BOLTS

FACE CLIP

SUPPORTING BEAM

Fig. 13.1d.1

DESIGN OF WOOD STRUCTURES

Face connections without the saddle feature shown in Fig. 13.1c may be used for light loads (Fig. 13.1d.1). However, construction is more difficult than with the saddle because the beams must be held in place while the bolts are installed. The connection at the end of the supported beam must be checked for shear in accordance with the procedures outlined in Example 12.5.

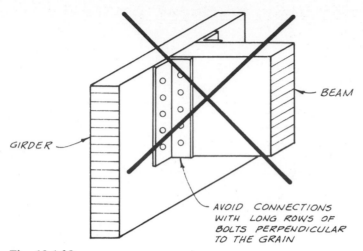

BEAM

GIRDER

AVOID CONNECTIONS WITH LONG ROWS OF BOLTS PERPENDICULAR TO THE GRAIN

Fig. 13.1d.2

Face clips should be limited to connections with light loads. Larger loads will require a long row of bolts perpendicular to the grain through steel side plates. Glulam timbers, although relatively dry at the time of manufacture, may shrink as the EMC is reached in service. The problems associated with cross-grain tension should be avoided by not using the type of connection in Fig. 13.1d.2.

Cantilever Beam Hinge Connections

PROBLEM

For vertical loads the suspended beam bears on the bottom of the hinge connector, and the hinge connector in turn bears on the top of the supporting (cantilever) member. See Fig. 13.1e.1. Cantilever systems may be subjected to lateral loads as well as bending loads. Some designers have used a tension connector with bolts lined up in both members near the bottom of the beam. The problem with this arrangement is that as the wood shrinks, the supporting member will permit the saddle to move downward. The bolts in the cantilever, however, will restrain this movement and a split may occur. The revised detail (Fig. 13.1e.2) avoids this problem.

SUGGESTED REVISION

This connection does not interfere with possible shrinkage in the members. It also allows a slight rotation in the joint. A positive tension tie can be provided with a separate tension strap at the mid-depth of the joint. This tension strap may also be an integral part of the hanger if the holes in the tension tie are slotted vertically in deep members. Also see Sec. 13.4.

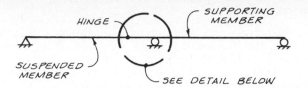

TYP CANTILEVER BEAM SYSTEM

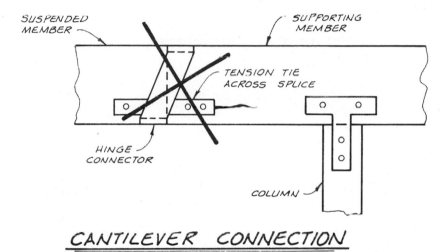

CANTILEVER CONNECTION

Fig. 13.1e.1

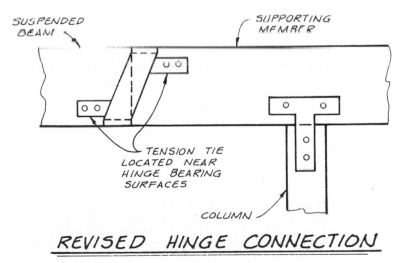

REVISED HINGE CONNECTION

Fig. 13.1e.2

DESIGN OF WOOD STRUCTURES

Beam Connection for Uplift

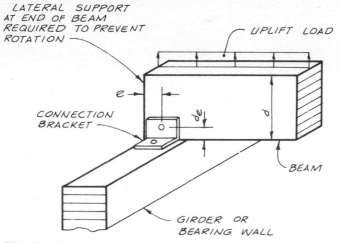

Fig. 13.1*f*

Where gravity loads are transferred by bearing perpendicular to grain, adequate fastenings must also be provided to carry horizontal and uplift loads. See Fig. 13.1*f*. These loads are usually of a transient nature and are of short duration. The fastener is usually placed toward the lower edge of the member. The distance d_e must be no less than the distance required for the perpendicular-to-grain edge spacing for the type of fastener used. It must also be deep enough so that the shear stresses caused by uplift (calculated in accordance with the methods of Example 12.5) are not excessive. The end distance *e* must also be adequate for the type of fastener used.

Beam Connection to Continuous Column

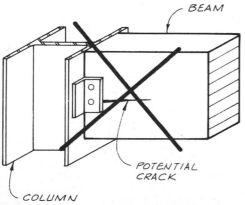

Fig. 13.1*g*.1

PROBLEM

Connections of the type shown in Fig. 13.1g.1 should not be used on large beams or girders because tension perpendicular to the grain and horizontal shear at the lower fastener tend to cause splitting of the member.

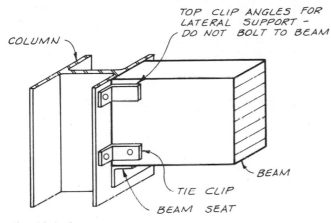

Fig. 13.1g.2

SUGGESTED REVISION

When the design is such that the bending member does not rest on a column, wall, or pilaster, but frames into another member, several methods can be used to support the ends. The preferred method is to transfer the end reaction by bearing perpendicular to grain. See Fig. 13.1g.2.

Beam with Notch in Tension Side

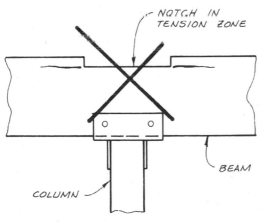

Fig. 13.1h

DESIGN OF WOOD STRUCTURES

PROBLEM

This detail (Fig. 13.1*h*) is critical in cantilever framing (negative moment at support). The top *tension* fibers have been cut to provide space for recessed connection hardware or for the passage of conduit or other elements over the top of the beam. This is particularly serious in glulam construction because the tension laminations are critical in the performance of the structure.

SUGGESTED REVISION

Notches in the tension zone are serious stress raisers and should be avoided. Detail the structure so that the cut in the beam is not necessary. Also see Sec. 6.2.

___

Beam with End Notch

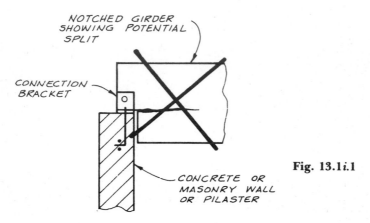

Fig. 13.1*i*.1

PROBLEM

An abrupt notch in the end of a wood member creates two problems (Fig. 13.1*i*.1). One is that the effective shear strength of the member is reduced because of the end notch. The next is that the exposure of end grain in the notch will permit a more rapid migration of moisture in the lower portion of the member causing the indicated split.

For horizontal shear calculations at notches at the end of a beam, see Sec. 6.4.

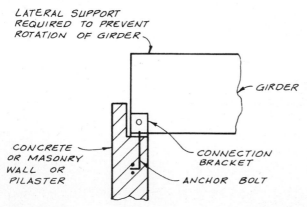

Fig. 13.1*i*.2

Where the height of the top of a beam is limited, the beam seat should be lowered to a pilaster or specially designed seat in the wall (Fig. 13.1i.2).

Inclined Beam Support Details

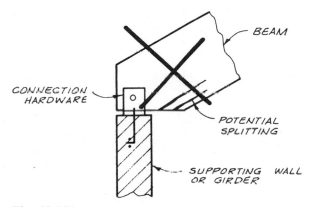

Fig. 13.1j.1

PROBLEM
This condition is similar to the problem in Fig. 13.1i, but it may not be as evident. The shear strength of the end of the member is reduced and the exposed end grain may result in splitting because of drying (Fig. 13.1j.1).

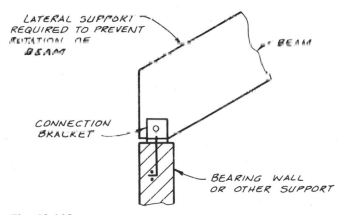

Fig. 13.1j.2

SUGGESTED REVISION
Where the end of the beam must be flush with outer wall, the beam seat should be lowered so that the tapered cut is loaded in bearing (Fig. 13.1j.2).

DESIGN OF WOOD STRUCTURES

Suspending Heavy Loads from Beams

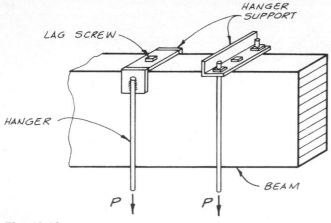

Fig. 13.1k

Heavy loads should preferably be suspended from the top of a beam or girder (Fig. 13.1k).

Suspending Loads from Beams

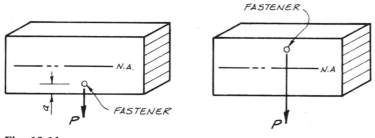

Fig. 13.1l

Isolated light loads may be suspended near the bottom face of a beam. The distance a must equal or exceed the required edge distance, and shear stresses must not exceed the limits given in Example 12.5.

It is good practice to locate the fastener above the neutral axis, and the detail in Figure 13.1k is recommended.

Suspending Multiple Loads

PROBLEM

The type of design shown in Fig. 13.1m.1 should be avoided. Tension stresses perpendicular to the grain tend to concentrate at the fastenings. These can interact with horizontal shear to cause splitting.

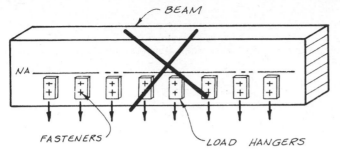

Fig. 13.1m.1

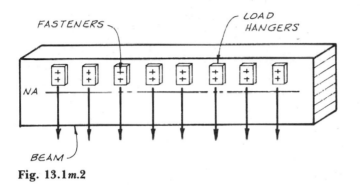

Fig. 13.1m.2

SUGGESTED REVISION

Fasteners should be located above the neutral axis (Fig. 13.1m.2). Where possible, the detail in Fig. 13.1k is recommended.

Eccentricity Considerations

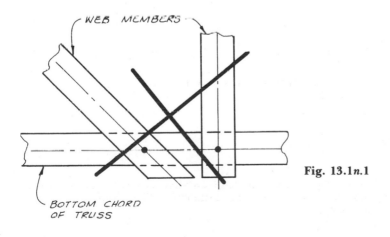

Fig. 13.1n.1

DESIGN OF WOOD STRUCTURES

PROBLEM

When the center lines of members do not line up at a common joint in a truss, considerable shear and moment may result in the bottom chord (Fig. 13.1n.1). When these stresses are combined with the presumably high tension stress in the member, failure may occur.

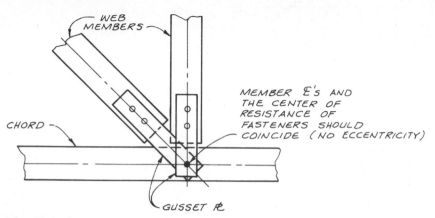

CHORD

WEB MEMBERS

MEMBER E's AND THE CENTER OF RESISTANCE OF FASTENERS SHOULD COINCIDE (NO ECCENTRICITY)

GUSSET ℄

Fig. 13.1n.2

SUGGESTED REVISION

For trusses with single-piece members (such as bowstring trusses with glulam upper and lower chords), the chords and web members should be in the same plane with straps or gusset plates used for the connection. See Fig. 13.1n.2.

Moisture Protection at Column Base

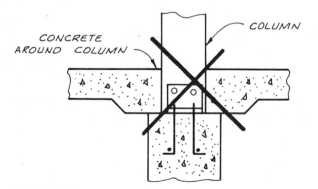

CONCRETE AROUND COLUMN

COLUMN

Fig. 13.1o.1

PROBLEM

Some designers try to conceal the base of a column or an arch by placing concrete around the connection (Fig. 13.1o.1). Moisture may migrate into the lower portion of the wood and may cause decay.

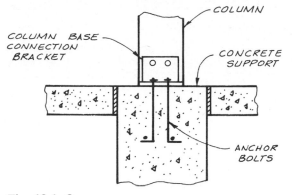

Fig. 13.1o.2

Detail the column base at the top of the floor level (Fig. 13.1o.2). Note that the base of the column is separated from concrete by a bottom bearing plate. Where subjected to splashing water, untreated columns should be supported by piers projecting up at least 2 in. above the finished floor.

13.3 Design Problem: Beam-to-Column Connection

Example 13.2 illustrates a typical beam-to-column connection. The beam is a 5⅛ × 27 glulam and the column is a 4-in. steel tube. Steel tubes and steel pipe columns are commonly used to support wood beams.

The design of this connection starts with a consideration of the required bearing length for the wood beam. Calculations of this nature were introduced in Chap. 6. The required overall length of the U bracket for bearing stresses is 9 in. In later calculations this length is shown to be less than the length required to satisfy the bolt spacing criteria.

The example concludes with a check of the bending stresses in the U bracket. Because of the small cantilever length, a bent plate is used for the bracket. For longer cantilever lengths a welded U bracket would be required. Complete details of the connection bracket should be included on the structural plans.

13.4 Cantilever Beam Hinge Connection

The type of connection used for an internal hinge in wood cantilever beam systems was introduced in Fig. 13.1e. There the detailing of the connection was considered to minimize the effects of shrinkage. In the design of this type of connection, the equilibrium of the hinge must also be considered. **443**

EXAMPLE 13.2 U-bracket Column Cap

Design a U bracket to connect the two glulam beams to the 4-in. steel tube column. Gravity loads are given in Fig. 13.2a. Uplift is zero. The lateral seismic load causes an axial force in the glulam beams of 6 k. Glulam is 24F Douglas fir which is initially dry and will remain dry in service.

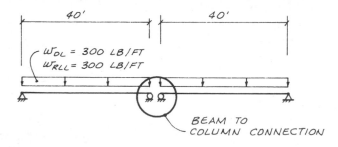

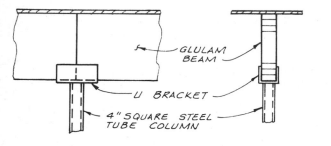

U-BRACKET CONNECTION

Fig. 13.2a

Vertical Load

Reaction from one beam:

$$R = \frac{wL}{2} = \frac{(600)(40)}{2} = 12,000 \text{ lb}$$

$$\text{Required bearing length} = \frac{R}{F_{c\perp} \times b}$$

$$= \frac{12,000}{450 \times 1.25 \times 5\frac{1}{8}} = 4.16 \text{ in.}$$

$$\text{Min. length of bracket} = 2(4.16) = 8.33$$
$$\textit{Say } 9 \text{ in. minimum}$$

Lateral Load

Assume ¾-in. bolts in double shear with steel side plates. From NDS Table 8.5A, *Bolt Design Values*, $(l = 5\frac{1}{8})$:

Species category 3
Tab. $p = 2860$ lb/bolt
Adj. p = tab. $p \times$ LDF $\times$ MPF $= 2860 \times 1.33 \times 1.25$
$= 4750$ lb/bolt

Required number of bolts $= \dfrac{6000}{4750}$

> *Use* two ¾-in. bolts each side of splice

Connection Details

There are a number of possible connection details that can be used. One possible design is included here (Fig. 13.2*b*). Assume lateral load is carried by U bracket.

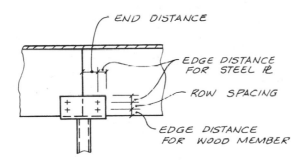

BOLT SPACINGS

Fig. 13.2*b*

Vertical load is carried by bearing. Spacing requirements are governed by the lateral load (parallel-to-grain loading criteria—Example 12.10).

HORIZONTAL SPACING:

End distance $= 7d = 7(\tfrac{3}{4}) = 5\tfrac{1}{4}$ in.
Steel plate edge distance $= 1\tfrac{1}{4}$ in.   (from Ref. 12)
Total bracket length $= (5\tfrac{1}{4} + 1\tfrac{1}{4})2 = 13$ in.
$13 > 9$
Spacing governs over bearing ($f_{c\perp}$)

VERTICAL SPACING:

Steel plate edge distance $= 1\tfrac{1}{4}$ in.
Row spacing $= 1\tfrac{1}{2}d = 1\tfrac{1}{2}(\tfrac{3}{4}) = 1\tfrac{1}{8}$ in.
    Say 1½ in.
Edge distance* $- 1\tfrac{1}{2}d = 1\tfrac{1}{8}$ in.
    or ½ row spacing $= \tfrac{3}{4}$ in. (the larger governs)
    Say 1½ in.

Minimum bracket height $= 1\tfrac{1}{4} + 2(1\tfrac{1}{2})$
$= 4\tfrac{1}{4}$ in.

---

* Uplift force was given as zero. If uplift occurs, the edge distance is governed by $4d$ (Example 12.11) or horizontal shear criteria (Example 12.5).

445

DESIGN OF WOOD STRUCTURES

PROPORTIONS OF U BRACKET:

To allow for construction tolerances, the overall length of the bracket will be taken as 13.5 in. (13 in. minimum). As a rule of thumb, a bracket height of one-third the length is reasonable.

$$\tfrac{1}{3} \times 13.5 = 4.5 \text{ in. } (4\tfrac{1}{4} \text{ in. minimum})$$

U-Bracket Stresses

The plan and elevation of the connection detail are shown in Fig. 13.2c. The bracket essentially cantilevers from the face of the column. [If a round column (e.g., a pipe column) is used, the cantilever length is figured from the face of an equivalent square column of equal area.]

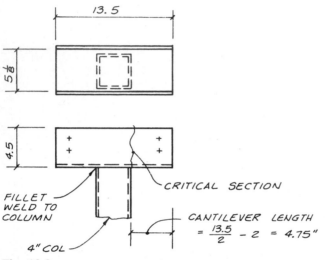

Fig. 13.2c

At this point it can be determined whether a bent plate bracket or welded plate bracket should be used (Fig. 13.2 d). If the cantilever length of the bracket is small in comparison with the distance between the vertical legs, a bracket of uniform thickness throughout can be used. This can be obtained by folding a plate into the desired shape. The bending stresses in the bracket caused by the cantilever action must be evaluated.

On the other hand, if the cantilever length is long in comparison with the distance between the vertical legs, the bottom plate may have to be thicker than the vertical legs. In this case the bottom plate of the bracket should first be designed for the bending stresses developed as it spans horizontally between the vertical plates. After the thickness of the bottom plate is determined, the composite section of the U bracket can be checked for cantilever stresses (as in the analysis of the stresses in the bent plate bracket). The weld between the vertical and horizontal plates must also be designed. The minimum weld size based on the plate thickness (Ref. 12) often governs the weld size.

In the problem at hand, a bent plate bracket will be used because the cantilever length of 4.75 in. is less than the distance between vertical plates (5⅛ in.). Check cantilever bending stresses using ¼-in. plate (Fig. 13.2 e).

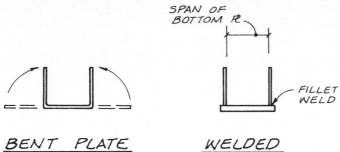

Fig. 13.2d

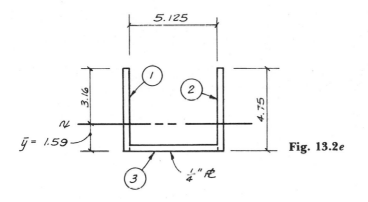

Fig. 13.2e

Section properties are conveniently calculated in a table from the formula

$$I_x = \Sigma I_g + \Sigma Ad^2$$

Element	A	y	Ay	$I_g = \dfrac{bh^3}{12}$	d	Ad^2
1	$4.75 \times \frac{1}{4} = 1.19$	2.38	2.82	2.23	0.79	0.73
2	1.19	2.38	2.82	2.23	0.79	0.73
3	$5.125 \times \frac{1}{4} = 1.28$	0.125	0.16	—	1.47	2.75
	$\Sigma\,3.66$		$\Sigma\,5.80$	$\Sigma\,4.47$		$\Sigma\,4.21$

$$\bar{y} = \frac{A_y}{A} = \frac{5.80}{3.66} = 1.59 \text{ in.}$$

$$I_x = \Sigma I_g + \Sigma A^2 = 4.47 + 4.21 = 8.68 \text{ in.}^4$$

Uniform load on bracket:

$$w = \frac{R}{A} = \frac{12{,}000 \times 2}{5.125 \times 13.5} = 347 \text{ psi}$$

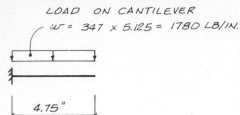

LOAD ON CANTILEVER

$w = 347 \times 5.125 = 1780$ LB/IN.

4.75"

Fig. 13.2*f*

$$M = \frac{wL^2}{2} = \frac{1780(4.75)^2}{2} = 20 \text{ in.-k}$$

$$f_b = \frac{Mc}{I} = \frac{20(3.16)}{8.68} = 7.3 \text{ ksi}$$

For A36 steel plate,

$$F_b = 22 \text{ ksi}* > 7.3 \qquad OK$$

Use ¼-in. plate U bracket

The U bracket must be welded to the top of the column. Without design loads, the weld size is governed by plate thickness requirements (Ref. 12).

See Fig. 13.3. The saddle hinge with tension tie is recommended for all types of loads. The tension ties are required to balance the eccentric moment generated by the vertical reactions.

Once the dimensions of the hinge connection have been established, the force on the tension ties can be calculated. This force is used to determine the bolt size, and the size and length of the weld between the tension tie and the main portion of the hinge connector. The spacing requirements for the bolts in the connection must also be considered. The top and bottom bearing plates for the hinge must be adequate to span horizontally between the vertical plates. The welds between the vertical and horizontal plates must also be designed.

In many cases there will be a tension force at the hinge due to an axial force in the beam. This is in addition to the tension force caused by the gravity loads. This typically occurs in buildings with concrete and masonry walls under seismic loading. Here the beams are designed as continuous cross ties (Chap. 14). A tension force can also result from diaphragm chord or strut action.

The additional tension force can be developed across the hinge by extending the basic tension tie system. See Fig. 13.4. The tension force due to

---

*The width-to-thickness ratios of plate elements in the compression zone are less than the limits given in Sec. 1.9 of Ref. 12.

448

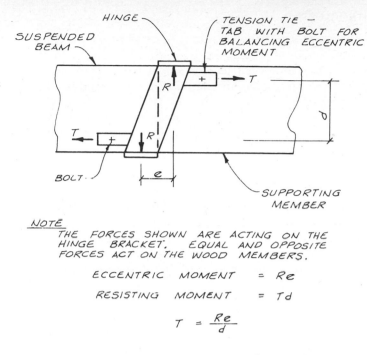

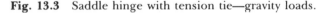

NOTE
THE FORCES SHOWN ARE ACTING ON THE HINGE BRACKET. EQUAL AND OPPOSITE FORCES ACT ON THE WOOD MEMBERS.

ECCENTRIC MOMENT = Re

RESISTING MOMENT = Td

$$T = \frac{Re}{d}$$

EQUILIBRIUM OF HINGE

Fig. 13.3 Saddle hinge with tension tie—gravity loads.

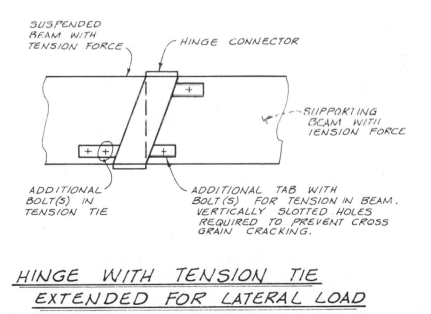

HINGE WITH TENSION TIE
EXTENDED FOR LATERAL LOAD

Fig. 13.4 Saddle hinge with general tension tie.

the gravity loads is taken in the usual manner. The additional tension is transferred through additional bolts near the bottom of the hinge. In order to minimize eccentricity, the additional tension tab is located opposite the normal tension tie. The holes in the additional tension tab should be slotted vertically to avoid the problem described in Fig. 13.1e. The bolt spacing requirements must be considered for the additional tie. Here the end distance may be more critical than with the other tension ties.

13.5   Prefabricated Connection Hardware

A number of manufacturers produce prefabricated connection hardware for use in wood construction. Catalogs (e.g., Refs. 21, 22, 23, and 31) are available from manufacturers or suppliers which show the configurations and dimensions of the hardware. In addition, the catalogs generally indicate allowable design loads for connections using the hardware. The designer should verify that the hardware and the allowable loads are accepted by the local building code authority. Catalogs usually indicate what building codes have accepted their design values.

In many cases prefabricated hardware will be the most economical solution

---

EXAMPLE 13.3   Connection Hardware

Framing Anchors

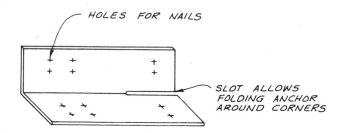

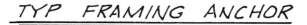

Fig. 13.5a.1

Framing anchors (Fig. 13.5a.1) are light-gauge steel connection brackets that are usually installed with nails. They have a variety of uses, especially where conventional nailed connections alone are inadequate. Framing anchors generally can be bent and nailed in a number of different ways (Fig. 13.5a.2). Allowable loads depend on the final configuration of the anchor and the direction of loading (different allowable loads apply to loadings A through E).

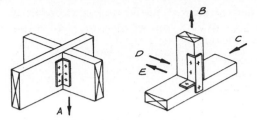

Fig. 13.5a.2 Framing anchor applications.

Diaphragm-to-Wall Anchors

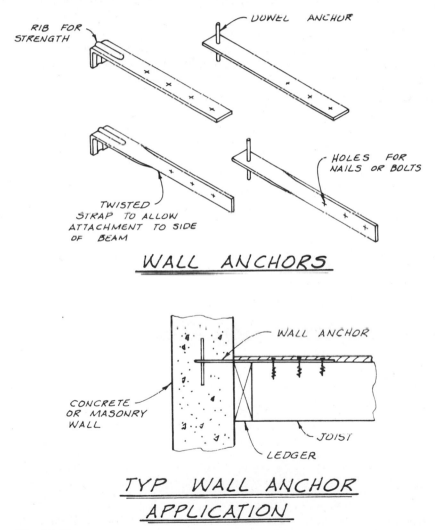

RIB FOR STRENGTH

DOWEL ANCHOR

TWISTED STRAP TO ALLOW ATTACHMENT TO SIDE OF BEAM

HOLES FOR NAILS OR BOLTS

WALL ANCHORS

WALL ANCHOR

CONCRETE OR MASONRY WALL

JOIST

LEDGER

TYP WALL ANCHOR APPLICATION

Fig. 13.5b

DESIGN OF WOOD STRUCTURES

Diaphragm-to-wall anchors (Fig. 13.5*b*) are known commercially by various names (purlin anchors, joist anchors, strap anchors, etc.). These anchors are used to tie concrete or masonry walls to horizontal diaphragms (Chap. 14). Their purpose is to prevent the ledger from being stressed in cross-grain bending (Example 6.1).

Tie Straps

TIE STRAP

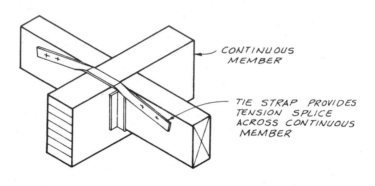

TYP TIE STRAP APPLICATION

Fig. 13.5*c*

Tie straps (Fig. 13.5*c*) are typically used to provide a connection across an intervening member. Ties of this type may be required if the beams are part of a horizontal diaphragm (Chap. 14).

to a connection problem. This is especially true if the hardware is a "shelf item." For example, light-gauge steel joist or purlin hangers (similar in configuration to the beam face hanger in Fig. 13.1*c*) are usually available as a stock item. In some cases allowable loads have been established through testing, and design values may be higher than can be justified by stress calculations. Obviously this type of hardware represents an efficient use of materials.

For connections involving larger loads, the appropriate hardware may be available from a manufacturer. However, the designer should have the ability to apply the fastener design principles of the previous chapters *and* basic structural steel design principles (Ref. 12), to develop appropriate connection details for a particular job. In some cases it may be more economical to fabricate special hardware, *or* there may not be suitable prefabricated hardware for a certain application.

Many manufacturers offer a number of variations of the hardware illustrated in Fig. 13.1. The fittings may range from light-gauge steel to ¼-in. and thicker steel plate. Allowable loads vary correspondingly. There are several other types of connection hardware that are generally available from manufacturers. See Example 13.3. These include

1. Framing anchors
2. Diaphragm-to-wall anchors
3. Tie straps

Some hardware that has been illustrated in previous chapters is also available commercially. For example, shearwall tie-down brackets (Fig. 10.10) are produced by a number of manufacturers. Metal knee braces (Fig. 6.20b) and metal bridging (mentioned in Example 6.4) are also available.

With some knowledge of structural steel design and the background provided in Chaps. 11, 12, and 13, the designer should be prepared to develop appropriate connections for the transfer of both vertical and lateral loads. Chapter 14 concludes the discussion of the lateral force design problem with final anchorage considerations and a brief introduction to some advanced problems in diaphragm design.

13.6 Problems

The design of connection hardware usually involves the use of structural steel design principles as well as wood design calculations. For the following problems, structural steel is A36, and design methods and allowable stresses should conform to the specifications contained in Ref. 12.

13.1 *Given:* The beam-to-column connection in Fig. 13.A. The two beams are 6 × 16 and the column is a 6 × 6. Lumber is No. 1 DF-larch. CUF = 1.0. Bolts in the connection are ⅞-in. diameter.

    Find: Design and detail a U-bracket connection for the following loads:
    a. An axial load in the column of DL = 5 k and snow = 15 k.
    b. The vertical load from part *a.* plus a tension force in the beam of 8 k. The tension force is a result of a seismic load.

13.2 Problem 13.2 is the same as Prob. 13.1 except that the lumber is Select Structural Hem-Fir.

13.3 *Given:* The beam-to-column connection in Fig. 13.A. The two beams are 6¾ × 31.5 DF-larch (combination 18F). CUF = 1.0. Bolts in the connection are ¾-in. diameter. The column is an 8-in. standard steel pipe column.

453

Find: Design and detail a U-bracket connection for the following loads:

    a. A total axial force in the column of DL + snow = 70 k.

    b. The vertical load from *a.* plus a tension force in the beam of 6 k. The tension force is a result of a seismic load.

    NOTE: The cantilever length of the U bracket is to be determined using the dimensions of an effective square column. The effective square column has an area equal to the area of a circle with a diameter of 8.625 in. (the outside diameter of a 8-in. standard pipe). Design the bottom plate of the U bracket for bending between the vertical supports.

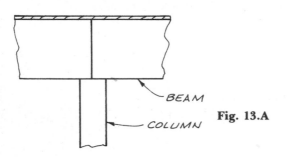

BEAM

COLUMN

Fig. 13.A

13.4 Problem 13.4 is the same as Prob. 13.3 except that lumber is 18F Southern pine glulam.

13.5 *Given:* The beam-to-girder connection in Fig. 13.B. The girder is a 6¾ × 30 glulam and the beams are 3⅛ × 15. Glulam is 20F DF-larch. CUF = 1.0.

Find: Design and detail the connection hardware if each beam transmits a reaction of 4.5 k to the girder. Load is DL + RLL.

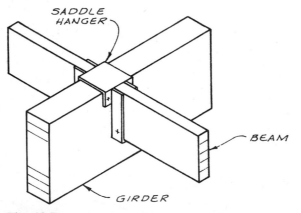

SADDLE HANGER

BEAM

GIRDER

Fig. 13.B

13.6 *Given:* The beam-to-girder connection in Fig. 13.C. The girder is a 5¾ × 25.5 glulam, and the beam is a 4 × 14 sawn member. DF-larch is used for both members. Bolts are ¾ in. diameter.

Find: Design and detail the connection hardware if the beam reaction on the girder is 2000 lb. Load is DL + RLL.

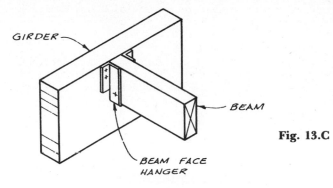

Fig. 13.C

13.7 *Given:* The hinge connector in Fig. 13.D. The girders are 6¾ × 31.5 DF-larch glulam (20F). CUF = 1.0. Bolts are ⅞-in. diameter.

    Find: Design and detail the hinge connector for the following loads:
        a. A vertical load of 16 k. Load is DL + snow.
        b. The vertical load from part *a.* plus a tension force in the girder of 8 k.

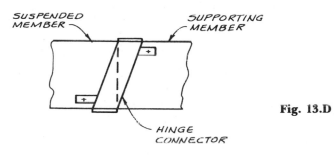

Fig. 13.D

13.8 *Given:* The hinge connector in Fig. 13.E. The supporting member is 5⅛ × 25.5 and the suspended member is a 5⅛ × 28.5 Southern pine glulam (20F). CUF = 1.0. Bolts are ¾-in. diameter.

    Find: Design and detail the hinge connector for the following loads:
        a. A vertical load of 12 k. Load is DL + RLL.
       b. The vertical load from part *a.* plus a tension force in the girder of 4 k.

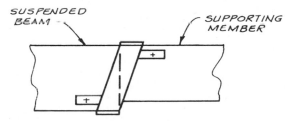

Fig. 13.E

13.9 Design and detail a shearwall anchor bracket similar to the one shown in Fig. 10.10. The lateral wind load causes a tension force in the chord of 4 k. The chord is composed of two 2 × 6 studs of No. 1 DF-larch. Use ¾-in. bolts for the connection to the chord, and choose the proper anchor bolt for the connection to the foundation from UBC Table 26G.

CHAPTER FOURTEEN

Diaphragm-to-Shearwall Anchorage

14.1 Introduction

Anchorage was defined previously as the tying together of the major elements of a building with an emphasis on the transfer of lateral loads. A systematic approach to anchorage involves a consideration of load transfer in the three principle directions of the building.

The anchorage requirements at the base of a shearwall are illustrated in Chap. 10. Chapter 14 concludes the anchorage problem with a detailed analysis of the connection between the horizontal diaphragm and the shearwalls. Several typical anchorage details are analyzed including connections to wood-frame shearwalls and concrete or masonry shearwalls.

After the general subject of diaphragm-to-wall anchorage, the subdiaphragm concept is introduced. This design technique was developed to ensure the integrity of a horizontal plywood diaphragm that supports seismic loads generated by *concrete or masonry walls*. Its purpose is to satisfy the Code requirement that "continuous ties" be provided to distribute these larger seismic loads into the diaphragm (UBC 2312, *Earthquake Regulations*). This is also an anchorage problem, but it is unique to buildings with concrete or masonry walls.

After these anchorage problems, the procedures for handling nonrectangular horizontal diaphragms are covered. The chapter concludes with a brief introduction to some recent developments in diaphragm research.

457

14.2 Anchorage Summary

Horizontal diaphragm anchorage refers to the design of connections between the horizontal diaphragm and the vertical elements of the building. These elements may support the horizontal diaphragm, or they may transfer a load to the diaphragm.

A systematic approach to anchorage is described in Example 10.6. This simply involves designing the connections between the various resisting elements for the following loads:

1. Vertical loads
2. Lateral loads parallel to a wall
3. Lateral loads perpendicular to a wall

In designing for these loads, it is important that a continuous path be developed for the transfer of forces. The path must take the load from its source, through the connections, and into the supporting element.

EXAMPLE 14.1   Anchorage Forces Parallel to Wall

Consider the anchorage forces *parallel* to the right end wall in the building shown in Fig. 14.1a. Three different forces are involved.

Unit Shear Transfer

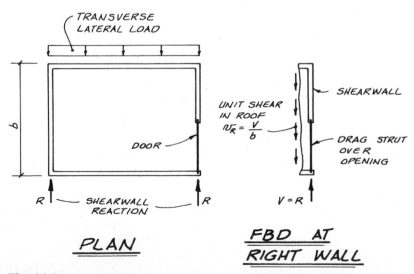

Fig. 14.1a

The unit shear in the roof diaphragm must be transferred to the supporting elements (drag strut and shearwall).

Drag Strut Connection to Shearwall

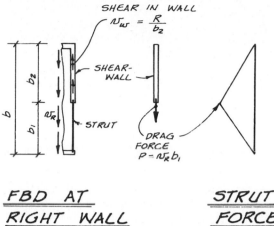

FBD AT
RIGHT WALL

STRUT
FORCE

Fig. 14.1b

The force in the drag strut must be transferred to the shearwall (Fig. 14.1b). The magnitude of the connection force may be read from the strut force diagram (Sec. 9.7).

Diaphragm Chord Connection

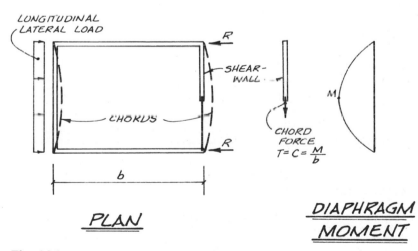

PLAN

DIAPHRAGM
MOMENT

Fig. 14.1c

The chord force in the transverse wall is developed by the longitudinal lateral load (Fig. 14.1c). The connection between the member over the door and the wall should be designed for the *chord force* or the *drag strut force*, whichever is larger.

The "path" in the progressive transfer of *vertical loads* is described in Example 10.7 and is not repeated here. Many of the connections for vertical loads can be made with the hardware described in Chap. 13.

There are several types of *lateral loads parallel* to a wall. See Example 14.1. These include

1. Basic unit shear transfer
2. Drag strut connection
3. Horizontal diaphragm chord force

The methods used to calculate drag strut forces and horizontal diaphragm chord forces were given in Chap. 9, and several connection designs were demonstrated in Chaps. 11 and 12. These loads are listed here simply to complete the anchorage design summary. However, the connection for the transfer of the unit shear has not been discussed previously, and it requires some additional consideration. The connections used to transfer this shear are different for different types of walls. Typical details of these connections and the calculations involved are illustrated in the following sections.

There are also several types of *lateral loads perpendicular* to a wall that must be considered when designing the anchorage of the horizontal diaphragm. These include

1. Wind load
2. Seismic force normal to the wall, F_p
3. In cases of masonry and concrete walls, an arbitrary Code minimum load of 200 lb/ft.

These are alternate design loads, and the anchorage must be designed for the maximum value. These loads act normal to the wall (either inward or outward) and attempt to separate the wall and the horizontal diaphragm. See Example 14.2.

Examples of these types of loads were also given in Examples 2.15 and 10.9. In the sketches in these examples the force labeled "roof reaction" corresponds to the anchorage force being discussed.

In wood-frame buildings the wind load usually controls the anchorage for perpendicular-to-wall loads. For masonry and concrete walls, however, the seismic force F_p often governs because of the large dead load of these walls. The Code minimum load of 200 lb/ft for these types of walls ensures a minimum factor of safety regardless of the calculated seismic loads. (Some designers also apply the 200 lb/ft minimum load parallel to the wall as a lower limit on the unit shear anchorage force—Example 14.1.)

There are a large number of connection details that can be used to anchor the horizontal diaphragm to the walls for both *parallel* and *perpendicular loads*. The type of wall framing (wood frame, concrete, or masonry) and the size and direction of the framing members for the horizontal diaphragm affect

EXAMPLE 14.2 **Anchorage Forces Perpendicular to Wall**

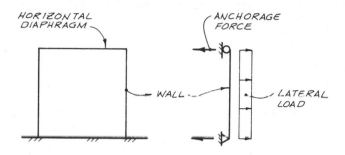

Fig. 14.2

Anchorage for normal-to-wall loads must be designed for the largest of the following:

1. Wind load tributary to horizontal diaphragm
2. Seismic load F_p tributary to horizontal diaphragm
3. For concrete and masonry walls, a separation force of 200 lb/ft at the diaphragm level

Anchorage must be provided for normal forces acting inward or outward (whichever is more critical).

---

the choice of the anchorage connection. Examples of two typical anchorage details are given in Secs. 14.3 and 14.4.

14.3   Connection Details—Horizontal Diaphragm to Wood-Frame Wall

A typical anchorage connection of a horizontal plywood diaphragm to a wood-frame wall is considered here. See Example 14.3. Two anchorage details are shown: one for the transverse wall and one for the longitudinal wall. In practice additional details will be required. For example, the connection detail where the girder ties into the transverse wall must also be included in the structural plans. This connection design must consider vertical reaction of the girder, and the horizontal diaphragm chord force for lateral loads in the longitudinal direction. The two connection details given in the sketches are intended to define the problem and be representative of the methods used for anchorage.

461

EXAMPLE 14.3   **Anchorage to Wood-Frame Walls**

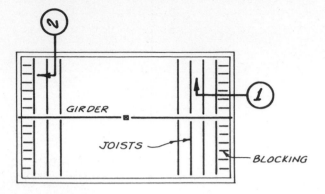

FRAMING PLAN

Fig. 14.3a

Different anchorage details are required for the diaphragm attachment to the transverse and longitudinal walls. More complicated roof or floor framing will require additional anchorage details.

The following details shown are representative only, and other properly designed connections are possible. The path for the transfer of anchorage forces is described for each connection.

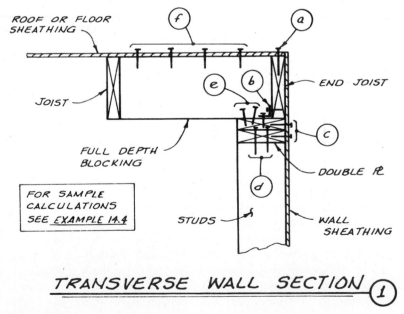

TRANSVERSE WALL SECTION ①

462   **Fig. 14.3b**

Parallel-to-Wall Load (diaphragm shear)

a. Diaphragm boundary nailing transfers shear into end joist (Fig. 14.3 *b*).

b. The shear must be transferred from the end joist into the wall plate. The top plate here serves as the diaphragm chord and strut. Various types of connections can be used for this transfer. Framing anchor is shown. Alternate connections are toenails (not allowed for concrete or masonry walls) or additional blocking as in Example 9.5.

c. Shearwall edge nailing transfers shear into wall sheathing.

Perpendicular-to-Wall Load (wind critical for most wood-frame walls)

d. Connect stud to double plate for reaction at top of stud. Standard connection from UBC Table 25P is two 16d nails into end grain of stud. Alternate connection for larger loads is framing anchor.

e. Full-depth blocking normal to wall is required to prevent the rotation of the joists under this type of loading. This blocking is necessary whether or not the horizontal diaphragm is a "blocked" diaphragm. Spacing of blocks depends on lateral load (2 to 4 ft is typical). Connect top plate to block for tributary lateral load. Toenails are shown (not allowed for concrete or masonry walls). Alternate connection is framing anchor.

f. Connect block to horizontal diaphragm sheathing for same load considered in *e*. For consistency, use the same size of nail as for diaphragm nailing. As long as the thickness of the plywood is sufficient to develop the nail capacity (Sec. 11.13), the allowable load per nail can be taken from NDS Table 8.8C, *Nails and Spikes—Lateral Load Design Values.*

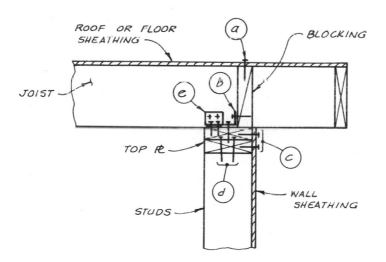

LONGITUDINAL WALL SECTION ②

Fig. 14.3c

DESIGN OF WOOD STRUCTURES

Parallel-to-Wall Load (diaphragm shear)

a. Diaphragm boundary nailing transfers shear into blocking (Fig. 14.3c).

b. Connection of blocking to double plate transfers shear into top plate. Framing anchor is shown. Alternate connections are toenails (not allowed for concrete or masonry walls) or additional blocking as in Example 9.5.

c. Shearwall edge nailing transfers shear from top plate to sheathing.

Perpendicular-to-Wall Load (wind critical for most wood-frame walls)

d. Same as transverse wall connection *d* (see page 463).

e. Connect top plate to joist for the tributary lateral load. Framing anchor is shown. Alternate connection is with toenails (not allowed for concrete or masonry walls). The connection of the joist to the horizontal diaphragm sheathing is provided directly by the diaphragm nailing. Specific nail design similar to the transverse wall connection *f* is not required.

As noted previously, the key to the anchorage problem is to provide a continuous *path* for the "flow" of forces in the connection. Each step in this flow is labeled on the sketch and a corresponding explanation is provided. Sample calculations are provided for the transverse wall anchorage detail. See Example 14.4.

EXAMPLE 14.4   Anchorage of Transverse Wood-Frame Wall

Design the connection for the anchorage of the transverse wall for the building in Example 14.3 (Section 1) using the following loads. The letter designations in this example correspond to the designations of the connection details in Fig. 14.3b.

Known Information

$$\text{Roof unit shear} = v = 160 \text{ lb/ft (parallel to wall)}$$

$$\text{Wind} = 20 \text{ psf} \qquad \text{(wall height} = 10 \text{ ft)}$$

$$\text{Separation force} = w = (20)\left(\frac{10}{2}\right)$$

$$= 100 \text{ lb/ft} \qquad \text{(perpendicular to wall)}$$

Assume allowable load on framing anchor.

$$\text{Allow. } P = 410 \text{ lb*} \qquad \text{(normal duration)}$$
$$= 545 \text{ lb} \qquad \text{(wind)}$$

Roof sheathing is ½-in. C-DX with 8d nails.

Parallel-to-Wall Loads

a. From UBC Table 25J, minimum allowable roof shear = 180 lb/ft > 160   *OK*
Minimum edge nailing (8d at 6 in. o.c.) will transfer roof shear into end joist.

---

    * In practice, the value is obtained from the manufacturer's catalog.

b. Connect end joist to plate with framing anchors (assume diaphragm width $= b$ $= 40$ ft)

$$V = vb = 160 \times 40 = 6400 \text{ lb}$$

$$\text{Number of anchors} = \frac{6400}{545} \approx 12$$

$$\text{Spacing of anchors} \approx \frac{40}{12} = 3.33 \text{ ft} = 3 \text{ ft-4 in.}$$

> *Use* 12 framing anchors total.
> Space anchors at approximately 3 ft-4 in. o.c.

Alternate calculation:

$$\text{Required spacing} = \frac{545}{160} = 3.41 \text{ ft} \qquad (\text{say } 3 \text{ ft-4 in. o.c.})$$

Alternate toenail connection: Assume 10d toenails (common) in DF lumber.

$$\text{Allow. } P = 94 \times 1.33 \times \tfrac{5}{6} = 104 \text{ lb/nail}$$

$$\text{Required spacing} = \frac{104}{160} - 0.65 \text{ ft} \qquad (\text{say } 6 \text{ in. o.c.})$$

Alternate connection *b*

> *Use* 10d common nails toenailed at 6 in. o.c.

c. Plywood shearwall edge nailing transfers shear from wall plate into shearwall (UBC Table 25K).

Perpendicular-to-Wall Loads

d. Check connection of stud to plate for separation force. Assume two 16d common nails in end grain of stud (UBC Table 25P). Lumber is DF.

$$\text{Allow. } P = 108 \times 1.33 \times \tfrac{2}{3} \times 2 = 191 \text{ lb}$$
$$P = 100 \text{ lb/ft} \times 1.33 \text{ ft} - 133 \text{ lb/stud} < 191 \qquad OK$$

$$\therefore \text{ Standard nailing OK}$$

e. Assume that blocking to stay wall is spaced 24 in. o.c.

$$\text{Load per block} = 100 \times 2 = 200 \text{ lb}$$

Assume 10d toenails.

$$\text{Allow. } P = 104 \text{ lb/nail} \qquad (\text{from } b)$$

$$\text{Number of toenails} = \frac{200}{104} \approx 2$$

> *Use* two 10d common toenails per block.
> Blocks at 24 in. o.c.

f. Attach blocking to roof sheathing.
Load per block $= 200$ lb   (from *e*)
8d nails in ½-in. plywood are used for roof diaphragm nailing
Determine number of 8d nails required per block.

$$\text{Allow. } P = 78 \times 1.33 = 104 \text{ lb/nail}$$

$$\text{Required number} = \frac{200}{104} \approx 2$$

Length of block will depend on spacing of joists (assume joists are 24 in. o.c.). If nails are spaced 6 in. o.c.,

$$N = \frac{24}{6} = 4 \text{ nails/block} > 2 \qquad OK$$

> *Use* 8d nails at 6 in. o.c. into blocking.

14.4 Connection Details—Horizontal Diaphragm to Concrete or Masonry Walls

The anchorage of concrete tilt-up and masonry (concrete block and brick) walls to horizontal plywood diaphragms has been the subject of considerable discussion in the design profession. The use of plywood diaphragms in buildings with concrete or masonry walls is very common, especially in one-story commercial and industrial buildings. Some failures of these connections occurred in the San Fernando earthquake, and revised design criteria have been included in the Code to strengthen these attachments.

The Code seismic loads that were introduced in Chaps. 2 and 3 are larger than the loads that were previously used for the design of these types of buildings. In addition, the Code has the following requirements for the anchorage of concrete and masonry walls for seismic loads:

1. Anchorage connections may not stress lumber in cross-grain bending or cross-grain tension.

2. Toenails and nails loaded in withdrawal are not allowed for anchorage in seismic zones 2, 3, and 4.

3. When the spacing of anchors exceeds 4 ft, the wall must be designed to resist bending between the anchors.

4. Continuous ties are to be provided between diaphragm chords in order to distribute the anchorage forces well into the diaphragm. Subdiaphragms (Sec. 14.5) may be used to transmit the anchorage forces to the main cross ties.

Two different anchorage problems are considered in this section, and the path used to transfer the forces is described. The first anchorage detail involves a ledger connection to a wall with a parapet. See Example 14.5. The vertical load must be transferred in addition to the lateral loads. The anchor bolts are designed to carry both the vertical load and the lateral load parallel to the wall. This requires that two different load combinations be checked, and the capacity of the bolts in the wall and in the ledger must be determined.

A special anchorage device is provided to transmit the load normal to the wall into the framing members. This is required in order to prevent cross-grain bending in the ledger. A numerical example is provided for the anchorage of a masonry wall with a ledger. See Example 14.6.

**EXAMPLE 14.5 Typical Ledger Anchorage to
Masonry Wall**

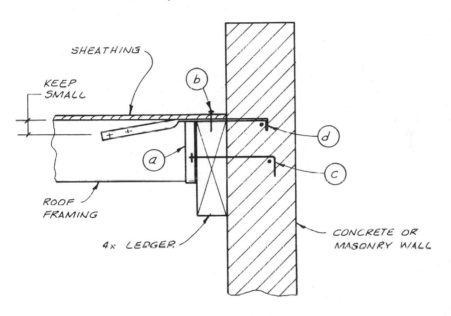

Fig. 14.4a

a. Vertical loads (DL + RLL) are transferred from roof framing members to ledger with prefabricated metal hangers.

b. Horizontal diaphragm unit shear parallel to the wall is transferred to ledger by diaphragm boundary nailing.

c. The loads from *a.* and *b.* are transferred from the ledger to the wall by the anchor bolts. The strength of the anchor bolts is governed by the capacity of the bolts in the

    1. *Wall*
        Concrete walls—UBC Table 26G
        Masonry walls—UBC Table 24G
    2. *Wood ledger* with bolt in single shear (Chap. 12)
        The following load cases must be considered to determine the bolt capacity in the wood ledger

467

(a) *Vertical load* (DL + RLL)

Single shear capacity perpendicular to grain, q

(b) *Vertical load plus lateral load parallel to wall* (DL + roof shear). RLL is not considered simultaneously with lateral load. Snow load (or floor LL), if applicable, must be included.

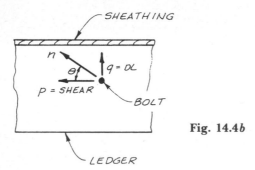

Fig. 14.4b

    i. Assume bolt size and spacing
   ii. Calculate tributary DL and tributary shear
  iii. Determine resultant n at angle θ
  iv. Compare to allowable n from the Hankinson formula

d. Loads perpendicular to the wall are carried by diaphragm-to-wall anchors (Fig. 13.5b) or similar-type hardware. If the strap is bent down to allow bolting through the framing member, the fold-down distance should be kept small to minimize eccentricity. However, the required edge distance for the fasteners should be provided. The anchors are embedded in the wall (preferably hooked around horizontal wall steel) and nailed, bolted, or lag screwed to the roof framing. The purpose of this separate anchorage connection for normal wall loads is to avoid cross-grain bending in the wood ledger (Example 6.1).

Subdiaphragm Problem

Once the perpendicular-to-wall load has been transferred into the framing members, it is carried into the sheathing. Because these normal wall loads may be large for concrete and masonry walls, it is important that the framing members extend back into the diaphragm (i.e., away from the wall) in order to make this transfer. The concept of subdiaphragms (Sec. 14.5) was developed to ensure that the framing is sufficiently anchored into the diaphragm.

EXAMPLE 14.6 Anchorage of Masonry Wall with Ledger

Design the connections for the ledger anchorage in Example 14.5. The letter designations in this example correspond to the designations of the connection details in Fig. 14.4.a. The following information is known.

Loads

        Roof DL = 60 lb/ft along wall
        Roof LL = 100 lb/ft

Roof diaphragm shear $= 300$ lb/ft parallel to wall
Separation force $= 190$ lb/ft < 200
$\therefore$ Use Code minimum of 200 lb/ft normal to wall

Construction

½-in. STR I plywood roof sheathing
Roof framing members are 2×6 at 24 in. o.c.
Ledger is 4×12 DF-larch
Walls are grouted concrete block

Anchorage

a.   Roof beams are 2 ft o.c. Reaction of roof beam on ledger:

$$R = (60 + 100)2 = 320 \text{ lb}$$

Choose prefabricated metal hanger from manufacturer's catalog.

b.   For ½-in. STR I plywood, Allow. $v = 320$ lb/ft (>300) for a blocked diaphragm with 10d nails at

6-in. o.c. edges
12-in. o.c. field

c.   Assume ¾-in. anchor bolts at 32 in. o.c.
Vertical loads:

$$DL + LL = (60 + 100) \frac{32}{12} = 427 \text{ lb/bolt}$$

Check bolt capacity in grouted masonry wall (UBC Table 24G).
Allow. $P = 1100$ lb/bolt > 427 *OK*

Check bolt capacity in ledger.
Load is perpendicular to grain, and bolt is in single shear (NDS Table 8.5A, *Bolt Design Values*).

$$\text{Allow. } q = \tfrac{1}{2} \text{ (tab. } q)(LDF) = \tfrac{1}{2}(1260)(1.25)$$
$$= 788 \text{ lb/bolt} > 427 \quad OK$$

DL + Lateral (roof LL may be omitted when lateral loads are checked):

$$DL = (60)\left(\frac{32}{12}\right) = 160 \text{ lb/bolt}$$

$$\text{Lateral} = 300 \times \frac{32}{12} = 800 \text{ lb/bolt}$$

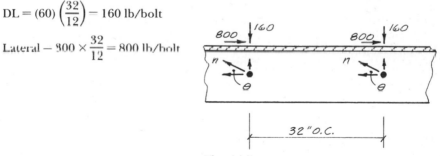

Fig. 14.5

Resultant load on bolt:

$$n = \sqrt{(800)^2 + (160)^2} = 816 \text{ lb/bolt}$$

Bolt capacity in grouted masonry wall (UBC Table 24G):
Allow. $P = 1100$ lb/bolt > 816 *OK*

469

Bolt capacity in ledger:
Load is at an angle θ to the grain

$$\theta = \tan^{-1}\left(\frac{160}{800}\right) = 11.3 \text{ degrees}$$

Allow. $p = \frac{1}{2}(2800) = 1400$

Allow. $q = \frac{1}{2}(1260) = 630$

Allow. $n = \dfrac{pq}{p \sin^2 \theta + q \cos^2 \theta} \times \text{LDF}$

$$= \frac{(1400)(630)1.33}{1400 \sin^2 (11.3°) + 630 \cos^2 (11.3°)}$$

$$= 1780 \text{ lb/bolt} > 816 \qquad OK$$

> Use ¾-in. anchor bolts at 32 in. o.c.

d. Provide wall anchors at every other roof framing member (i.e., 4 ft-0 in. o.c.) Perpendicular-to-wall load:

$$P = 200 \times 4 = 800 \text{ lb}$$

Choose an anchor from manufacturer's catalog with an allowable load greater than 800 lb.

The second anchorage detail is for a diaphragm that connects to the top of a concrete or masonry wall. See Example 14.7. The anchor bolts in the wall are used to transfer the lateral loads both parallel and perpendicular to the wall. These loads are not checked concurrently, and the bolt values of p and q in the wood plate are used independently. The strength of the bolt in the wall must also be checked.

The problem in this connection centers around the need to avoid cross-grain tension in the wood plate when loads normal to the wall are considered. Two possible connections for preventing cross-grain tension are presented.

Calculations for the anchorage of this type of connection are similar in many respects to those for the ledger connection (Example 14.6), and a numerical example is not provided.

14.5   Subdiaphragm Anchorage of Concrete and Masonry Walls

The anchorage requirements for concrete and masonry walls were outlined in Sec. 14.4. There it was noted that the Code requires *continuous ties* between the diaphragm chords in order to distribute the anchorage forces perpendicular to a wall into the diaphragm. The concept of subdiaphragms was developed to satisfy this Code requirement. The concept was first presented in Ref.

17.

EXAMPLE 14.7   Anchorage at Top of Masonry Wall

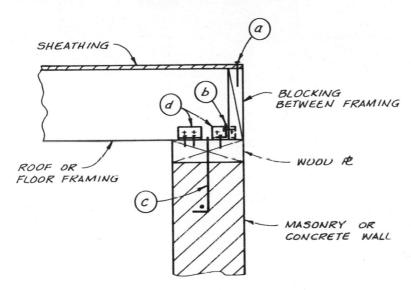

SHEATHING

BLOCKING
BETWEEN FRAMING

ROOF OR
FLOOR FRAMING

WOOD PL

MASONRY OR
CONCRETE WALL

TYPICAL CONNECTION TO
TOP OF WALL

Fig. 14.6a

Parallel to Wall Load (diaphragm shear)

a.   Boundary nailing transfers diaphragm shear from sheathing into blocking.

b.   Connection for transferring shear from blocking to plate. Framing anchor is shown. Alternate connection is addition blocking as in Example 9.5. Toenails are not allowed for anchorage of concrete or masonry walls.

c1.   Shear is transferred from plate to wall by anchor bolts. Bolt strength is governed by wall capacity (UBC Table 24G or 26G) or by the allowable load parallel to grain of the bolt in the wood plate, *p*.

Perpendicular-to-Wall Load (wind, seismic, or Code minimum)

c2.   Lateral load normal to wall is transferred from wall to plate by anchor bolts. Bolt strength is governed by wall capacity (UBC Table 24G or 26G) or by the allowable load perpendicular to grain of the bolt in the wood plate, *q*.

d.   The attachment of the wall plate to the framing presents a problem. The Code requires that anchorage of concrete and masonry walls be accomplished without cross-grain bending or *cross-grain tension*.

The connection in Fig. 14.6b is shown with framing anchors on both the right and left sides of the wall plate. In this way cross-grain tension in the plate is not being relied upon for load transfer. The framing anchors shown with solid lines are those that stress the plate in compression.

471

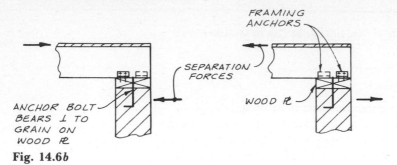

Fig. 14.6*b*

For a given direction of loading, the effective anchors cause the wood plate to bear on the anchor bolt in compression. Without fasteners on both sides of the anchor bolt, cross-grain tension would be developed in the plate for one of the directions of loading. The shear in the plate should be checked in accordance with the procedures given in Example 12.5.

There are other connection details that can be used to avoid cross-grain tension. One alternative, shown in Fig. 14.6*c*, avoids the problem by tying the framing member directly to the anchor bolt. Use of this type of anchorage requires close field coordination. The anchor bolts must be set accurately when the wall is constructed to match the future location of the roof or floor framing.

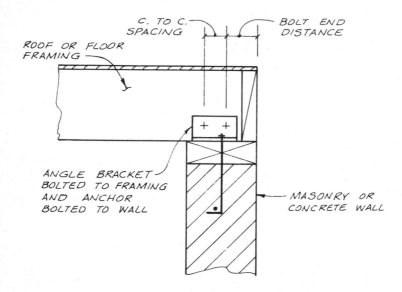

PARTIAL CONNECTION DETAIL
SHOWING ALT ANCHORAGE FOR
NORMAL LOADS

472 **Fig. 14.6***c*

Subdiaphragm Anchorage

For a discussion of the subdiaphragm anchorage of concrete and masonry walls see Example 14.5 and Sec. 14.5.

If the framing members in a horizontal diaphragm are continuous members from one side of the building to the other, the diaphragm would naturally have continuous ties between the chords. See Example 14.8. This is possible, however, only in very small buildings, or in buildings with closely spaced trusses. Even if continuous framing is present in one direction, it would normally not be present in the other direction.

EXAMPLE 14.8   Diaphragm with Continuous Framing

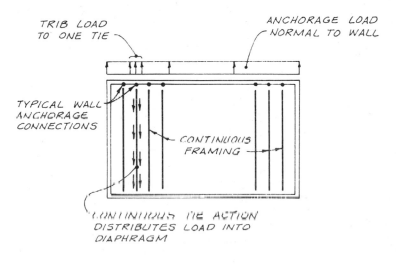

PLAN

Fig. 14.7

Continuous framing in the transverse direction would provide a continuous tie between diaphragm chords for transverse lateral loads. For this loading the chords are parallel to the longitudinal walls. Continuous ties distribute the anchorage load back into the diaphragm. The action of a continuous tie is similar in concept to a drag strut. The tie drags the wall force into the diaphragm.

If the spacing of wall anchors exceeds 4 ft, the wall must be designed to resist bending between the anchors. For lateral loads in the longitudinal direction, no intermediate ties are shown perpendicular to the continuous framing (Fig. 14.7). Ties   **473** must be designed and spliced across the transverse framing, or the transverse wall must be designed for bending (at the diaphragm level) between the longitudinal shearwalls.

The majority of buildings do not have continuous framing. In a typical panelized roof system, for example, the subpurlins span between purlins, purlins span between girders, and girders span between columns. See Example 14.9. In addition, the tops of the beams are kept at the same elevation, and the diaphragm sheathing is nailed directly to these members. In order for this type of construction to be used, the lighter beams are not continuous across supporting members. Lighter beams are typically suspended from heavier beams with metal hangers (Chap. 13).

If continuous cross ties are to be provided between diaphragm chords in both directions of the building, a large number of *continuity ties* will be required. In order to meet the continuous tie requirement literally, each beam would have to be spliced for an axial drag-type force across the members to which it frames.

EXAMPLE 14.9   Diaphragm Without Continuous Framing

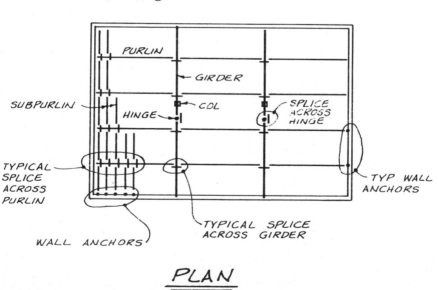

PLAN

Fig. 14.8

In a building without continuous framing between the diaphragm chords, a large number of "continuity ties" or splices would be required (Fig. 14.8).

a. Subpurlins spliced across purlins
b. Purlins spliced across girders
c. Girders spliced across hinges

Splices must be capable of transmitting the lateral load tributary to a given member (Fig. 14.7). With subdiaphragms, the continuous tie requirement may be satisfied without connection splices at every beam crossing.

The idea of a *subdiaphragm* was developed to ensure the proper anchorage of the wall seismic forces without requiring a tie splice at every beam crossing. With this method the designer selects a portion of the total diaphragm (known as the subdiaphragm) which must be designed as a separate diaphragm. See Example 14.10. The unit *shear* capacity (plywood thickness and nailing) must be sufficient to carry the tributary wall lateral loads, and the subdiaphragm must have its own *chords* and *continuous cross ties* between the chords.

The subdiaphragm reactions are in turn carried by cross ties which must be continuous across the full horizontal diaphragm. Use of the subdiaphragm avoids the requirement for continuous ties across the full width of the diaphragm except at the subdiaphragm boundaries.

EXAMPLE 14.10   Subdiaphragm Analysis

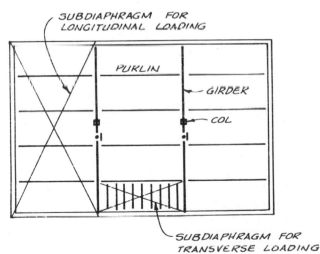

PLAN

Fig. 14.9a

The building from Figure 14.8 is redrawn here (Fig. 14.9a) showing the subdiaphragms which will be used for the wall anchorage. The assumed action of the subdiaphragms for resisting the lateral loads is shown in the following diagrams. The subdiaphragm concept is only a computational device that is used to develop adequate anchorage of concrete and masonry walls. It is recognized that the subdiaphragms are actually part of the total horizontal diaphragm, and they do not deflect independently as shown.

Subdiaphragm proportions (span/width ratio) are limited to a maximum of 4:1 (the same as for the total horizontal diaphragm).

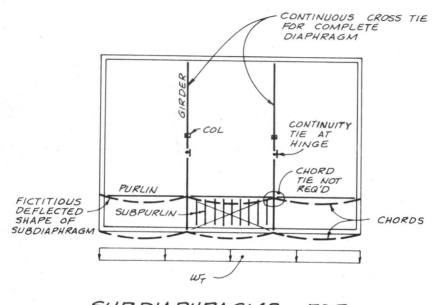

SUBDIAPHRAGMS FOR ANCHORING FRONT WALL

Fig. 14.9b

Three subdiaphragms for the anchorage of the front wall are shown in Fig. 14.9b. Three similar subdiaphragms will also be used for the anchorage of the rear wall. Note that the transverse anchorage force can act in either direction.

The purlin and wall serve as the chords for the subdiaphragms. The purlin must be designed for combined bending and axial forces. Because the subdiaphragms are treated as separate diaphragms, the chord forces are zero at the girder. The purlins, therefore, need not be tied across the girders.

The subpurlins serve as continuous cross ties between the subdiaphragm chords and do not have to be spliced across the purlin. The girders act as continuous cross ties for the complete diaphragm, and a continuity splice at the hinge is required. The splice must carry the reactions from two subdiaphragms.

The subdiaphragm for anchorage of the left wall is shown in Fig. 14.9c. A similar subdiaphragm is required for anchoring the right wall. The girder and wall serve as chords for the subdiaphragm. The girder must be designed for combined bending and axial loads.

476

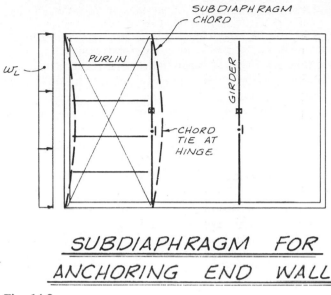

SUBDIAPHRAGM FOR ANCHORING END WALL

Fig. 14.9c

The splice at the girder hinge must be designed for the subdiaphragm chord force. This is an alternate load to the continuity tie force caused by the transverse lateral load. The purlins serve as continuous ties between subdiaphragm chords and do not have to be spliced across the girder.

The designer has some flexibility in the choice of what portions of the complete diaphragm are used as subdiaphragms. See Example 14.11. Figure 14.10a in this example shows that a number of small subdiaphragms are used to anchor the transverse wall. In Fig. 14.10b, one large subdiaphragm is used to anchor the same wall force between the longitudinal walls only.

The flexibility permitted in the choice of subdiaphragm arrangements results from the fact that the subdiaphragm concept is a computational device only. The independent deflection of the subdiaphragms is not possible, and the analysis is used arbitrarily to ensure adequate anchorage connections.

Various types of connections can be used to provide the continuity ties or chord ties for the subdiaphragms. Because most beams frame into the sides of the supporting members, compression forces are normally assumed to be taken by bearing. Thus the connections are usually designed for tension. This tension tie can be incorporated as part of the basic connection hardware (e.g., the extended tension splice on the hinge connection in Fig. 13.4), or a separate tension strap may be provided (Fig. 13.5c). The connections may be specially designed using the principles of Chaps. 11, 12, and 13, or some form of prefabricated connection hardware may be used.

477

14.6 Design Problem: Subdiaphragm

In Example 14.12 the wall anchorage requirements for a large one-story build-ing are analyzed. Exterior walls are of brick masonry construction.

In order to check the unit shear in the subdiaphragms, the construction of the entire roof diaphragm must be known. For this reason the unit shear diagrams for the complete diaphragm are shown along with the nailing and the allowable unit shears. Two different nailing patterns are used for the roof diaphragm. The heavier nailing occurs near the ends of the building in areas of high diaphragm shear. This different nailing is of importance when the unit shears in the subdiaphragms are checked.

Along the longitudinal wall a series of small subdiaphragms is used to anchor the wall. Subdiaphragm 1 spans between girders which serve as contin-uous cross ties for the entire roof diaphragm in the transverse direction. Because subdiaphragm 1 occurs in both areas of diaphragm nailing, the unit

---

**EXAMPLE 14.11   Alternate Subdiaphragm
Configurations**

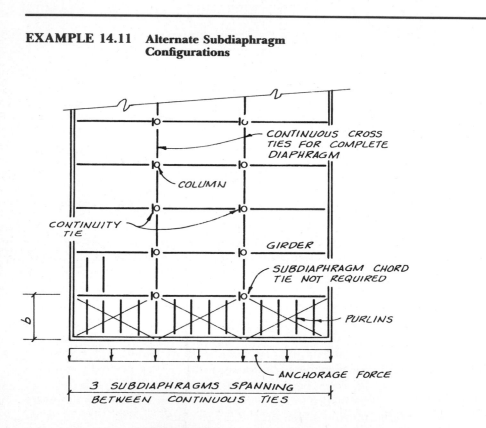

PLAN

Fig. 14.10a

One subdiaphragm arrangement (Fig. 14.10a) is to provide small subdiaphragms to span between continuous cross ties. Continuity tie splices are required across the complete diaphragm for these main cross ties only. Subdiaphragm chords need not be spliced at the columns because the chord force is zero at these locations.

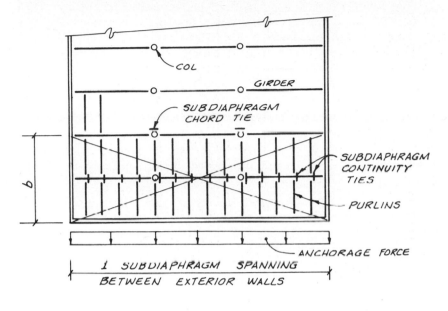

PLAN

Fig. 14.10b

In the second subdiaphragm layout (Fig. 14.10b), one large subdiaphragm is used to span between the exterior walls. Continuous cross-tie splices are required for the purlins across the first girder only. The subdiaphragm chord formed by the second line of girders must be tied across the columns for the appropriate chord force.

The choice of the subdiaphragm configuration in practice will depend on the required number of connections when both the transverse and longitudinal loads are considered.

EXAMPLE 14.12   Subdiaphragm Anchorage for Masonry Walls

The complete roof diaphragm for the building in Example 14.11 has been designed for seismic loads in both the transverse and longitudinal directions. The unit shears and plywood nailing are summarized below. A subdiaphragm analysis is to be performed using subdiaphragm 1 (Fig. 14.11b) for the longitudinal walls and subdiaphragm 2 (Fig. 14.11c) for the transverse walls. The calculation of subdiaphragm unit shears and tie forces is demonstrated.

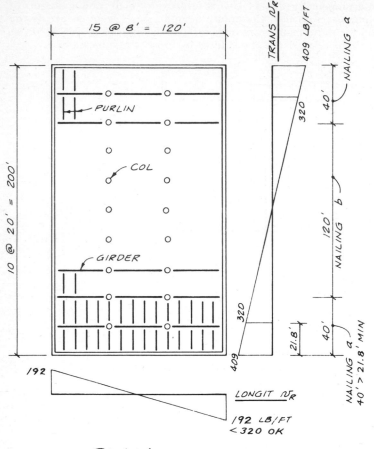

PLAN

Fig. 14.11a

Known Information

Dead loads:
    Roof DL = 10 psf
    Wall DL = 90 psf
Seismic load:
    $w = 0.186\ W$
    Transverse $w = 491$ lb/ft
    Longitudinal $w = 640$ lb/ft
Roof unit shears:
    See unit shear diagrams.

Plywood sheathing:
    ½-in. STR I plywood.
    All edges supported.
    Load cases 2 and 4.
Nailing *a*:
    10d at 4-in. o.c. boun.
        6-in. o.c. edges
        12-in. o.c. field
    Allow. $v = 425$ lb/ft > 409
Nailing *b*:
    10d at 6-in. o.c. edges
        12-in. o.c. field
    Allow. $v = 320$ lb/ft

Longitudinal Wall Anchorage
Subdiaphragm 1 (8 ft × 20 ft):

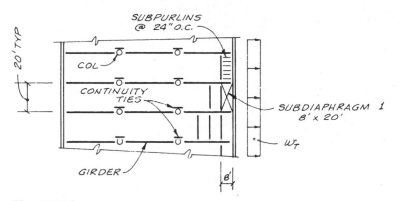

Fig. 14.11*b*

Wall anchorage force:

Trib. wall height = 8 ft

$w = ZIC_p W_p$

$\quad = (1)\ (1)\ (0.3)90 = 27$ psf > wind of 20 psf

$\quad = (27)\ 8 = 216$ lb/ft > Code minimum of 200

Span-to-width ratio $= \dfrac{L}{b} = \dfrac{20}{8} = 2.5 < 4 \qquad OK$

Subdiaphragm shear:

$$V = \frac{w*L}{2} = \frac{(216)(20)}{2} = 2160 \text{ lb}$$

$$v = \frac{V}{b} = \frac{2160}{8} = 270 \text{ lb/ft} < 320 \text{ lb/ft} \qquad OK$$

∴ Minimum diaphragm nailing (i.e., nailing *b*)
is adequate for subdiaphragm 1

Girder—continuity tie force at each column:

$$T = (216)(20) = 4320 \text{ lb}$$

Subpurlin—anchor wall to subpurlin at 4 ft-0 in. o.c.
(wall need not be designed for bending between anchors spaced 4 ft-0 in.
o.c. maximum)

$$T = (216)(4) = 864 \text{ lb/wall anchor}$$

Chord force in purlin:

$$T = C = \frac{M}{b} = \frac{wL^2}{8b} = \frac{(216)(20)^2}{8 \times 8} = 1350 \text{ lb}$$

No chord tie required across girder.

481

Transverse Wall Anchorage
Subdiaphragm 2 (40 ft × 120 ft):

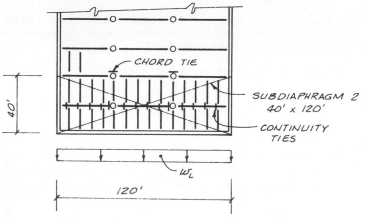

Fig. 14.11c

Wall anchorage force (from subdiaphragm 1):

$$w = 216 \text{ lb/ft}$$

Span-to-width ratio:

$$\frac{L}{b} = \frac{120}{40} = 3 < 4 \qquad OK$$

Subdiaphragm shear:

$$V = \frac{w^*L}{2} = \frac{(216)(120)}{2} = 12{,}960 \text{ lb}$$

$$v = \frac{V}{b} = \frac{12{,}960}{40} = 324 \text{ lb/ft} < 425 \qquad OK$$

Diaphragm nailing *a* is adequate for subdiaphragm 2.
   Purlin—continuity tie force across first girder:

$$T = (216)\ (8) = 1728 \text{ lb}$$

Purlin—anchor wall to purlin at 8 ft-0 in. o.c. for the same force:

$$T = 1728 \text{ lb}$$

---

  * The subdiaphragm load suggested in Ref. 17 is made up of the wall anchorage force (demonstrated here) *plus* the seismic force generated by the weight of the subdiaphragm. This additional uniform load can be calculated as the seismic base shear coefficient times the unit DL of the roof times the length of the subdiaphragm parallel to the direction of loading. Because subdiaphragms are basically used to provide *wall* anchorage, the roof seismic force is not included in this example. The designer, however, may choose to include it, and it would be conservative to do so.

(Wall must be designed for bending when spacing of anchors exceeds 4 ft-0 in. o.c.)
Chord force in girder:

$$T = C = \frac{M}{b} = \frac{wL^2}{8b} = \frac{(216)(120)^2}{8(40)} = 9720 \text{ lb}$$

The subdiaphragm chord along the second girder should be tied across the column for the chord force. An alternate for subdiaphragm 2 is to use several smaller subdiaphragms similar to the first arrangement in Example 14.11.

---

shear in the subdiaphragm must be checked against the lower allowable shear for the complete diaphragm.

The transverse wall is anchored by a large subdiaphragm that spans between the longitudinal walls. Continuous cross ties are required only between the subdiaphragm chords. Because subdiaphragm 2 coincides with the area of heavy nailing for the complete diaphragm, the unit shear can be checked against larger shear capacity.

If the unit shear in the subdiaphragms exceeds the shear capacity of the complete diaphragm, several solutions are possible. One obvious answer would be to provide heavier nailing (with an increased shear capacity) for the subdiaphragm. An alternate solution is to change the dimensions of the subdiaphragm. This can be used to lower the shear in the subdiaphragm, but the continuity or chord tie requirements will also change.

It should be noted that subdiaphragms are assumed to be separate small diaphragms for the purpose of determining the nailing requirements for the sheathing. If plywood edges do not occur at the subdiaphragm boundaries, the standard nailing in the "field" of the plywood will not be adequate to provide the required unit shear capacity. Special nailing can be designed for this, or two rows of edge nailing can simply be specified at the boundaries of the subdiaphragm.

14.7 Lateral Analysis of Nonrectangular Buildings

The buildings that have been considered thus far have been rectangular buildings with exterior and possibly interior shearwalls. These types of buildings have been used to illustrate basic concepts about horizontal diaphragm design.

In practice, however, buildings are often not rectangular in plan view. Obviously, an unlimited number of plan configurations is possible. However, a simple L-shaped building can be used to illustrate how the lateral forces on a nonrectangular building are resisted. See Example 14.13.

The building is divided into rectangles which are designed as separate horizontal diaphragms. The shearwall that is common to both diaphragms receives a portion of its load from the drag strut. The drag strut collects **483** the roof diaphragm unit shear from the unsupported segments of the two horizontal diaphragms. A numerical example using this same building illustrates the method of calculating the design forces in both directions of loading. See Example 14.14.

EXAMPLE 14.13   Lateral Analysis of Nonrectangular Buildings

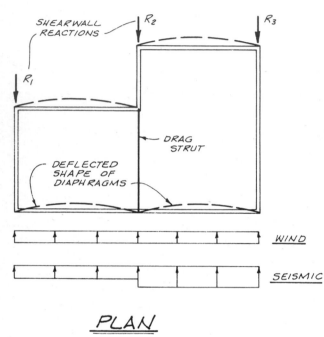

PLAN

TYPICAL L-SHAPED BUILDING

Fig. 14.12

A building with a nonrectangular plan may use rectangular horizontal diaphragms. Reactions to the shearwalls are calculated on a tributary width basis. The nonuniform seismic load results from the distribution of the dead load (Example 2.14). A similar analysis is used for lateral loads in the other direction.

EXAMPLE 14.14   L-Shaped Building

Determine the design unit roof shears, chord forces, and maximum drag strut connection forces for the one-story building in Fig. 14.13a. Consider loads in both the N-S and E-W directions.

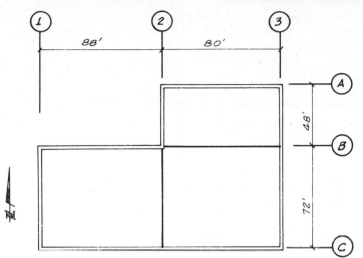

PLAN

Fig. 14.13 a

Loads

Roof DL = 20 psf
Wall DL = 75
Wind = 20 psf
Seismic = 0.186 W
Trib. height to roof level = 12 ft

Loads in N-S Direction

Diaphragm loads (Fig. 14.13 b):

$$\text{Wind} = 20 \times 12 = 240 \text{ lb/ft}$$
$$\text{Seismic } w_1 = 0.186 \left[(20 \times 72) + 2(75 \times 12)\right]$$
$$= 602 \text{ lb/ft}$$
$$w_2 = 0.186 \left[(20 \times 120) + 2(75 \times 12)\right]$$
$$= 781 \text{ lb/ft} \quad \text{seismic governs}$$

Roof shears:
  Line 1 and west of line 2:

$$v = \frac{V}{b} = \frac{wL}{2b} = \frac{(602)\ (88)}{2(72)} = 368 \text{ lb/ft}$$

  Line 3 and east of line 2:

$$v = \frac{wL}{2b} = \frac{(781)\ (80)}{2(120)} = 260 \text{ lb/ft}$$

Chord forces:
  Lines B and C between lines 1 and 2:

$$T = C = \frac{M}{b} = \frac{wL^2}{8b} = \frac{602(88)^2}{8(72)} = 8.1 \text{ k}$$

485

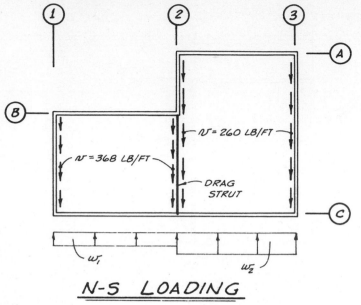

N-S LOADING

Fig. 14.13b

Lines *A* and *C* between lines 2 and 3:

$$T = C = \frac{wL^2}{8b} = \frac{781(80)^2}{8(120)} = 5.21 \text{ k}$$

Drag tie force:

Connection on line 2 at line *B*. Drag strut collects unsupported roof shear from both diaphragms.

$$T = (368 + 260)72 = 45.2 \text{ k}$$

Loads in E-W Direction

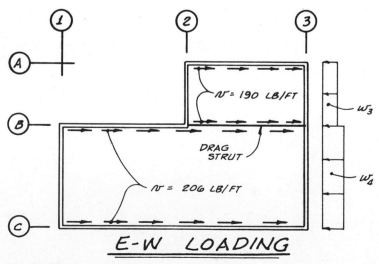

E-W LOADING

Fig. 14.13c

Diaphragm loads (Fig. 14.13c):

Wind does not govern (dng)

Seismic
$$w_3 = 0.186 \, [(20 \times 80) + 2(75 \times 12)]$$
$$= 632 \text{ lb/ft}$$
$$w_4 = 0.186 \, [(20 \times 168) + 2(75 \times 12)]$$
$$= 960 \text{ lb/ft}$$

Roof shears:

Line A and north of line B:

$$v = \frac{V}{b} = \frac{wL}{2b} = \frac{632(48)}{2(80)} = 190 \text{ lb/ft}$$

Line C and south of line B:

$$v = \frac{wL}{2b} = \frac{960(72)}{2(168)} = 206 \text{ lb/ft}$$

Chord forces:

Lines 2 and 3 between lines A and B:

$$T = C = \frac{M}{b} = \frac{wL^2}{8b} = \frac{632(48)^2}{8(80)} = 2.28 \text{ k}$$

Lines 1 and 3 between lines B and C:

$$T = C = \frac{wL^2}{8b} = \frac{960(72)^2}{8(168)} = 3.7 \text{ k}$$

Drag tie force:

Connection on line B at line 2.

$$T = (190 + 206) \, (80) = 31.7 \text{ k}$$

14.8 Recent Developments in Plywood Diaphragms

The lateral loads in areas of high seismic risk have increased substantially in recent years. In 1977 the American Plywood Association (APA) conducted tests on diaphragms that were designed to develop high allowable unit shears. These diaphragms are referred to as *high-load* diaphragms, and initial test results were published in Ref. 15.

One type of diaphragm used two layers of plywood sheathing. The first layer was attached to the framing members with conventional nailing. The top layer was attached to the first layer with 14-gauge $\times$ 1¾-in. staples. The panel edges for the second layer were offset from the edges of the first layer, and the staples were purposely not driven into the framing members (i.e., staples penetrated the first layer of plywood only).

Another method of obtaining high-load diaphragms was with the use of relatively thick plywood (⅝ in. and ¾ in.) and closely spaced fasteners. Tests indicated that closely spaced 10d and 16d nails often caused the framing lumber to split. However, very closely spaced (1 in. o.c.) pneumatically driven

487

wire staples were found to not cause splitting. High unit shears were obtained with these diaphragms.

Among other objectives, these diaphragm tests also were designed to measure the effects of openings in horizontal diaphragms. The forces developed at the corners of the openings were larger than expected. A method of analyzing these forces is under investigation.

Formal publication of high-load diaphragm design procedures and other design considerations relating to the 1977 diaphragm tests is still pending. Additional information may be obtained from APA.

14.9 Problems

14.1 *Given:* The roof framing of Fig. 9.A (Chap. 9), and the typical section of the same building in Fig. 14.A. The lateral wind pressure is 20 psf. Roof DL = 15 psf and RLL = 20 psf.

     Find: Design and sketch the roof to wall anchorage details for:
- *a.* The rear longitudinal wall at the 25-ft and 20-ft headers.
- *b.* The longitudinal bearing walls.
- *c.* The right transverse wall. Consider one case with full-height wall studs, and a second case with 10-ft studs to a double plate and a filler truss over.

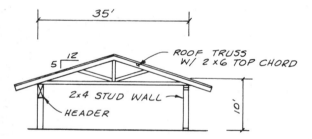

Fig. 14.A

14.2 *Given:* The roof framing plan of Fig. 9.A (Chap. 9) and the typical section of the same building in Fig. 14.B. Roof DL = 20 psf; roof LL = 20 psf; wall DL = 75 psf. Seismic base shear coefficient = 0.186.

     Find: Design and sketch the roof to wall anchorage details for:
- *a.* The longitudinal bearing walls.
- *b.* The transverse walls, assuming continuous full-height masonry from foundation to roof level. Include cross ties at 4 ft-0 in. o.c.

14.3 *Given:* The roof framing plan of Fig. 9.C (Chap. 9). Roof height is 14 ft above finish floor, and wall height is 16.5 ft (2.5-ft parapet). Roof DL = 14 psf, and roof LL = 20 psf. Seismic base shear coefficient = 0.186. Wind = 20 psf.

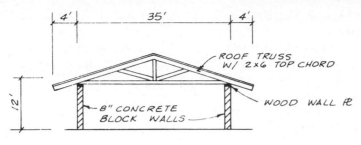

SECTION

Fig. 14.B

Find: *a.* Design and sketch the roof-to-wall anchorage connection where the purlins attach to the wall.

  b. Perform a subdiaphragm analysis for both directions. Calculate the magnitude of all continuity tie and chord tie forces.

  c. Suggest appropriate connection details for developing the tie forces in *b*.

14.4 *Given:* The plan of the U-shaped building in Fig. 14.C. Exterior walls serve as shearwalls. Roof DL = 20 psf, and wall DL = 60 psf. Seismic coefficient − 0.186.

Find: *a.* Unit shear in the horizontal diaphragms and shearwalls.

  b. The magnitude of the drag strut force at the point where these members connect to the shearwalls.

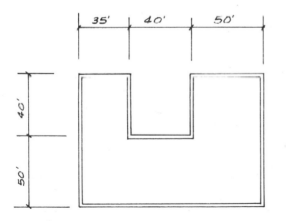

PLAN

Fig. 14.C

APPENDIX A

Equivalent Uniform Weights of Wood Framing

Weights are for Douglas fir-larch lumber (S4S) used at an equilibrium moisture content of 15 percent (the maximum found in most covered structures). The unit weight of Douglas fir-larch equals or exceeds the unit weight of most softwood species, and the dead loads given below are conservative for most designs.

Nominal size	Spacing					
	12 in. o.c.		16 in. o.c.		24 in. c.c.	
	Weight, psf	Board feet* per ft²	Weight, psf	Board feet* per ft²	Weight, psf	Board feet* per ft²
2 × 3	0.9	0.50	0.7	0.38	0.5	0.25
2 × 4	1.3	0.67	1.0	0.50	0.7	0.34
2 × 6	2.1	1.00	1.6	0.75	1.0	0.5
2 × 8	2.7	1.33	2.1	1.00	1.4	0.07
2 × 10	3.5	1.67	2.6	1.25	1.7	0.84
2 × 12	4.3	2.00	3.2	1.50	2.1	1.00
3 × 6	3.5	1.50	2.6	1.13	1.7	0.75
3 × 8	4.6	2.00	3.4	1.50	2.3	1.00
3 × 10	5.8	2.50	4.4	1.88	2.9	1.25
3 × 12	7.1	3.00	5.3	2.25	3.5	1.50
3 × 14	8.3	3.50	6.3	2.63	4.2	1.75
4 × 8	6.4	2.67	4.8	2.00	3.2	1.34
4 × 10	8.2	3.33	6.1	2.50	4.1	1.67
4 × 12	9.9	4.00	7.4	3.00	5.0	2.00
4 × 14	11.7	4.67	8.8	3.50	5.8	2.34
4 × 16	13.5	5.33	10.1	4.00	6.7	2.67

* Lumber is ordered and priced by the board foot. A board foot is a volume of lumber corresponding to 1 in. thick by 12 in. wide by 1 ft long. Nominal dimensions are used to calculate board measure.

SOURCE: Table from Ref. 4. Courtesy of Western Wood Products Association.

Weights of Building Materials

Loads given in Appendix B are typical values. Specific products may have weights which differ considerably from those shown, and manufacturer's catalogs should be consulted for actual loads.

Roof dead loads			
Material	Weight, psf		
Lumber sheathing, 1 in. nominal	2.5		
Plywood, per inch of thickness	3.0		
Timber decking (MC = 15%):	2 in. nom.	3 in. nom.	4 in. nom.
DF-larch	4.5	7.6	10.6
DF (south)	4.1	6.8	9.5
Hem-Fir	4.0	6.6	9.3
Mtn. hemlock—Hem-Fir	4.0	6.6	9.3
Subalpine fir	2.9	4.9	6.9
Engelmann spruce	3.1	5.1	7.2
Lodgepole pine	3.7	6.2	8.6
Ponderosa pine—Sugar pine	3.7	6.2	8.6
Idaho white pine	3.5	5.8	8.1
Western cedars	3.3	5.5	7.7
Aluminum (including laps):	Flat	Corrugated (1½ and 2½ in.)	
12 American or B&S gage	1.2	. . .	
14	0.9	1.1	
16	0.7	0.9	
18	0.6	0.7	
20	0.5	0.6	
22	. . .	0.4	

Roof dead loads		
Material	Weight, psf	
Galvanized steel (including laps):	Flat	Corrugated (2½ and 3 in.)
12 U.S. std. gage	4.5	4.9
14	3.3	3.6
16	2.7	2.9
18	2.2	2.4
20	1.7	1.8
22	1.4	1.5
24	1.2	1.3
26	0.9	1.0
Other types of decking (per inch of thickness):		
Concrete plank	6.5	
Insulrock	2.7	
Petrical	2.7	
Porex	2.7	
Poured gypsum	6.5	
Tectum	2.0	
Vermiculite concrete	2.6	
Corrugated asbestos (¼ in.)	3.0	
Felt:		
3-ply	1.5	
3-ply with gravel	5.5	
5-ply	2.5	
5-ply with gravel	6.5	
Insulation (per inch of thickness):		
Expanded polystyrene	0.2	
Fiber glass, rigid	1.5	
Loose	0.5	
Roll roofing	1.0	
Shingles:		
Asphalt (¼ in. approx.)	2.0	
Book tile (2 in.)	12.0	
Book tile (3 in.)	20.0	
Cement asbestos (⅜ in. approx.)	4.0	
Clay tile (for mortar add 10 psf)	9.0 to 14.0	
Ludowici	10.0	
Roman	12.0	
Slate (¼ in.)	10.0	
Spanish	19.0	
Wood (1 in.)	3.0	

Ceiling dead loads	
Material	Weight, psf
Acoustical fiber tile	1.0
Channel-suspended system	1.0
For gypsum wallboard and plaster, see *Wall and partition dead loads*	

Floor dead loads	
Material	Weight, psf
Hardwood (1 in. nominal)	4.0
Plywood (per inch of thickness)	3.0
Asphalt mastic (per inch of thickness)	12.0
Cement finish (per inch of thickness)	12.0
Ceramic and quarry tile (¾ in.)	10.0
Concrete (per inch of thickness):	
Lightweight	6.0 to 10.0
Reinforced (normal weight)	12.5
Stone	12.0
Cork tile (1/16 in.)	0.5
Flexicore (6-in. slab)	46.0
Linoleum (¼ in.)	1.0
Terrazo finish (1½ in.)	19.0
Vinyl tile (⅛ in.)	1.4

Wall and Partition Dead Loads	
Material	Weight, psf
Wood paneling (1 in.)	2.5
Wood studs (2 × 4 DF-larch):	
12 in. o.c.	1.3
16 in. o.c.	1.0
24 in. o.c.	0.7
Glass block (4 in.)	18.0
Glass (¼-in. plate)	3.3
Glazed tile	10.0
Marble or marble wainscoting	15.0
Masonry (per 4 in. of thickness):	
Brick	38.0
Concrete block	30.0
Cinder concrete block	20.0
Hollow clay tile, load bearing	23.0
Hollow clay tile, non-load-bearing	18.0
Hollow gypsum block	13.0
Limestone	55.0
Terra-cotta tile	25.0
Stone	55.0
(The average weights of completed reinforced and grouted concrete block and brick walls can be found in Ref. 13.)	

Wall and Partition Dead Loads	
Material	Weight, psf
Plaster (1 in.)	8.0
Plaster (1 in.) on wood lath	10.0
Plaster (1 in.) on metal lath	8.5
Gypsum wallboard (1 in.)	5.0
Porcelain-enameled steel	3.0
Stucco (⅞ in.)	10.0
Windows (glass, frame, and sash)	8.0

SOURCE: Weights from Ref. 4. Courtesy Western Wood Product Association.

Section Properties of Glulam Members

Weights are for Douglas fir-larch at MC = 15 percent (the maximum MC found in most covered structures). This corresponds to a minimum unit weight of 34.1 pcf which is heavier than the unit weight for most species of wood. Weights are rounded up to the nearest lb/ft.

							Weight
No. of lams			Size		Section	Moment of	per
		Depth d,	factor	Area A,	modulus	inertia I,	lineal foot,
1½ in.	¾ in.	in.	C_F	in.²	S, in.³	in.⁴	lb/ft
2	4	3.00	1.00	9.4	4.7	7.0	3
	5	3.75	1.00	11.7	7.3	13.7	3
3	6	4.50	1.00	14.1	10.5	23.7	4
	7	5.25	1.00	16.4	14.4	37.7	4
4	8	6.00	1.00	18.8	18.8	56.2	5
	9	6.75	1.00	21.1	23.7	80.1	5
5	10	7.50	1.00	23.4	29.3	109.9	6
	11	8.25	1.00	25.8	35.4	146.2	7
6	12	9.00	1.00	28.1	42.2	189.8	7
	13	9.75	1.00	30.5	49.5	241.4	8
7	14	10.50	1.00	32.8	57.4	301.5	8
	15	11.25	1.00	35.2	65.9	370.8	9
8	16	12.00	1.00	37.5	75.0	450.0	9
	17	12.75	0.99	39.8	84.7	539.8	10
9	18	13.50	0.99	42.2	94.9	640.7	10
	19	14.25	0.98	44.5	105.8	753.6	11

3⅛-in.-wide glulams

3⅛-in.-wide glulams

No. of lams		Depth d, in.	Size factor C_F	Area A, in.²	Section modulus S, in.³	Moment of inertia I, in.⁴	Weight per lineal foot, lb/ft
1½ in.	¾ in.						
10	20	15.00	0.98	46.9	117.2	878.9	12
	21	15.75	0.97	49.2	129.2	1017.4	12
11	22	16.50	0.97	51.6	141.8	1169.8	13
	23	17.25	0.96	53.9	155.0	1336.7	13
12	24	18.00	0.96	56.3	168.8	1518.8	14
	25	18.75	0.95	58.6	183.1	1716.6	14
13	26	19.50	0.95	60.9	198.0	1931.0	15
	27	20.25	0.94	63.3	213.6	2162.4	15
14	28	21.00	0.94	65.6	229.7	2411.7	16
	29	21.75	0.94	68.0	246.4	2679.5	17
15	30	22.50	0.93	70.3	263.7	2966.3	17
	31	23.25	0.93	72.7	281.5	3272.9	18
16	32	24.00	0.93	75.0	300.0	3600.0	18

5⅛-in.-wide glulams

No. of lams		Depth d, in.	Size factor C_F	Area A, in.²	Section modulus S, in.³	Moment of inertia I, in.⁴	Weight per lineal foot, lb/ft
1½ in.	¾ in.						
3	6	4.50	1.00	23.1	17.3	38.9	6
	7	5.25	1.00	26.9	23.5	61.8	7
4	8	6.00	1.00	30.8	30.8	92.2	8
	9	6.75	1.00	34.6	38.9	131.3	9
5	10	7.50	1.00	38.4	48.0	180.2	10
	11	8.25	1.00	42.3	58.1	239.8	10
6	12	9.00	1.00	46.1	69.2	311.3	11
	13	9.75	1.00	50.0	81.2	395.8	12
7	14	10.50	1.00	53.8	94.2	494.4	13
	15	11.25	1.00	57.7	108.1	608.1	14
8	16	12.00	1.00	61.5	123.0	738.0	15
	17	12.75	0.99	65.3	138.9	885.2	16
9	18	13.50	0.99	69.2	155.7	1050.8	17
	19	14.25	0.98	73.0	173.4	1235.8	18
10	20	15.00	0.98	76.9	192.2	1441.4	19
	21	15.75	0.97	80.7	211.9	1668.6	20
11	22	16.50	0.97	84.6	232.5	1918.5	20
	23	17.25	0.96	88.4	254.2	2192.2	21
12	24	18.00	0.96	92.3	276.8	2490.8	22
	25	18.75	0.95	96.1	300.3	2815.2	23
13	26	19.50	0.95	99.9	324.8	3166.8	24
	27	20.25	0.94	103.8	350.3	3546.4	25

5⅛-in.-wide glulams

No. of lams		Depth d, in.	Size factor C_F	Area A, in.²	Section modulus S, in.³	Moment of inertia I, in.⁴	Weight per lineal foot, lb/ft
1½ in.	¾ in.						
14	28	21.00	0.94	107.6	376.7	3955.2	26
	29	21.75	0.94	111.5	404.1	4394.3	27
15	30	22.50	0.93	115.3	432.4	4864.7	28
	31	23.25	0.93	119.2	461.7	5367.6	29
16	32	24.00	0.93	123.0	492.0	5904.0	30
	33	24.75	0.92	126.8	523.2	6475.0	30
17	34	25.50	0.92	130.7	555.4	7081.6	31
	35	26.25	0.92	134.5	588.6	7725.0	32
18	36	27.00	0.91	138.4	622.7	8406.3	33
	37	27.75	0.91	142.2	657.8	9126.4	34
19	38	28.50	0.91	146.1	693.8	9886.6	35
	39	29.25	0.91	149.9	730.8	10,687.8	36
20	40	30.00	0.90	153.8	768.8	11,531.3	37
	41	30.75	0.90	157.6	807.7	12,417.9	38
21	42	31.50	0.90	161.4	847.5	13,348.9	39
	43	32.25	0.90	165.3	888.4	14,325.2	40
22	44	33.00	0.89	169.1	930.2	15,348.1	40
	45	33.75	0.89	173.0	972.9	16,418.5	41
23	46	34.50	0.89	176.8	1016.7	17,537.6	42
	47	35.25	0.89	180.7	1061.4	18,706.4	43
24	48	36.00	0.88	184.5	1107.0	19,926.0	44

6¾-in.-wide glulams

No. of lams		Depth d, in.	Size factor C_F	Area A, in.²	Section modulus S, in.³	Moment of inertia I, in.⁴	Weight per lineal foot, lb/ft
1½ in.	¾ in.						
4	8	6.00	1.00	40.5	40.5	121.5	10
	9	6.75	1.00	45.6	51.3	173.0	11
5	10	7.50	1.00	50.6	63.3	237.3	12
	11	8.25	1.00	55.7	76.6	315.9	14
6	12	9.00	1.00	60.8	91.1	410.1	15
	13	9.75	1.00	65.8	106.9	521.4	16
7	14	10.50	1.00	70.9	124.0	651.2	17
	15	11.25	1.00	75.9	142.4	800.9	18
8	16	12.00	1.00	81.0	162.0	972.0	20
	17	12.75	0.99	86.1	182.9	1165.9	21
9	18	13.50	0.99	91.1	205.0	1384.0	22
	19	14.25	0.98	96.2	228.4	1627.7	23
10	20	15.00	0.98	101.3	253.1	1898.4	24
	21	15.75	0.97	106.3	279.1	2197.7	26

499

6¾-in.-wide glulams							
No. of lams		Depth d, in.	Size factor C_F	Area A, in.²	Section modulus S, in.³	Moment of inertia I, in.⁴	Weight per lineal foot, lb/ft
1½ in.	¾ in.						
11	22	16.50	0.97	111.4	306.3	2526.8	27
	23	17.25	0.96	116.4	334.8	2887.3	28
12	24	18.00	0.96	121.5	364.5	3280.5	29
	25	18.75	0.95	126.6	395.5	3707.9	30
13	26	19.50	0.95	131.6	427.8	4170.9	32
	27	20.25	0.94	136.7	461.3	4670.9	33
14	28	21.00	0.94	141.8	496.1	5209.3	34
	29	21.75	0.94	146.8	532.2	5787.6	35
15	30	22.50	0.93	151.9	569.5	6407.2	36
	31	23.25	0.93	156.9	608.1	7069.5	38
16	32	24.00	0.93	162.0	648.0	7776.0	39
	33	24.75	0.92	167.1	689.1	8528.0	40
17	34	25.50	0.92	172.1	731.5	9327.0	41
	35	26.25	0.92	177.2	775.2	10,174.4	42
18	36	27.00	0.91	182.3	820.1	11,071.7	44
	37	27.75	0.91	187.3	866.3	12,020.2	45
19	38	28.50	0.91	192.4	913.8	13,021.4	46
	39	29.25	0.91	197.4	962.5	14,076.7	47
20	40	30.00	0.90	202.5	1012.5	15,187.5	48
	41	30.75	0.90	207.6	1063.8	16,355.3	50
21	42	31.50	0.90	212.6	1116.3	17,581.4	51
	43	32.25	0.90	217.7	1170.1	18,867.4	52
22	44	33.00	0.89	222.8	1225.1	20,214.6	53
	45	33.75	0.89	227.8	1281.4	21,624.4	54
23	46	34.50	0.89	232.9	1339.0	23,098.3	56
	47	35.25	0.89	237.9	1397.9	24,637.7	57
24	48	36.00	0.88	243.0	1458.0	26,244.0	58
	49	36.75	0.88	248.1	1519.4	27,918.7	59
25	50	37.50	0.88	253.1	1582.0	29,663.1	60
	51	38.25	0.88	258.2	1645.9	31,478.7	62
26	52	39.00	0.88	263.3	1711.1	33,366.9	63
	53	39.75	0.88	268.3	1777.6	35,329.2	64
27	54	40.50	0.87	273.4	1845.3	37,366.9	65
	55	41.25	0.87	278.4	1914.3	39,481.6	66
28	56	42.00	0.87	283.5	1984.5	41,674.5	68
	57	42.75	0.87	288.6	2056.0	43,947.2	69
29	58	43.50	0.87	293.6	2128.8	46,301.0	70
	59	44.25	0.87	298.7	2202.8	48,737.4	71
30	60	45.00	0.86	303.8	2278.1	51,257.8	72
	61	45.75	0.86	308.8	2354.7	53,863.7	74
31	62	46.50	0.86	313.9	2432.5	56,556.4	75
	63	47.25	0.86	318.9	2511.6	59,337.3	76
32	64	48.00	0.86	324.0	2592.0	62,208.0	77

8¾-in.-wide glulams

No. of lams		Depth d, in.	Size factor C_F	Area A, in.²	Section modulus S, in.³	Moment of inertia I, in.⁴	Weight per lineal foot, lb/ft
1½ in.	¾ in.						
6	12	9.0	1.00	78.8	118.1	531.6	19
	13	9.75	1.00	85.3	138.6	675.8	21
7	14	10.5	1.00	91.9	160.8	844.1	22
	15	11.25	1.00	98.4	184.6	1038.2	24
8	16	12.00	1.00	105.0	210.0	1260.0	25
	17	12.75	0.99	111.6	237.1	1511.3	27
9	18	13.50	0.99	118.1	265.8	1794.0	28
	19	14.25	0.98	124.7	296.1	2109.9	30
10	20	15.00	0.98	131.3	328.1	2460.9	32
	21	15.75	0.97	137.8	361.8	2848.8	33
11	22	16.50	0.97	144.4	397.0	3275.5	35
	23	17.25	0.96	150.9	433.9	3742.8	36
12	24	18.00	0.96	157.5	472.5	4252.5	38
	25	18.75	0.95	164.1	512.7	4806.5	39
13	26	19.50	0.95	170.6	554.5	5406.7	41
	27	20.25	0.94	177.2	598.0	6054.8	42
14	28	21.00	0.94	183.8	643.1	6752.8	44
	29	21.75	0.94	190.3	689.9	7502.5	46
15	30	22.50	0.93	196.9	738.3	8305.7	47
	31	23.25	0.93	203.4	788.3	9164.2	49
16	32	24.00	0.93	210.0	840.0	10,080.0	50
	33	24.75	0.92	216.6	893.3	11,054.8	52
17	34	25.50	0.92	223.1	948.3	12,090.6	53
	35	26.25	0.92	229.7	1004.9	13,189.1	55
18	36	27.00	0.91	236.3	1063.1	14,352.2	56
	37	27.75	0.91	242.8	1123.0	15,581.7	58
19	38	28.50	0.91	249.4	1184.5	16,879.6	60
	39	29.25	0.91	255.9	1247.7	18,247.5	61
20	40	30.00	0.90	262.5	1312.5	19,687.5	63
	41	30.75	0.90	269.1	1378.9	21,201.3	64
21	42	31.50	0.90	275.6	1447.0	22,790.7	66
	43	32.25	0.90	282.2	1516.8	24,457.7	67
22	44	33.00	0.89	288.8	1588.1	26,204.1	69
	45	33.75	0.89	295.3	1661.1	28,031.6	70
23	46	34.50	0.89	301.9	1735.8	29,942.2	72
	47	35.25	0.89	308.4	1812.1	31,937.7	73
24	48	36.00	0.88	315.0	1890.0	34,020.0	75
	49	36.75	0.88	321.6	1969.6	36,190.9	77
25	50	37.50	0.88	328.1	2050.8	38,452.1	78
	51	38.25	0.88	334.7	2133.6	40,805.7	80
26	52	39.00	0.88	341.3	2218.1	43,253.4	81
	53	39.75	0.88	347.8	2304.3	45,797.1	83
27	54	40.50	0.87	354.4	2392.0	48,438.6	84
	55	41.25	0.87	360.9	2481.4	51,179.8	86

501

8¾-in.-wide glulams							

No. of lams		Depth d, in.	Size factor C_F	Area A, in.²	Section modulus S, in.³	Moment of inertia I, in.⁴	Weight per lineal foot, lb/ft
1½ in.	¾ in.						
28	56	42.00	0.87	367.5	2572.5	54,022.5	87
	57	42.75	0.87	374.1	2665.2	56,968.6	89
29	58	43.50	0.87	380.6	2759.5	60,019.8	91
	59	44.25	0.87	387.2	2855.5	63,178.1	92
30	60	45.00	0.86	393.8	2953.1	66,445.3	94
	61	45.75	0.86	400.3	3052.4	69,823.3	95
31	62	46.50	0.86	406.9	3153.3	73,313.8	97
	63	47.25	0.86	413.4	3255.8	76,918.8	98
32	64	48.00	0.86	420.0	3360.0	80,640.0	100
	65	48.75	0.86	426.6	3465.8	84,479.4	101
33	66	49.50	0.85	433.1	3573.3	88,438.7	103
	67	50.25	0.85	439.7	3682.4	92,519.9	105
34	68	51.00	0.85	446.3	3793.1	96,724.7	106
	69	51.75	0.85	452.8	3905.5	101,055	108
35	70	52.50	0.85	459.4	4019.5	105,513	109
	71	53.25	0.85	465.9	4135.2	110,100	111
36	72	54.00	0.85	472.5	4252.5	114,818	112
	73	54.75	0.85	479.1	4371.4	119,668	114
37	74	55.50	0.84	485.6	4492.0	124,654	115
	75	56.25	0.84	492.2	4614.3	129,776	117
38	76	57.00	0.84	498.8	4738.1	135,037	119
	77	57.75	0.84	505.3	4863.6	140,437	120
39	78	58.50	0.84	511.9	4990.8	145,980	122
	79	59.25	0.84	518.4	5119.6	151,667	123
40	80	60.00	0.84	525.0	5250.0	157,500	125
	81	60.75	0.84	531.6	5382.1	163,480	126
41	82	61.50	0.83	538.1	5515.8	169,610	128
	83	62.25	0.83	544.7	5651.1	175,892	129
42	84	63.00	0.83	551.3	5788.1	182,326	131

10¾-in.-wide glulams							

No. of lams		Depth d, in.	Size factor C_F	Area A, in.²	Section modulus S, in.³	Moment of inertia I, in.⁴	Weight per lineal foot, lb/ft
1½ in.	¾ in.						
7	14	10.5	1.00	112.9	197.5	1037.0	27
	15	11.25	1.00	120.9	226.8	1275.5	29
8	16	12.00	1.00	129.0	258.0	1548.0	31
	17	12.75	0.99	137.1	291.3	1856.8	33
9	18	13.50	0.99	145.1	326.5	2204.1	35
	19	14.25	0.98	153.2	363.8	2592.2	37
10	20	15.00	0.98	161.3	403.1	3023.4	39
	21	15.75	0.97	169.3	444.4	3500.0	41

10¾-in.-wide glulams

No. of lams		Depth d, in.	Size factor C_F	Area A, in.²	Section modulus S, in.³	Moment of inertia I, in.⁴	Weight per lineal foot, lb/ft
1½ in.	¾ in.						
11	22	16.50	0.97	177.4	487.8	4024.2	42
	23	17.25	0.96	185.4	533.1	4598.3	44
12	24	18.00	0.96	193.5	580.5	5224.5	46
	25	18.75	0.95	201.6	629.9	5905.2	48
13	26	19.50	0.95	209.6	681.3	6642.5	50
	27	20.25	0.94	217.7	734.7	7438.8	52
14	28	21.00	0.94	225.8	790.1	8296.3	54
	29	21.75	0.94	233.8	847.6	9217.3	56
15	30	22.50	0.93	241.9	907.0	10,204.1	58
	31	23.25	0.93	249.9	968.5	11,258.9	60
16	32	24.00	0.93	258.0	1032.0	12,384.0	62
	33	24.75	0.92	266.1	1097.5	13,581.7	63
17	34	25.50	0.92	274.1	1165.0	14,854.1	65
	35	26.25	0.92	282.2	1234.6	16,203.7	67
18	36	27.00	0.91	290.3	1306.1	17,632.7	69
	37	27.75	0.91	298.3	1379.7	19,143.3	71
19	38	28.50	0.91	306.4	1455.3	20,737.8	73
	39	29.25	0.91	314.4	1532.9	22,418.4	75
20	40	30.00	0.90	322.5	1612.5	24,187.5	77
	41	30.75	0.90	330.6	1694.1	26,047.3	79
21	42	31.50	0.90	338.6	1777.8	28,000.1	81
	43	32.25	0.90	346.7	1863.4	30,048.1	83
22	44	33.00	0.89	354.8	1951.1	32,193.6	84
	45	33.75	0.89	362.8	2040.8	34,438.8	86
23	46	34.50	0.89	370.9	2132.5	36,786.2	88
	47	35.25	0.89	378.9	2226.3	39,237.8	90
24	48	36.00	0.88	387.0	2322.0	41,796.0	92
	49	36.75	0.88	395.1	2419.8	44,463.0	94
25	50	37.50	0.88	403.1	2519.5	47,241.2	96
	51	38.25	0.88	411.2	2621.3	50,132.8	98
26	52	39.00	0.88	419.3	2725.1	53,139.9	100
	53	39.75	0.88	427.3	2830.9	56,265.0	102
27	54	40.50	0.87	435.4	2938.8	59,510.3	104
	55	41.25	0.87	443.4	3048.6	62,878.1	105
28	56	42.00	0.87	451.5	3160.5	66,370.5	107
	57	42.75	0.87	459.6	3274.4	69,989.9	109
29	58	43.50	0.87	467.6	3390.3	73,738.6	111
	59	44.25	0.87	475.7	3508.2	77,618.8	113
30	60	45.00	0.86	483.8	3628.1	81,632.8	115
	61	45.75	0.86	491.8	3750.1	85,782.9	117
31	62	46.50	0.86	499.9	3874.0	90,071.2	119
	63	47.25	0.86	507.9	4000.0	94,500.2	121
32	64	48.00	0.86	516.0	4128.0	99,072.0	123
	65	48.75	0.86	524.1	4258.0	103,789	125
33	66	49.50	0.85	532.1	4390.0	108,653	126

503

10¾-in.-wide glulams							
No. of lams		Depth d, in.	Size factor C_F	Area A, in.²	Section modulus S, in.³	Moment of inertia I, in.⁴	Weight per lineal foot, lb/ft
1½ in.	¾ in.						
	67	50.25	0.85	540.2	4524.1	113,667	128
34	68	51.00	0.85	548.3	4660.1	118,833	130
	69	51.75	0.85	556.3	4798.2	124,153	132
35	70	52.50	0.85	564.4	4938.3	129,630	134
	71	53.25	0.85	572.4	5080.4	135,265	136
36	72	54.00	0.85	580.5	5224.5	141,062	138
	73	54.75	0.85	588.6	5370.6	147,021	140
37	74	55.50	0.84	596.6	5518.8	153,146	142
	75	56.25	0.84	604.7	5668.9	159,439	144
38	76	57.00	0.84	612.8	5821.1	165,902	146
	77	57.75	0.84	620.8	5975.3	172,537	147
39	78	58.50	0.84	628.9	6131.5	179,347	149
	79	59.25	0.84	636.9	6289.8	186,334	151
40	80	60.00	0.84	645.0	6450.0	193,500	153
	81	60.75	0.84	653.1	6612.3	200,847	155
41	82	61.50	0.83	661.1	6776.5	208,378	157
	83	62.25	0.83	669.2	6942.8	216,095	159
42	84	63.00	0.83	677.3	7111.1	224,000	161
	85	63.75	0.83	685.3	7281.4	232,096	163
43	86	64.50	0.83	693.4	7453.8	240,384	165
	87	65.25	0.83	701.4	7628.1	248,868	167
44	88	66.00	0.83	709.5	7804.5	257,549	168
	89	66.75	0.83	717.6	7982.9	266,429	170
45	90	67.50	0.83	725.6	8163.3	275,511	172
	91	68.25	0.82	733.7	8345.7	284,797	174
46	92	69.00	0.82	741.8	8530.1	294,289	176
	93	69.75	0.82	749.8	8716.6	303,990	178
47	94	70.50	0.82	757.9	8905.0	313,902	180
	95	71.25	0.82	765.9	9095.5	324,027	182
48	96	72.00	0.82	774.0	9288.0	334,368	184
	97	72.75	0.82	782.1	9482.5	344,926	186
49	98	73.50	0.82	790.1	9679.0	355,704	188
	99	74.25	0.82	798.2	9877.6	366,705	189
50	100	75.00	0.82	806.3	10,078.1	377,930	191

Selected Tables
from the
Uniform
Building Code,
1979 Edition *

Tables for wood design values *(cont.)*	UBC table no.
17. Allowable shears for wind or seismic loading on vertical diaphragms of fiberboard sheathing board construction for Type V construction only	25-O
18. Nailing schedule	25-P
19. Allowable spans for lumber floor and roof sheathing	25-Q
20. Allowable spans for plywood subfloor and roof sheathing continuous over two or more spans and face grain perpendicular to supports	25-R-1
21. Allowable loads for plywood roof sheathing continuous over two or more spans and face grain parallel to supports	25-R-2
22. Allowable span for plywood combination subfloor-underlayment	25-S
23. Allowable spans for two-inch tongue-and-groove decking	25-T
24. Allowable shear for wind or seismic forces in pounds per foot for vertical diaphragms of lath and plaster or gypsum board frame wall assemblies	47-I

Tables for determining allowable loads on anchor bolts in concrete and masonry	UBC table no.
25. Allowable shear on bolts for all masonry except gypsum and unburned clay units	24-G
26. Allowable shear and tension on bolts [in concrete]	26-G

Note: For a discussion of other Code and wood design criteria used in this book, see Chapter 1.

TABLE NO. 23-A—UNIFORM AND CONCENTRATED LOADS

USE OR OCCUPANCY		UNIFORM LOAD[1]	CONCEN-TRATED LOAD
CATEGORY	DESCRIPTION		
1. Armories		150	0
2. Assembly areas[4] and auditoriums and balconies therewith	Fixed seating areas	50	0
	Movable seating and other areas	100	0
	Stage areas and enclosed platforms	125	0
3. Cornices, marquees and residential balconies		60	0
4. Exit facilities[5]		100	0[8]
5. Garages	General storage and/or repair	100	[3]
	Private pleasure car storage	50	[3]
6. Hospitals	Wards and rooms	40	1000[2]
7. Libraries	Reading rooms	60	1000[2]
	Stack rooms	125	1500[2]
8. Manufacturing	Light	75	2000[2]
	Heavy	125	3000[2]
9. Offices		50	2000[2]
10. Printing plants	Press rooms	150	2500[2]
	Composing and linotype rooms	100	2000[2]
11. Residential[6]		40	0[8]
12. Rest rooms[7]			
13. Reviewing stands, grand stands and bleachers		100	0
14. Roof deck	Same as area served or for the type of occupancy accommodated		
15. Schools	Classrooms	40	1000[2]
16. Sidewalks and driveways	Public access	250	[3]
17. Storage	Light	125	
	Heavy	250	
18. Stores	Retail	75	2000[2]
	Wholesale	100	3000[2]

[1]See Section 2306 for live load reductions.
[2]See Section 2304 (c), first paragraph, for area of load application.
[3]See Section 2304 (c), second paragraph, for concentrated loads.
[4]Assembly areas include such occupancies as dance halls, drill rooms, gymnasiums, play-

grounds, plazas, terraces and similar occupancies which are generally accessible to the public.

[2]Exit facilities shall include such uses as corridors serving an occupant load of 10 or more persons, exterior exit balconies, stairways, fire escapes and similar uses.

[4]Residential occupancies include private dwellings, apartments and hotel guest rooms.

[7]Rest room loads shall be not less than the load for the occupancy with which they are associated, but need not exceed 50 pounds per square foot.

[8]Individual stair treads shall be designed to support a 300-pound concentrated load placed in a position which would cause maximum stress. Stair stringers may be designed for the uniform load set forth in the table.

TABLE NO. 23-B—SPECIAL LOADS[1]

USE		VERTICAL LOAD	LATERAL LOAD
CATEGORY	DESCRIPTION	(Pounds per Square Foot Unless Otherwise Noted)	
1. Construction, public access at site (live load)	Walkway See Sec. 4406	150	
	Canopy See Sec. 4407	150	
2. Grandstands, reviewing stands and bleachers (live load)	Seats and foot-boards	120[2]	See Footnote 3
3. Stage accessories, see Sec. 3902 (live load)	Gridirons and fly galleries	75	
	Loft block wells[4]	250	250
	Head block wells and sheave beams[4]	250	250
4. Ceiling framing (live load)	Over stages	20	
	All uses except over stages	10[5]	
5. Partitions and interior walls, see Sec. 2309 (live load)			5
6. Elevators and dumbwaiters (dead and live load)		2 x Total loads[6]	
7. Mechanical and electrical equipment (dead load)		Total loads	
8. Cranes (dead and live load)[7]	Total load including impact increase	1.25 x Total load[7]	0.10 x Total load[8]
9. Balcony railings, guard rails and handrails	Exit facilities serving an occupant load greater than 50		50[9]
	Other		20[9]
10. Storage racks	Over 8 feet high	Total loads[10]	See Table No. 23-J

(Footnotes on following page)

FOOTNOTES FOR TABLE NO. 23-B

[1]The tabulated loads are minimum loads. Where other vertical loads required by this code or required by the design would cause greater stresses they shall be used.

[2]Pounds per lineal foot.

[3]Lateral sway bracing loads of 24 pounds per foot parallel and 10 pounds per foot perpendicular to seat and footboards.

[4]All loads are in pounds per lineal foot. Head block wells and sheave beams shall be designed for all loft block well loads tributary thereto. Sheave blocks shall be designed with a factor of safety of five.

[5]Does not apply to ceilings which have sufficient total access from below, such that access is not required within the space above the ceiling. Does not apply to ceilings if the attic areas above the ceiling are not provided with access. This live load need not be considered acting simultaneously with other live loads imposed upon the ceiling framing or its supporting structure.

[6]Where Appendix Chapter 51 has been adopted, see reference standard cited therein for additional design requirements.

[7]The impact factors included are for cranes with steel wheels riding on steel rails. They may be modified if substantiating technical data acceptable to the building official is submitted. Live loads on crane support girders and their connections shall be taken as the maximum crane wheel loads. For pendant-operated traveling crane support girders and their connections, the impact factors shall be 1.10.

[8]This applies in the direction parallel to the runway rails (longitudinal). The factor for forces perpendicular to the rail is 0.20 × the transverse traveling loads (trolley, cab, hooks and lifted loads). Forces shall be applied at top of rail and may be distributed among rails of multiple rail cranes and shall be distributed with due regard for lateral stiffness of the structures supporting these rails.

[9]A load per lineal foot to be applied horizontally at right angles to the top rail.

[10]Vertical members of storage racks shall be protected from impact forces of operating equipment or racks shall be designed so that failure of one vertical member will not cause collapse of more than the bay or bays directly supported by that member.

TABLE NO. 23-C—MINIMUM ROOF LIVE LOADS[1]

ROOF SLOPE	METHOD 1			METHOD 2		
	TRIBUTARY LOADED AREA IN SQUARE FEET FOR ANY STRUCTURAL MEMBER			UNIFORM LOAD[2]	RATE OF REDUC-TION r (Percent)	MAXIMUM REDUC-TION R (Percent)
	0 to 200	201 to 600	Over 600			
1. Flat or rise less than 4 inches per foot. Arch or dome with rise less than one-eighth of span	20	16	12	20	.08	40
2. Rise 4 inches per foot to less than 12 inches per foot. Arch or dome with rise one-eighth of span to less than three-eighths of span	16	14	12	16	.06	25
3. Rise 12 inches per foot and greater. Arch or dome with rise three-eighths of span or greater	12	12	12	12	No Reductions Permitted	
4. Awnings except cloth covered[3]	5	5	5	5		
5. Greenhouses, lath houses and agricultural buildings[4]	10	10	10	10		

[1]Where snow loads occur, the roof structure shall be designed for such loads as determined by the building official. See Section 2305 (d). For special purpose roofs, see Section 2305 (e).

[2]See Section 2306 for live load reductions. The rate of reduction r in Section 2306 Formula (6-1) shall be as indicated in the table. The maximum reduction R shall not exceed the value indicated in the table.

[3]As defined in Section 4506.

[4]See Section 2305 (e) for concentrated load requirements for greenhouse roof members.

510

TABLE NO. 23-D—MAXIMUM ALLOWABLE DEFLECTION FOR STRUCTURAL MEMBERS[1]

TYPE OF MEMBER	MEMBER LOADED WITH LIVE LOAD ONLY (L.L.)	MEMBER LOADED WITH LIVE LOAD PLUS DEAD LOAD (L.L. + K D.L.)
Roof Member Supporting Plaster or Floor Member	$L/360$	$L/240$

[1]Sufficient slope or camber shall be provided for flat roofs in accordance with Section 2305 (f).

$L.L.$ = Live load
$D.L.$ = Dead load
K = Factor as determined by Table No. 23-E
L = Length of member in same units as deflection

TABLE NO. 23-E—VALUE OF "K"

WOOD		REINFORCED CONCRETE[2]	STEEL
Unseasoned	Seasoned[1]		
1.0	0.5	$[2 - 1.2 (A'_x/A_x)] \geqq 0.6$	0

[1]Seasoned lumber is lumber having a moisture content of less than 16 percent at time of installation and used under dry conditions of use such as in covered structures.

[2]See also Section 2609.

A'_s = Area of compression reinforcement.
A_s = Area of nonprestressed tension reinforcement.

TABLE NO. 23-F—WIND PRESSURES FOR VARIOUS HEIGHT ZONES ABOVE GROUND[1]

HEIGHT ZONES (in feet)	WIND-PRESSURE MAP AREAS (pounds per square foot)						
	20	25	30	35	40	45	50
Less than 30	15	20	25	25	30	35	40
30 to 49	20	25	30	35	40	45	50
50 to 99	25	30	40	45	50	55	60
100 to 499	30	40	45	55	60	70	75
500 to 1199	35	45	55	60	70	80	90
1200 and over	40	50	60	70	80	90	100

[1]See Figure No. 4. Wind pressure column in the table should be selected which is headed by a value corresponding to the minimum permissible, resultant wind pressure indicated for the particular locality.

The figures given are recommended as minimum. These requirements do not provide for tornadoes.

TABLE NO. 23-I—HORIZONTAL FORCE FACTOR *K* FOR BUILDINGS OR OTHER STRUCTURES[1]

TYPE OR ARRANGEMENT OF RESISTING ELEMENTS	VALUE[2] OF *K*
1. All building framing systems except as hereinafter classified	1.00
2. Buildings with a box system as specified in Section 2312 (b)	1.33
3. Buildings with a dual bracing system consisting of a ductile moment-resisting space frame and shear walls or braced frames using the following design criteria: a. The frames and shear walls shall resist the total lateral force in accordance with their relative rigidities considering the interaction of the shear walls and frames b. The shear walls acting independently of the ductile moment-resisting portions of the space frame shall resist the total required lateral forces c. The ductile moment-resisting space frame shall have the capacity to resist not less than 25 percent of the required lateral force	0.80
4. Buildings with a ductile moment-resisting space frame designed in accordance with the following criteria: The ductile moment-resisting space frame shall have the capacity to resist the total required lateral force	0.67
5. Elevated tanks plus full contents, on four or more cross-braced legs and not supported by a building.	2.5[3]
6. Structures other than buildings and other than those set forth in Table No. 23-J	2.00

[1]Where wind load as specified in Section 2311 would produce higher stresses, this load shall be used in lieu of the loads resulting from earthquake forces.

[2]See Figures Nos. 1, 2 and 3 in this chapter and definition of *Z* as specified in Section 2312 (c).

[3]The minimum value of *KC* shall be 0.12 and the maximum value of *KC* need not exceed 0.25.

The tower shall be designed for an accidental torsion of 5 percent as specified in Section 2312 (e) 5. Elevated tanks which are supported by buildings or do not conform to type or arrangement of supporting elements as described above shall be designed in accordance with Section 2312 (g) using $C_p = .3$.

TABLE NO. 23-J—HORIZONTAL FORCE FACTOR C_p FOR ELEMENTS OF STRUCTURES AND NONSTRUCTURAL COMPONENTS

PART OR PORTION OF BUILDINGS	DIRECTION OF HORIZONTAL FORCE	VALUE OF C_p[1]
1. Exterior bearing and nonbearing walls, interior bearing walls and partitions, interior nonbearing walls and partitions —see also Section 2312 (j) 3 C. Masonry or concrete fences over 6 feet high	Normal to flat surface	0.3[6]
2. Cantilever elements: a. Parapets	Normal to flat surfaces	0.8
b. Chimneys or stacks	Any direction	
3. Exterior and interior ornamentations and appendages	Any direction	0.8
4. When connected to, part of, or housed within a building: a. Penthouses, anchorage and supports for chimneys and stacks and tanks, including contents b. Storage racks with upper storage level at more than 8 feet in height, plus contents c. All equipment or machinery	Any direction	0.3[2 3]
5. Suspended ceiling framing systems (applies to Seismic Zones Nos. 2, 3 and 4 only)	Any direction	0.3[4]
6. Connections for prefabricated structural elements other than walls, with force applied at center of gravity of assembly	Any direction	0.3[5]

[1]C_p for elements laterally self-supported only at the ground level may be two-thirds of value shown.

[2]W_p for storage racks shall be the weight of the racks plus contents. The value of C_p for racks over two storage support levels in height shall be 0.24 for the levels below the top two levels. In lieu of the tabulated values steel storage racks may be designed in accordance with U.B.C. Standard No. 27-11.

Where a number of storage rack units are interconnected so that there are a minimum of four vertical elements in each direction on each column line designed to resist horizontal forces, the design coefficients may be as for a building with K values from Table No. 23-1, $CS - 0.2$ for use in the formula $V = ZIKCSW$ and W equal to the total dead load plus 50 percent of the rack-rated capacity. Where the design and rack configurations are in accordance with this paragraph, the design provisions in U.B.C. Standard No. 27-11 do not apply.

[3]For flexible and flexibly mounted equipment and machinery, the appropriate values of C_p shall be determined with consideration given to both the dynamic properties of the equipment and

(Continued)

machinery and to the building or structure in which it is placed but shall be not less than the listed values. The design of the equipment and machinery and their anchorage is an integral part of the design and specification of such equipment and machinery.

For essential facilities and life safety systems, the design and detailing of equipment which must remain in place and be functional following a major earthquake shall consider drifts in accordance with Section 2312 (k).

⁵Ceiling weight shall include all light fixtures and other equipment which is laterally supported by the ceiling. For purposes of determining the lateral force, a ceiling weight of not less than 4 pounds per square foot shall be used.

⁶The force shall be resisted by positive anchorage and not by friction.

⁶See also Section 2309 (b) for minimum load and deflection criteria for interior partitions.

TABLE NO. 23-K
VALUES FOR OCCUPANCY IMPORTANCE FACTOR I

TYPE OF OCCUPANCY	I
Essential Facilities¹	1.5
Any building where the primary occupancy is for assembly use for more than 300 persons (in one room)	1.25
All others	1.0

¹See Section 2312 (k) for definition and additional requirements for essential facilities.

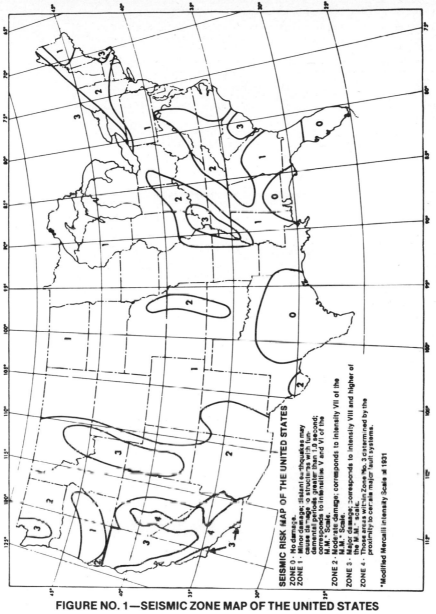

SEISMIC RISK MAP OF THE UNITED STATES

ZONE 0 - No damage.
ZONE 1 - Minor damage; distant earthquakes may cause damage to structures with fundamental periods greater than 1.0 second; corresponds to intensities V and VI of the M.M.* Scale.

ZONE 2 - Moderate damage; corresponds to intensity VII of the M.M.* Scale.

ZONE 3 - Major damage; corresponds to intensity VIII and higher of the M.M.* scale.

ZONE 4 - Those areas within Zone No. 3 determined by the proximity to certain major fault systems.

*Modified Mercalli Intensity Scale of 1931

FIGURE NO. 1—SEISMIC ZONE MAP OF THE UNITED STATES

515

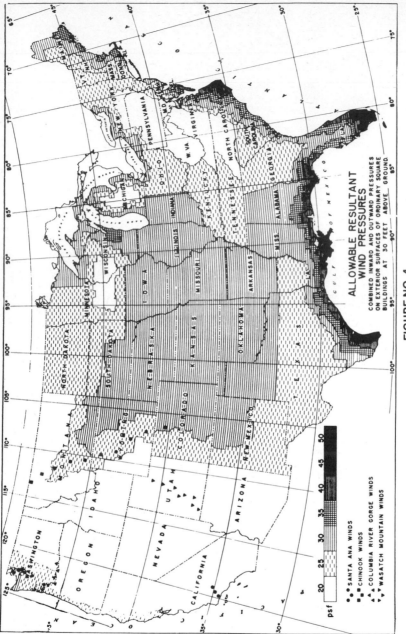

ALLOWABLE RESULTANT
WIND PRESSURES

COMBINED INWARD AND OUTWARD PRESSURES
ON EXTERIOR SURFACES OF ORDINARY SQUARE
BUILDINGS AT 30 FEET ABOVE GROUND

FIGURE NO. 4

•• SANTA ANA WINDS
■■ CHINOOK WINDS
▲ COLUMBIA RIVER GORGE WINDS
▼ WASATCH MOUNTAIN WINDS

psf   20 25 30 35 40 45 50

TABLE NO. 25-K—ALLOWABLE SHEAR FOR WIND OR SEISMIC FORCES IN POUNDS PER FOOT FOR PLYWOOD SHEAR WALLS WITH FRAMING OF DOUGLAS FIR-LARCH OR SOUTHERN PINE[1]

PLYWOOD GRADE	MINIMUM NOMINAL PLYWOOD THICKNESS (inches)	MINIMUM NAIL PENE-TRATION IN FRAMING (inches)	NAIL SIZE (Common or Galvanized Box)	PLYWOOD APPLIED DIRECT TO FRAMING — Nail Spacing at Plywood Panel Edges				NAIL SIZE (Common or Galvanized Box)	PLYWOOD APPLIED OVER ½-INCH GYPSUM SHEATHING — Nail Spacing at Plywood Panel Edges			
				6	4	2½	2		6	4	2½	2
STRUCTURAL I	5/16	1¼	6d	200	300³	450³	510³	8d	200	300	450	510
	3/8	1½	8d	230³	360³	530³	610³	10d	280	430	640²	730²
	1/2	1⅝	10d	340	510	770²	870²	—	—	—	—	—
C-D, C-C, STRUCTURAL II and other grades covered in U.B.C. Standard No. 25-9	5/16	1¼	6d	180	270	400	450	8d	180	270	400	450
	3/8	1½	8d	220³	320³	470³	530³	10d	260	380	570²	640²
	1/2	1⅝	10d	310	460	690²	770²	—	—	—	—	—
			NAIL SIZE (Galvanized Casing)					NAIL SIZE (Galvanized Casing)				
Plywood Panel Siding in Grades Covered in U.B.C. Standard No. 25-9	5/16	1¼	6d	140	210	320	360	8d	140	210	320	360
	3/8	1½	8d	130³	200³	300³	340³	10d	160	240	410	410

[1]All panel edges backed with 2-inch nominal or wider framing. Plywood installed either horizontally or vertically. Space nails at 6 inches on center along intermediate framing members for ⅜-inch plywood installed with face grain parallel to studs spaced 24 inches on center and 12 inches on center for other conditions and plywood thicknesses. These values are for short time loads due to wind or earthquake and must be reduced 25 percent for normal loading.

Allowable shear values for nails in framing members of other species set forth in Table No. 25-17-J of U.B.C. Standards shall be calculated for all grades by multiplying the values for common and galvanized box nails in STRUCTURAL I and galvanized casing nails in other grades by the following factors: Group III, 0.82 and Group IV, 0.65.

[2]Reduce tabulated allowable shears 10 percent when boundary members provide less than 3-inch nominal nailing surface.

[3]The values for ⅜-inch-thick plywood applied direct to framing may be increased 20 percent, provided studs are spaced a maximum of 16 inches on center or plywood is applied with face grain across studs or if the plywood thickness is increased to ½ inch or greater.

TABLE NO. 25-J—ALLOWABLE SHEAR IN POUNDS PER FOOT FOR HORIZONTAL PLYWOOD DIAPHRAGMS WITH FRAMING OF DOUGLAS FIR-LARCH OR SOUTHERN PINE[1]

PLYWOOD GRADE	Common Nail Size	Minimum Nominal Penetration In Framing (In Inches)	Minimum Nominal Plywood Thickness (In Inches)	Minimum Nominal Width of Framing Member (In Inches)	BLOCKED DIAPHRAGMS — Nail spacing at diaphragm boundaries (all cases), at continuous panel edges parallel to load (Cases 3 and 4) and at all panel edges (Cases 5 and 6) — 6	4	2½	2	UNBLOCKED DIAPHRAGM — Load perpendicular to unblocked edges and continuous panel joints (Case 1)	Other configurations (Cases 2, 3 & 4)
					Nail spacing at other plywood panel edges: 6	6	4	3		
STRUCTURAL I	6d	1¼	5/16	2 / 3	185 / 210	250 / 280	375 / 420	420 / 475	165 / 185	125 / 140
	8d	1½	3/8	2 / 3	270 / 300	360 / 400	530 / 600	600 / 675	240 / 265	180 / 200
	10d	1⅝	1/2	2 / 3	320 / 360	425 / 480	640² / 720	730² / 820	285 / 320	215 / 240
C-D, C-C, STRUCTURAL II and other grades covered in U.B.C. Standard No. 25-9	6d	1¼	5/16	2 / 3	170 / 190	225 / 250	335 / 380	380 / 430	150 / 170	110 / 125
	6d	1¼	3/8	2 / 3	185 / 210	250 / 280	375 / 420	420 / 475	165 / 185	125 / 140
	8d	1½	3/8	2 / 3	240 / 270	320 / 360	480 / 540	545 / 610	215 / 240	160 / 180
	8d	1½	1/2	2 / 3	270 / 300	360 / 400	530 / 600	600 / 675	240 / 265	180 / 200
	10d	1⅝	1/2	2 / 3	290 / 325	385 / 430	575² / 650	655² / 735	255 / 290	190 / 215
	10d	1⅝	5/8	2 / 3	320 / 360	425 / 480	640² / 720	730² / 820	285 / 320	215 / 240

¹These values are for short time loads due to wind or earthquake and must be reduced 25 percent for normal loading. Space nails 10 inches on center for floors and 12 inches on center for roofs along intermediate framing members.

Allowable shear values for nails in framing members of other species set forth in Table No. 25-17-J of U.B.C. Standards shall be calculated for all grades by multiplying the values for nails in STRUCTURAL I by, the following factors: Group III, 0.82 and Group IV, 0.65.

²Reduce tabulated allowable shears 10 percent when boundary members provide less than 3-inch nominal nailing surface.

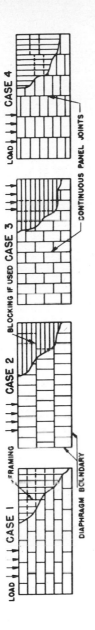

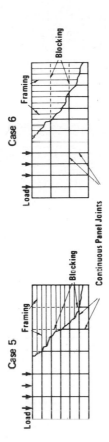

NOTE: Framing may be located in either direction for blocked diaphragms.

TABLE NO. 25-I—MAXIMUM DIAPHRAGM DIMENSION RATIOS

MATERIAL	HORIZONTAL DIAPHRAGMS Maximum Span-Width Ratios	VERTICAL DIAPHRAGMS Maximum Height-Width Ratios
1. Diagonal sheathing, conventional	3:1	2:1
2. Diagonal sheathing, special	4:1	$3\frac{1}{2}$:1
3. Plywood, nailed all edges	4:1	$3\frac{1}{2}$:1
4. Plywood, blocking omitted at intermediate joints	4:1	2:1

TABLE NO. 25-M—EXPOSED PLYWOOD PANEL SIDING

MINIMUM THICKNESS[1]	MINIMUM NO. OF PLYS	STUD SPACING (INCHES) PLYWOOD SIDING APPLIED DIRECT TO STUDS OR OVER SHEATHING
$\frac{3}{8}$''	3	16[2]
$\frac{1}{2}$''	4	24

[1]Thickness of grooved panels is measured at bottom of grooves.

[2]May be 24 inches if plywood siding applied with face grain perpendicular to studs or over one of the following: (a) 1-inch board sheathing; (b) $\frac{1}{2}$-inch plywood sheathing, (c) $\frac{3}{8}$-inch plywood sheathing with face grain of sheathing perpendicular to studs.

TABLE NO. 25-N—PLYWOOD WALL SHEATHING[1]
(Not Exposed to the Weather, Face Grain Parallel or Perpendicular to Studs)

Minimum Thickness	Panel Identification Index	STUD SPACING (Inches) Siding Nailed to Studs	Sheathing Under Coverings Specified in Section 2517 (g) 4 Sheathing Parallel To Studs	Sheathing Perpendicular to Studs
5/16	12/0, 16/0, 20/0	16	—	16
3/8	16/0, 20/0 24/0	24	16	24
1/2	24/0, 32/16	24	24	24

[1]In reference to Section 2518 (g) 5, blocking of horizontal joints is not required.

TABLE NO. 25-O—ALLOWABLE SHEARS FOR WIND OR SEISMIC LOADING ON VERTICAL DIAPHRAGMS OF FIBERBOARD SHEATHING BOARD CONSTRUCTION FOR TYPE V CONSTRUCTION ONLY[1]

SIZE AND APPLICATION	NAIL SIZE	SHEAR VALUE 3-INCH NAIL SPACING AROUND PERIMETER AND 6-INCH AT INTERMEDIATE POINTS
$\frac{7}{16}$" x 4' x 8'	No. 11 ga. gal. roofing nail 1½" long, $\frac{7}{16}$" head	125[2]
$\frac{25}{32}$" x 4' x 8'	No. 11 ga. gal. roofing nail 1¾" long, $\frac{7}{16}$" head	175

[1]Fiberboard sheathing diaphragms shall not be used to brace concrete or masonry walls.

[2]The shear value may be 175 for ½-inch x 4 foot x 8 foot fiberboard nailbase sheathing.

TABLE NO. 25-P—NAILING SCHEDULE

CONNECTION	NAILING[1]
1. Joist to sill or girder, toenail	3-8d
2. Bridging to joist, toenail each end	2-8d
3. 1" x 6" subfloor or less to each joist, face nail	2-8d
4. Wider than 1" x 6" subfloor to each joist, face nail	3-8d
5. 2" subfloor to joist or girder, blind and face nail	2-16d
6. Sole plate to joist or blocking, face nail	16d at 16" o.c.
7. Top plate to stud, end nail	2-16d
8. Stud to sole plate	4-8d, toenail or 2-16d, end nail
9. Doubled studs, face nail	16d at 24" o.c.
10. Doubled top plates, face nail	16d at 16" o.c.
11. Top plates, laps and intersections, face nail	2-16d
12. Continuous header, two pieces	16d at 16" o.c. along each edge
13. Ceiling joists to plate, toenail	3-8d
14. Continuous header to stud, toenail	4-8d
15. Ceiling joists, laps over partitions, face nail	3-16d
16. Ceiling joists to parallel rafters, face nail	3-16d
17. Rafter to plate, toenail	3-8d
18. 1" brace to each stud and plate, face nail	2-8d
19. 1" x 8" sheathing or less to each bearing, face nail	2-8d
20. Wider than 1" x 8" sheathing to each bearing, face nail	3-8d
21. Built-up corner studs	16d at 24" o.c.

CONNECTION	NAILING[1]
22. Built-up girder and beams	20d at 32″ o.c. at top and bottom and staggered 2-20d at ends and at each splice
23. 2″ planks	2-16d at each bearing

CONNECTION	NAILING[1]
24. Particleboard:[5]	
Wall Sheathing (to framing):	
⅜″-½″	6d[3]
⅝″-¾″	8d[3]
25. Plywood:[5]	
Subfloor, roof and wall sheathing (to framing):	
½″ and less	6d[2]
⅝″-¾″	8d[3] or 6d[4]
⅞″-1″	8d[2]
1⅛″-1¼″	10d[3] or 8d[4]
Combination Subfloor-underlayment (to framing):	
¾″ and less	6d[4]
⅞″-1″	8d[4]
1⅛″-1¼″	10d[3] or 8d[4]
26. Panel Siding (to framing)	
½″ or less	6d[6]
⅝″	8d[6]
27. Fiberboard Sheathing:[7]	
½″	No. 11 ga.[8] 6d[3]
	No. 16 ga.[9]
²⁵⁄₃₂″	No. 11 ga.[8] 8d[3]
	No. 16 ga.[9]

[1]Common or box nails may be used except where otherwise stated.
[2]Common or deformed shank.
[3]Common.
[4]Deformed shank.
[5]Nails spaced at 6 inches on center at edges, 12 inches at intermediate supports (10 inches at intermediate supports for floors), except 6 inches at all supports where spans are 48 inches or more. For nailing of plywood diaphragms and shear walls refer to Section 2514 (c). Nails for wall sheathing may be common, box or casing.
[6]Corrosion-resistant siding and casing nails.
[7]Fasteners spaced 3 inches on center at exterior edges and 6 inches on center at intermediate supports.
[8]Galvanized roofing nails with ⁷⁄₁₆-inch-diameter head and 1½-inch length for ½-inch sheathing and 1¾ inch for ²⁵⁄₃₂-inch sheathing.
[9]Galvanized staple with ⁷⁄₁₆-inch crown and 1⅛-inch length for ½-inch sheathing and 1½-inch length for ²⁵⁄₃₂-inch sheathing.

TABLE NO. 25-Q—ALLOWABLE SPANS FOR LUMBER FLOOR AND ROOF SHEATHING[3]

SPAN (Inches)	MINIMUM NET THICKNESS (Inches) OF LUMBER PLACED			
	PERPENDICULAR TO SUPPORTS		DIAGONALLY TO SUPPORTS	
	Surfaced Dry[2]	Surfaced Unseasoned	Surfaced Dry[2]	Surfaced Unseasoned
FLOORS				
24	¾	$\frac{25}{32}$	¾	$\frac{25}{32}$
16	⅝	$\frac{11}{16}$	⅝	$\frac{11}{16}$
ROOFS				
24	⅝	$\frac{11}{16}$	¾	$\frac{25}{32}$

[1]Installation details shall conform to Sections 2518 (e) 1 and 2518 (h) 7 for floor and roof sheathing, respectively.

[2]Maximum 19 percent moisture content.

[3]Floor or roof sheathing conforming with this table shall be deemed to meet the design criteria of Section 2517.

SHEATHING LUMBER SHALL MEET THE FOLLOWING MINIMUM GRADE REQUIREMENTS: BOARD GRADE

SOLID FLOOR OR ROOF SHEATHING	SPACED ROOF SHEATHING	U.B.C. STANDARD NUMBER
1. Utility	Standard	25-2, 25-3 or 25-4
2. 4 Common, or Utility	3 Common, or Standard	25-2, 25-3, 25-4 25-5 or 25-8
3. No. 3	No. 2	25-6
4. Merchantable	Construction Common	25-7

TABLE NO. 25-R-1—ALLOWABLE SPANS FOR PLYWOOD SUBFLOOR AND ROOF SHEATHING CONTINUOUS OVER TWO OR MORE SPANS AND FACE GRAIN PERPENDICULAR TO SUPPORTS[1][9]

PANEL IDENTIFICATION INDEX[3]	PLYWOOD THICKNESS (Inch)	ROOF[2]				FLOOR MAXIMUM SPAN[4] (In Inches)
		MAXIMUM SPAN (In Inches)		LOAD (IN POUNDS PER SQUARE FOOT)		
		Edges Blocked	Edges Unblocked	Total Load	Live Load	
12/0	⁵⁄₁₆	12		155	150	0
16/0	⁵⁄₁₆, ⅜	16		95	75	0
20/0	⁵⁄₁₆, ⅜	20		75	65	0
24/0	⅜	24	16	65	50	0
24/0	½	24	24	65	50	0
30/12	⅝	30	26	70	50	12[5]
32/16	½, ⅝	32	28	55	40	16[7]
36/16	¾	36	30	55	50	16[7]
42/20	⅝, ¾, ⅞	42	32	40[6]	35[6]	20[7][8]
48/24	¾, ⅞	48	36	40[6]	35[6]	24

[1]These values apply for C-C, C-D, Structural I and II grades only. Spans shall be limited to values shown because of possible effect of concentrated loads.

[2]Uniform load deflection limitations: 1/180th of the span under live load plus dead load, 1/240th under live load only. Edges may be blocked with lumber or other approved type of edge support.

[3]Identification index appears on all panels in the construction grades listed in Footnote No. 1.

[4]Plywood edges shall have approved tongue-and-groove joints or shall be supported with blocking, unless ¼-inch minimum thickness underlayment is installed, or finish floor is ²⁵⁄₃₂-inch wood strip. Allowable uniform load based on deflection of 1/360 of span is 165 pounds per square foot.

[5]May be 16-inch if ²⁵⁄₃₂-inch wood strip flooring is installed at right angles to joists.

[6]For roof live load of 40 pounds per square foot or total load of 55 pounds per square foot, decrease spans by 13 percent or use panel with next greater identification index.

[7]May be 24 inch if ²⁵⁄₃₂-inch wood strip flooring is installed at right angles to joists.

[8]May be 24 inches where a minimum of 1½ inches of approved cellular or lightweight concrete is placed over the subfloor and the plywood sheathing is manufactured with exterior glue.

[9]Floor or roof sheathing conforming with this table shall be deemed to meet the design criteria of Section 2517.

TABLE NO. 25-R-2—ALLOWABLE LOADS FOR PLYWOOD ROOF SHEATHING CONTINUOUS OVER TWO OR MORE SPANS AND FACE GRAIN PARALLEL TO SUPPORTS[1][2]

	THICKNESS	NO. OF PLIES	SPAN	TOTAL LOAD	LIVE LOAD
STRUCTURAL I	½	4	24	35	25
	½	5	24	55	40
Other grades covered in U.B.C. Standard No. 25-9	½	5	24	30	25
	⅝	4	24	40	30
	⅝	5	24	60	45

[1]Uniform load deflection limitations: 1/180 of span under live load plus dead load, 1/240 under live load only. Edges shall be blocked with lumber or other approved type of edge supports.

[2]Floor or roof sheathing conforming with this table shall be deemed to meet the design criteria of Section 2517.

TABLE NO. 25-S—ALLOWABLE SPAN FOR PLYWOOD COMBINATION SUBFLOOR-UNDERLAYMENT[1]
Plywood Continuous over Two or More Spans and Face Grain Perpendicular to Supports

SPECIES GROUPS[2]	MAXIMUM SPACING OF JOISTS		
	16″	20″	24″
1	½″	⅝″	¾″
2, 3	⅝″	¾″	⅞″
4	¾″	⅞″	1″

[1]Applicable to Underlayment Grade, C-C (plugged) and all grades of sanded Exterior-type plywood. Spans limited to values shown because of possible effect of concentrated loads. Allowable uniform load based on deflection of 1/360 of span is 125 pounds per square foot. Plywood edges shall have approved tongue and groove joints or shall be supported with blocking, unless ¼- inch minimum thickness underlayment is installed, or finish floor is 25/32-inch wood strip. If wood strips are perpendicular to supports, thicknesses shown for 16- and 20-inch spans may be used on 24-inch span. Except for ½ inch, Underlayment Grade and C-C (plugged) panels may be of nominal thicknesses 1/32 inch thinner than the nominal thicknesses shown when marked with the reduced thickness.

[2]See U.B.C. Standard No. 25-9 for plywood species groups.

TABLE NO. 25-T—ALLOWABLE SPANS FOR TWO-INCH TONGUE-AND-GROOVE DECKING

SPAN[1] (In Feet)	LIVE LOAD	DEFLECTION LIMIT	*f* (psi)	*E* (psi)
ROOFS				
4	20	1/240 1/360	160	170,000 256,000
	30	1/240 1/360	210	256,000 384,000
	40	1/240 1/360	270	340,000 512,000
4.5	20	1/240 1/360	200	242,000 305,000
	30	1/240 1/360	270	363,000 105,000
	40	1/240 1/360	350	484,000 725,000
5.0	20	1/240 1/360	250	332,000 500,000
	30	1/240 1/360	330	495,000 742,000

SPAN [1] (In Feet)	LIVE LOAD	DEFLECTION LIMIT	f (psi)	E (psi)
	40	1/240 1/360	420	660,000 1,000,000
5.5	20	1/240 1/360	300	442,000 660,000
	30	1/240 1/360	400	662,000 998,000
	40	1/240 1/360	500	884,000 1,330,000
6.0	20	1/240 1/360	360	575,000 862,000
	30	1/240 1/360	480	862,000 1,295,000
	40	1/240 1/360	600	1,150,000 1,730,000
6.5	20	1/240 1/360	420	595,000 892,000
	30	1/240 1/360	560	892,000 1,340,000
	40	1/240 1/360	700	1,190,000 1,730,000
7.0	20	1/240 1/360	490	910,000 1,360,000
	30	1/240 1/360	650	1,370,000 2,000,000
	40	1/240 1/360	810	1,820,000 2,725,000
7.5	20	1/240 1/360	560	1,125,000 1,685,000
	30	1/240 1/360	750	1,685,000 2,530,000
	40	1/240 1/360	930	2,250,000 3,380,000
8.0	20	1/240 1/360	640	1,360,000 2,040,000
	30	1/240 1/360	850	2,040,000 3,060,000
FLOORS				
4 4.5 5.0	40	1/360	840 950 1060	1,000,000 1,300,000 1,600,000

[1]Spans are based on simple beam action with 10 pounds per square foot dead load and provisions for a 300-pound concentrated load on a 12-inch width of floor decking. Random lay-up permitted in accordance with the provisions of Section 2518 (e) 3 or 2518 (h) 8. Lumber thickness assumed at 1½ inches, net.

TABLE NO. 47-I—ALLOWABLE SHEAR FOR WIND OR SEISMIC FORCES IN POUNDS PER FOOT FOR VERTICAL DIAPHRAGMS OF LATH AND PLASTER OR GYPSUM BOARD FRAME WALL ASSEMBLIES[1]

TYPE OF MATERIAL	THICKNESS OF MATERIAL	WALL CONSTRUCTION	NAIL SPACING MAXIMUM (in inches)	SHEAR VALUE	MINIMUM NAIL SIZE
1. Expanded metal, or woven wire lath and portland cement plaster	⅞"	Unblocked	6	180	No. 11 gauge, 1½" long, 7/16" head; No. 16 gauge staple, ⅞" legs
2. Gypsum lath, plain or perforated	⅜" Lath and ½" Plaster	Unblocked	5	100	No. 13 gauge, 1⅛" long, 19/64" head, plasterboard blued nail.
3. Gypsum sheathing board	½" x 2' x 8'	Unblocked	4	75	No. 11 gauge, 1¾" long, 7/16" head, diamond-point, galvanized.
	½" x 4' ½" x 4'	Blocked Unblocked	4 7	175 100	
4. Gypsum wallboard or veneer base	½"	Unblocked	7	100	5d cooler nails.
		Unblocked	4	125	
		Blocked	7	125	
		Blocked	4	150	
	⅝"	Blocked	4	175	6d cooler nails.
		Blocked Two-ply	Base ply 9 Face ply 7	250	Base ply—6d cooler nails. Face ply—8d cooler nails.

[1]These vertical diaphragms shall not be used to resist loads imposed by masonry or concrete construction. See Section 4713 (b). Values are for short-time loading due to wind or earthquake and must be reduced 25 percent for normal loading.

[2]Applies to nailing at all studs, top and bottom plates and blocking.

TABLE NO. 24-G—ALLOWABLE SHEAR ON BOLTS FOR ALL MASONRY EXCEPT GYPSUM AND UNBURNED CLAY UNITS

DIAMETER OF BOLT (Inches)	EMBEDMENT[1] (Inches)	SOLID MASONRY (Shear in Pounds)	GROUTED MASONRY (Shear in Pounds)
½	4	350	550
⅝	4	500	750
¾	5	750	1100
⅞	6	1000	1500
1	7	1250	1850[2]
1⅛	8	1500	2250[2]

[1]An additional 2 inches of embedment shall be provided for anchor bolts located in the top of columns for buildings located in Seismic Zones Nos. 2, 3 and 4.

[2]Permitted only with not less than 2500 pounds per square inch units.

TABLE NO. 26-G
ALLOWABLE SHEAR AND TENSION ON BOLTS [in concrete] (In Pounds)[1][2]

DIAMETER (In Inches)	MINIMUM[3] EMBEDMENT (In Inches)	MINIMUM CONCRETE STRENGTH (In psi)		
		SHEAR[4]		TENSION[5]
		2000	3000	2000 to 5000
¼	2½	500	500	200
⅜	3	1100	1100	500
½	4	2000	2000	950
⅝	4	2750	3000	1500
¾	5	2940	3560	2250
⅞	6	3580	4150	3200
1	7	3580	4150	3200
1⅛	8	3580	4500	3200
1¼	9	3580	5300	3200

NOTES:

[1]Values are for natural stone aggregate concrete and bolts of at least A307 quality. Bolts shall have a standard bolt head or an equal deformity in the embedded portion.

[2]Values are based upon a bolt spacing of 12 diameters with a minimum edge distance of 6 diameters. Such spacing and edge distance may be reduced 50 percent with an equal reduction in value. Use linear interpolation for intermediate spacings and edge margins.

[3]An additional 2 inches of embedment shall be provided for anchor bolts located in the top of columns for buildings located in Seismic Zones Nos. 2, 3 and 4.

[4]Values shown are for work with or without special inspection.

[5]Values shown are for work without special inspection. Where special inspection is provided values may be increased 100 percent.

SI Metric Units

Introduction

In 1960 the Eleventh General Conference of Weights and Measures adopted the name International System of Units (accepted abbreviation SI, from Système International d'Unites) for a practical and consistent set of units of measure. Rules for common usage, notation, and abbreviation were adopted. Most industrial nations in the world have adopted and converted to SI. SI is not the old metric system (cgs or MKS).

A large number of code and industry design tables are required to carry out a timber structural design. Until these tables are converted to SI metric units by the appropriate agencies, there will be little practical wood design done in metric units.

However, the future trend will be toward the use of SI units. The following brief introduction to SI is included to aid the structural engineer in conversion between U.S. Customary (USC) units and SI units. Additional information can be found in Refs. 37 and 38.

Notation

SI is made up of seven base units, two supplementary units, and many consistent derived units. The following units are pertinent to structural design.

	Quantity	Unit	SI symbol
Base unit	Length	meter	m
	Mass	kilogram	kg
	Time	second	s
Supplementary unit	Plane angle	radian	rad
Derived unit	Density (weight)		N/m^3
	Density (mass)		kg/m^3
	Energy	joule	$J (N \cdot m)$
	Force	newton	$N (kg \cdot m/s^2)$
	Frequency	hertz	$Hz (1/s)$
	Moment of force		$N \cdot m (kg \cdot m^2/s^2)$
	Moment of inertia		m^4
	Stress, pressure	pascal	$Pa (N/m^2)$

Prefixes

The following are SI approved prefixes and their abbreviations used to denote very small or very large quantities

Multiplication factor	Name	Symbol
$0.000\ 000\ 001 = 10^{-9}$	nano	n
$0.000\ 001 = 10^{-6}$	micro	μ
$0.001 = 10^{-3}$	milli	m
$1000 = 10^3$	kilo	k
$1\ 000\ 000 = 10^6$	mega	M
$1000\ 000\ 000 = 10^9$	giga	G

Conversion Factors

The following is a list of structurally pertinent conversions between USC units and SI units.

Area

$$1 \text{ in.}^2 = 645.2 \text{ mm}^2$$
$$1 \text{ ft}^2 = 92.90 \times 10^{-3} \text{ m}^2$$

Bending Moment or Torque

$$1 \text{ in.-lb} = 0.1130 \text{ N} \cdot \text{m}$$
$$1 \text{ ft-lb} = 1.356 \text{ N} \cdot \text{m}$$
$$1 \text{ ft-k} = 1.356 \text{ kN} \cdot \text{m}$$

Lengths or Displacements

$$1 \text{ in.} = 25.40 \text{ mm}$$
$$1 \text{ ft} = 0.3048 \text{ m}$$

Loads

$$1 \text{ lb} = 4.448 \text{ N}$$
$$1 \text{ k} = 4.448 \text{ kN}$$
$$1 \text{ lb/ft} = 14.59 \text{ N/m}$$
$$1 \text{ k/ft} = 14.59 \text{ kN/m}$$

Moment of Inertia

$$1 \text{ in.}^4 = 0.4162 \times 10^6 \text{ mm}^4$$
$$1 \text{ ft}^4 = 8.631 \times 10^{-3} \text{ m}^4$$

Section Modulus or Volume

$$1 \text{ in.}^3 = 16.39 \times 10^3 \text{ mm}^3$$
$$1 \text{ ft}^3 = 28.32 \times 10^{-3} \text{ m}^3$$

Stress and Modulus of Elasticity

$$1 \text{ psi} = 6.895 \text{ kPa}$$
$$1 \text{ ksi} = 6.895 \text{ MPa}$$
$$1 \text{ psf} = 47.88 \text{ Pa}$$

Unit Weight, Density

$$1 \text{ lb/ft}^3 = 0.157 \text{ kN/m}^3$$

APPENDIX F

1. *Uniform Building Code,* 1979 Edition, International Conference of Building Officials (5360 South Workman Mill Road, Whittier, CA 90601).

2. *National Design Specification for Wood Construction* 1977 Edition, National Forest Products Association (1619 Massachusetts Avenue, N.W., Washington, DC 20036).

3. American Institute of Timber Construction (333 West Hampden Avenue, Englewood, CO 80110): *Timber Construction Manual,* 2d ed., Wiley, 1974.

4. *Western Woods Use Book,* Western Woods Products Association (1500 Yeon Building, Portland, OR 97204), 1973.

5. *Wood Handbook: Wood as an Engineering Material,* Agricultural Handbook No. 72, 1974, Forest Products Laboratory, U.S. Department of Agriculture, Washington, DC.

6. *Wood Handbook,* Agriculture Handbook No. 72, 1955, Forest Products Laboratory, U.S. Department of Agriculture, Washington, DC.

7. Gurfinkel, German: *Wood Engineering,* 1973, Southern Forest Products Association (Box 52468, New Orleans, LA 70152).

8. Hoyle, Robert J.: *Wood Technology in the Design of Structures,* 3d ed., Mountain Press Publishing Company (287 West Front Street, Missoula, MT 59801).

9. American Society of Civil Engineers (345 East 47th Street, New York, NY 10017): *Wood Structures,* 1975.

10. *Seismic Design for Buildings,* 1973, Technical Manual 5-809-10, NAV FAC P-355, Air Force Manual No. 88-3, Chap. 13, Departments of the Army, Navy, and Air Force, Washington, DC.

11. Goers, Ralph W., and Associates: ATC-4, *A Methodology for Seismic Design and Construction of Single Family Dwellings,* 1976, Applied Technology Council (480 California Ave., Suite 205, Palo Alto, CA 94306).

533

12. American Institute of Steel Construction, Inc. (1221 Avenue of the Americas, New York, NY 10020): *Manual of Steel Construction,* 7th ed.

13. Amrhein, James E.: *Reinforced Masonry Engineering Handbook,* 1st ed., Masonry Institute of America (2550 Beverly Boulevard, Los Angeles, CA 90057).

14. Publications of the American Plywood Association (1119 A Street, P.O. Box 2277, Tacoma, WA 98401):
  14.1 "Plywood Design Specification," 1976.
  14.2 "Plywood Design Specification," Supplements 1–4, 1974.
  14.3 "Guide to Plywood Grades," 1978.
  14.4 "Plywood Roof Systems," 1978.
  14.5 "Plywood Wall Systems," 1978.
  14.6 "Plywood Systems: Floors," 1978.
  14.7 "APA Sturd-I-Floor," 1978.
  14.8 "Plywood Encyclopedia."
  14.9 "APA Laboratory Report 105—Plywood Shearwalls," 1966.
  14.10 "APA Laboratory Report 106—1966, Horizontal Plywood Diaphragm Tests," 1966.
  14.11 "U.S. Product Standard PS1-74 for Construction and Industrial Plywood," reproduction by APA, 1974.
  14.12 "Plywood Diaphragm Construction," 1976.

15. Tissell, John R.: "Plywood Diaphragms," 1977, American Society of Civil Engineers, Preprint 3076. (ASCE, 345 East 47th Street, New York, NY 10017).

16. Bower, Warren H.: "Lateral Analysis of Plywood Diaphragms," *Journal of the Structural Division,* ASCE, Vol. 100, No. ST4, Proc. Paper 10494, April, 1974, pp. 759–772. (ASCE, 345 East 47th Street, New York, NY 10017).

17. Nosse, John: "Anchorage of Concrete or Masonry Walls," *Building Standards,* Part 1, November–December 1975, International Conference of Building Officials (5360 South Workman Mill Road, Whittier, CA 90601).

18. "Pneumatic and Mechanically Driven Building Construction Fasteners," Manual No. 19–73, 1973, Industrial Stapling and Nailing Technical Association, (435 N. Michigan Avenue, Suite 1717, Chicago, IL 60611).

19. Senco Products, Inc. (8485 Broadwell Road, Cincinnati, OH 45244): "Portable Air-Driven Nailing and Stapling System, SENCO Fastening Systems."

20. Stern, E. George: "Nails—Definitions and Sizes, a Handbook for Nail Users," No. 61, 1967, Virginia Polytechnic Institute Research Division, Blacksburg, Virginia).

21. Simpson Company (1450 Doolittle Drive, P.O. Box 1568, San Leandro, CA 94577): "Handbook of Structural Designs and Load Values," Catalog No. 79 H-1, 1978.

22. KC Metal Products, Inc. (1960 Hartog Drive, San Jose, CA 95131): "Rough Carpentry, Wood Framing Systems," catalog, 1978.

23. TECO (5530 Wisconsin Avenue, Washington, DC 20015): "Structural Wood Fasteners," No. 101.

24. TECO (5530 Wisconsin Avenue, Washington, DC 20015): "Design Manual for TECO Timber Connector Construction," No. 109, 1973.

25. American Institute of Timber Construction (333 West Hampden Avenue, Englewood, CO 80110): "Connections in Glued Laminated Timber," Technical Note No. 3, 1977.

26. Structural Engineers Association of California (171 Second Street, San Francisco, CA 94105): *Recommended Lateral Force Requirements and Commentary*, 1975.

27. American Wood Preservers Institute (1651 Old Meadow Road, McLean, VA): "Pressure-Treated Wood—Unlimited Versatility in Design."

28. American Institute of Timber Construction (333 West Hampden Avenue Englewood, CO 80110): "Glulam Systems," published yearly.

29. American Society for Testing and Materials (1916 Race Street, Philadelphia, PA 19103): *ASTM Annual Book of Standards: Part 22—Wood; Adhesives.*

30. Federal Specification FF-N-105B, "Nails, Brads, Staples, and Spikes: Wire, Cut and Wrought" (General Services Administration, Federal Supply Service, Washington, DC 20406).

31. Easy Arch Rib (6612 Columbus Street, Riverside, CA 92504): "Easy Arch Rib," catalog.

32. Building Officials and Code Administrators International, Inc. (BOCA) (17926 South Halsted, Homewood, IL 60430): *Basic Building Code 1978*, 7th ed.

33. Southern Building Code Congress International, Inc. (SBCC) (5200 Montclair Road, Birmingham, AL 35213): *Standard Building Code*, 1976 ed. with 1977 and 1978 revisions.

34. American National Standards Institute (ANSI) (1430 Broadway, New York, NY 10018): ANSI A58.1 *Building Code Requirements for Minimum Design Loads in Buildings and Other Structures*, 1972.

35. Truss Plate Institute, Inc. (Washington, DC): "Design Specification for Metal Plate Connected Wood Trusses," TPI-78, 1978 (7411 Riggs Road, Hyattsville, MD 20783).

36. Douglas Fir Plywood Association (Tacoma, WA): Parkins, Nelson S.: *Plywood Properties Design and Construction* (out of print).

37. National Bureau of Standards (Washington, DC): *The International System of Units (SI)*, NBS Special Publication 330, 1974.

38. American Society for Testing and Materials (1916 Race Street, Philadelphia, PA 19103): *Metric Practice Guide*, E380-74.

Index

539